分布式数据库架构及企业实践

基于Mycat中间件

/ 周继锋 冯钻优 陈胜尊 左越宗　著 /

電子工業出版社
Publishing House of Electronics Industry
北京•BEIJING

内 容 简 介

本书由资深 Mycat 专家及一线架构师、DBA 编写而成。全书总计 8 章，首先简单介绍了分布式系统和分布式数据库的需求，然后讲解了分布式数据库的实现原理，并对市场上存在的各种分布式数据库中间件进行了对比，再围绕着如何利用 Mycat 实现分布式数据库而展开。本书对 Mycat 从入门到进阶、从高级技术实践到架构剖析、从网络通信协议解析到系统工作原理的方方面面进行了详细讲解，并剖析了 Mycat 的 SQL 路由、跨库联合查询、分布式事务及原生 MySQL、PostgreSQL 协议等核心技术。通过本书不仅可以了解 Mycat 的基本概念，掌握 Mycat 配置等技术，还能感受到 Mycat 的架构设计之美，了解 Mycat 2.0 的未来规划。

无论是对于软件工程师、测试工程师、运维工程师、软件架构师、技术经理，还是对于资深 IT 人士来说，本书都极具参考价值。

图书在版编目（CIP）数据

分布式数据库架构及企业实践：基于 Mycat 中间件 / 周继锋等著. —北京：电子工业出版社，2016.11
ISBN 978-7-121-30287-9

Ⅰ. ①分… Ⅱ. ①周… Ⅲ. ①分布式数据库－数据库系统 Ⅳ. ①TP311.133.1

中国版本图书馆 CIP 数据核字（2016）第 268198 号

策划编辑：张国霞
责任编辑：徐津平
印　　刷：北京盛通商印快线网络科技有限公司
装　　订：北京盛通商印快线网络科技有限公司
出版发行：电子工业出版社
　　　　　北京市海淀区万寿路 173 信箱　邮编　100036
开　　本：787×980　1/16　印张：20　　字数：450 千字
版　　次：2016 年 11 月第 1 版
印　　次：2021 年 2 月第 7 次印刷
定　　价：79.00 元

凡所购买电子工业出版社图书有缺损问题，请向购买书店调换。若书店售缺，请与本社发行部联系，联系及邮购电话：（010）88254888，88258888。

质量投诉请发邮件至 zlts@phei.com.cn，盗版侵权举报请发邮件至 dbqq@phei.com.cn。

本书咨询联系方式：010-51260888-819　faq@phei.com.cn。

推荐序 1

随着大数据时代的到来，海量数据存储、并行计算、异构数据互联等一系列新技术在市场上不断地涌现。相信数据库行业的很多从业者都对传统关系型数据库的单点故障及容量问题头疼不已，而“分库分表”也早已成为解决这类问题的基础，此时，Mycat应运而生。

Mycat 是一款面向企业级应用的开源数据库中间件产品，它目前支持数据库集群、分布式事务与 ACID，被普遍视为基于 MySQL 技术的集群分布式数据库解决方案，在一些互联网、金融、运营商客户中用来替代昂贵的 Oracle。

Mycat 不仅可以轻松对接 MySQL、SQL Server 等传统关系型数据库，也融合了内存缓存、NoSQL、HDFS 等新兴大数据技术，是一款非常优秀的数据库中间件。

在如今的大数据时代，分布式架构已经成为企业级数据应用的标配，传统的关系型数据库产品已经面临一个真正的拐点：一方面，关系型数据库自身难以实现分布式，这大大限制了其数据存储能力及整体的性能表现；另一方面，商业化的传统数据库产品的成本和性价比在分布式架构崛起的状况下毫无优势可言。因此，无论是从底层全新实现分布式计算存储的 NoSQL、Hadoop，还是使用 Mycat 这样的分库分表工具，对关系型数据库大刀阔斧地进行“改装”都是大势所趋。

作为一名专注于数据库领域多年的从业者，我认为 Mycat 从中间件工具的角度成功地弥补了 MySQL 的诸多局限。

- 分布式存储：通过 Mycat，MySQL 可以实现集群化与分布式管理，使数据库容量与处理能力大大改善。
- 性能加速：通过分布式集群及 Mycat Booster 对 MySQL 数据库在集群环境下的加速，Mycat 大大提升了 MySQL 集群的性能。
- 异构数据互联互通：除了 MySQL，Mycat 同时支持如 SequoiaDB、MongoDB 这样的 NoSQL 数据库及 HDFS 分布式文件系统，实现了对非结构化数据、半结构化数据及结构化数据的存储及互联。

- 多样化的数据库工具：Mycat 为用户提供了丰富的管理工具，可以帮助用户更好地管理数据库系统。

本书非常适合作为 Mycat 的入门及进阶参考读物，它非常全面地阐述了分库分表的基本原理、实现机制及实践经验。本书的作者有着丰富的行业经验及技术底蕴，能够把业界非常前沿的知识用深入浅出的语言传授给各位读者。

最后，作为 SequoiaDB 的联合创始人，我十分钦佩 Mycat 团队的技术及勇气。虽然基础软件的开发难度很大，但是我们都敢于去挑战一个个技术难点，并填补国内基础软件产品的巨大空白。因此，我在这里衷心地祝愿 Mycat 前程似锦！

巨杉数据库联合创始人　王涛

推荐序 2

随着分布式系统的发展，应用的分布式由于无状态的特性，可以利用消息机制相对简单地进行拆分，计算的分布式也可以通过 Map、Reduce 等相关算法来解决。但是随着业务压力和并发压力的增加，我们急需一种分布式数据库解决方案来支持数据库的水平扩展，通过简单地增加服务器及线性地提升数据库的并发访问能力，为闯过分布式系统的最后一道难关铺平道路。

从阿里巴巴的 Cobar 到开源社区的 Mycat，从 Cobar 的架构师贺贤懋、朱海清、邱硕到 Mycat 的核心人员南哥、冰风影，作为一名专注于 MySQL 数据库十多年的从业者，我见证了分布式数据库的从无到有到百花齐放，在收到本书的序言邀请时，我感到非常荣幸。

现在的分布式数据库产品越来越多。YouTube 公司提供的 Vitness 功能强大，在 YouTube 的生产环境下支撑了大量的业务访问；360 公司的 Altas 基于 MySQL Proxy 开发而成，最初主要在应用层进行透明的读写分离，于 2013 年引入了分库分表；陈菲在离开 360 公司后在 WPS 云平台用 Go 语言编写了 Kingshard；楼方鑫（黄忠）在离开支付宝后编写了 OneProxy；腾讯互娱的 DBA 团队基于 Spider 打造了自己的分布式数据库平台；淘宝在内部将 TDDL 的客户端工具作为了分库分表中间件；阿里巴巴的 B2B 开源了支撑其内部业务生产环境 3 年的 Cobar，为开源社区提供了一大助力；而基于 Cobar 开发的 Mycat 及其各种分支由于其易用性，将分布式数据库进一步推广到互联网和传统行业的各个业务领域。

Mycat 无疑是这些中间件中的佼佼者，支持百亿级别的数据分片和并行计算，支持高可用和 MySQL 的读写分离，并随着版本的更新进一步支持 Oracle、DB2、MongoDB 等后端数据库，随着周边产品的进一步成熟，在越来越多的分布式或者非分布式（仅用它的读写分离或者高可用）生产环境中得到部署，受到越来越多的企业的关注。本书恰逢其会，由 Mycat 核心开发人员撰写而成，详细讲述了 Mycat 的由来、架构特点、核心模块、实际使用案例和企业实践，是一本不可多得的好书。

沃趣科技 MySQL 负责人　李春

推荐序 3

作为国产开源数据库中间件——Mycat 的发起者，我不得不为本书作序。

这是一本由众多技术精英合著的数据库+中间件领域的专业书籍，这些人包括 Mycat Commiter、Mycat 志愿者及资深 DBA，大家在工作之余抽出大量时间来编写和完善此书，历经一年完成了本书的编写工作，实属不易。

数据库中间件是新兴的重要的互联网中间件，目前业界仍然缺乏一本系统性介绍相关领域的软件产品、常用技术、架构等的纸质书籍。本书围绕 Mycat 开源中间件，从基础入门到架构原理，从运行机制到源码实现，从系统运维到应用实践，讲解得详尽而又完善。本书内容丰富、图文并茂、由浅入深，对数据库中间件的基本原理阐述清晰，对程序源码分析透彻，对实践经验讲解深刻。

从内容上讲，本书从一个使用者的角度去理解、分析和解决问题，通过大量的实例操作和源码解析，帮助读者深入理解 Mycat 的各种概念。读者对其中的案例只要稍做修改，再结合实际的业务需求，就可以在正在开发的项目中应用，达到事半功倍的效果。并且，通过学习书中应用实战方面的内容，不仅可以直接提高开发技能，还可以解决在实践过程中经常遇到的各种关键问题。另外，本书中的所有观点和经验均是作者们在多年建设、维护大型应用系统的过程中积累形成的，非常值得借鉴和推广。

希望 Mycat 在大家的支持下走得更远，成为中国软件的骄傲。

Leader-us

前言

随着移动互联网的兴起和大数据的蓬勃发展，系统的数据量正呈几何倍数增长，系统的压力也越来越大，这时最容易出现的问题就是服务器繁忙，我们可以通过增加服务器及改造系统来缓解压力，然后采用负载均衡、动静分离、缓存系统来提高系统的吞吐量。然而，当数据量的增长达到一定程度的时候，增加应用服务器并不能明显地提高系统的效率，因为所有压力都会传导到数据库层面，而大多数系统都是用一个数据库来存储和管理系统数据的。这时，Mycat应运而生。

谈到 Mycat 就不得不谈谈 Cobar，Cobar 是阿里巴巴开源的数据库中间件，由于其存在使用限制及一些比较严重的问题，Leader-us 在其基础上于 2013 年年底实现了 Mycat 1.0 版本，Mycat一经发布便引起了很多人的关注。之后 Mycat 社区对 Cobar 的代码进行了彻底重构，使用 NIO重构了网络模块，并且优化了 Buffer 内核，增强了聚合、Join 等基本特性，同时兼容了绝大多数数据库，使之成为通用的数据库中间件。Mycat 在 1.4 版本以后完全脱离了 Cobar 内核，同时采用了 Mycat 集群管理、自动扩容及智能优化，成为了高性能的数据库中间件。Mycat 从诞生至今已有三年多了，一直在坚持做最好的开源数据库中间件产品。

本书总计 8 章，涵盖了 Mycat 入门、进阶、高级技术实战、企业运维、架构剖析、核心技术分析、多数据库支持原理与实现等内容，内容详尽、图文并茂，几乎囊括了 Mycat 所涉及的方方面面，无论是对于软件工程师、测试工程师、运维工程师、软件架构师、技术经理，还是对于资深 IT 人士来说，本书都极具参考价值。

第 1 章：介绍了分布式系统和分布式数据库系统的原理，介绍 Mycat 的起源和发展状况，并对各种数据库中间件做了简要介绍和对比。

第 2 章：讲解了 Mycat 的入门知识，介绍了 Mycat 的安装环境、核心概念和分库分表的原理，以及 Mycat 源码开发调试的过程。

第 3 章：讲解了 Mycat 的进阶知识，主要介绍 Mycat 的各种配置和分片算法。

第 4 章：讲解了 Mycat 和 MySQL 实战案例，由拥有丰富的 Mycat 线上实战经验的专家和 DBA 共同编写而成，有很高的参考价值。

第 5 章：简要介绍了用于 Mycat 性能监控的工具——Mycat-web，详细讲解了 Mycat 和 MySQL 的优化技术，是 DBA 的亲身总结和经验之谈。

第 6 章：重点阐述了 Mycat 的架构，包括网络、线程、连接池、内存管理及缓存实现等，是了解 Mycat 框架的基础。

第 7 章：介绍了 Mycat 的核心技术，包括分布式事务的实现、跨库 Join 的三种实现方式等，介绍了多节点数据汇聚和排序的原理，并详细阐述了在 Mycat 1.6 版本中实现的一致性分布式事务的功能。

第 8 章：介绍了 MySQL 和 PostgreSQL 的通信协议及 Mycat 对这些通信协议的实现，然后介绍了 Mycat 对 JDBC 及多种数据库的支持，例如 Oracle、SQL Server、MongoDB 等。

本书的编写和校对历经一年，参与编写的作者都是 Mycat 开源项目中参与度比较高、提交过不少代码或有丰富的实战经验的资深人士。非常感谢参与本书编写、指导或校对的专家：Leader-us、南哥（曹宗南）、从零开始（宋伟）、小张哥（张超）、yuanfang（杨鹏飞）、顽石神（张治春）、冰麒麟（杨峰）、望舒（胡雅辉）、明明 Ben（朱阿明）、零（章爱国）、little-pan（潘自朋）、CrazyPig（陈建欣）、毛茸茸的逻辑（王成瑞）、海王星（林志强）、石头狮子（林晁）、HanSenJ（姬文刚）、武（王灯武）、战狼（刘胡波）、KK（刘军）、董海雄（易班网）、arx（李秋伟）、正能量（王金剑）、吉光（李伟）。

由于作者的写作水平有限，书中难免会有不妥或者疏漏之处，欢迎读者批评指正。

冰风影

Mycat 社区负责人

2016 年 11 月 6 日于广州番禺

目录

第 1 章 数据库中间件与分布式数据库的实现

从第一台计算机产生至今的半个多世纪里，计算机已经深入人们的生活，在各个领域得到广泛应用，并不断地改变人们的生活方式。从电子管到晶体管，从集成电路到超大规模集成电路，从单个 CPU、存储器、外设和一些终端在内的集中式计算到大量 CPU 通过高速网络连接组成的并行计算，计算机的发展有着惊人的进步。

1.1 什么是分布式系统

分布式系统是指其组件分布在网络上，组件之间通过传递消息进行通信和动作协调的系统。它的核心理念是让多台服务器协同工作，完成单台服务器无法处理的任务，尤其是高并发或者大数据量的任务。它的主要特点如下。

- **透明性**：分布式系统对于用户来说是透明的，一个分布式系统在用户面前的表现就像一个传统的单处理机分时系统，可让用户不必了解其内部结构就能使用。
- **扩展性**：分布式系统的最大特点是可扩展性，它能够根据需求的增加而扩展，可以通过横向扩展使集群的整体性能得到线性提升，也可以通过纵向扩展单台服务器的性能（上升空间有限，使用不多）使服务器集群的性能得到提升。
- **可靠性**：分布式系统不允许单点失效的问题存在，它的基本思想是，如果一台机器坏了，则其他机器能够接替它进行工作，具有持续服务的特性。

- **高性能**：高性能是人们设计分布式系统的一个初衷，如果建立了一个透明、灵活、可靠的分布式系统，但是它运行起来像蜗牛一样慢，那么这个分布式系统是失败的。

分布式系统在拥有众多优点的同时自然有其缺点。

- 在节点通信部分的开销比较大，线程安全问题也变得复杂，需要在保证数据完整性的同时兼顾性能。
- 过分依赖网络，网络信息的丢失或饱和将会抵消分布式系统的大部分优势。
- 有潜在的数据安全和网络安全等安全性问题。

下面讲讲分布式系统与集中式系统的区别。首先，简单地讲，分布式系统采用并行计算，而集中式系统采用串行计算，这是二者本质上的区别；其次，分布式系统的性价比高，对于处理大规模数据而言，最节约成本的办法是在一个系统中使用集中在一起的大量的廉价 CPU；然后，分布式系统的可靠性高，通过把负载分散到众多机器上，单个芯片发生故障时最多会使一台机器停止运行，而其他机器不会受到影响；另外，分布式系统比集中式系统的可扩展性强，可根据业务量的增长逐渐扩展系统的计算能力；最后，从性能上讲，分布式系统的计算能力比单个大型主机更强。

1.2 为什么需要分布式数据库

随着计算机和信息技术的迅猛发展和普及，行业应用系统的规模迅速扩大，行业应用所产生的数据量呈爆炸式增长，动辄达到数百 TB 甚至数百 PB 的规模，已远远超出现有的传统计算技术和信息系统的处理能力，而集中式数据库面对大规模数据处理逐渐表现出其局限性。因此，人们希望寻找一种能快速处理数据和及时响应用户的访问的方法，也希望对数据进行集中分析、管理和维护。这已成为现实世界的迫切需求。

分布式数据库是在集中式数据库的基础上发展起来的，是计算机技术和网络技术结合的产物。分布式数据库是指数据在物理上分布而在逻辑上集中管理的数据库系统。物理上分布是指分布式数据库的数据分布在物理位置不同并由网络连接的节点或站点上；逻辑上集中是指各数据库节点之间在逻辑上是一个整体，并由统一的数据库管理系统管理。不同的节点分布可以跨不同的机房、城市甚至国家。

分布式数据库的主要特点如下。

- **透明性**：用户不必关心数据的逻辑分区和物理位置分布的细节，也不必关心重复副本

（冗余数据）的一致性问题，同时不必关心在局部场地上数据库支持哪种数据模型。对于系统开发工程师而言，当数据从一个场地移到另一个场地时不必改写应用程序，使用起来如同一个集中式数据库。

- **数据冗余性：**分布式数据库通过冗余实现系统的可靠性、可用性，并改善其性能。多个节点存储数据副本，当某一节点的数据遭到破坏时，冗余的副本可保证数据的完整性；当工作的节点受损害时，可通过心跳等机制进行切换，系统整体不被破坏。还可以通过热点数据的就近分布原则减少网络通信的消耗，加快访问速度，改善性能。
- **易于扩展性：**在分布式数据库中能够方便地通过水平扩展提高系统的整体性能，也能够通过垂直扩展来提高性能，扩展并不需要修改系统程序。
- **自治性：**各节点上的数据由本地的 DBMS 管理，具有自治处理能力，完成本场地的应用或局部应用。

分布式数据库还具有经济、性能优越、响应速度更快、灵活的体系结构、易于集成现有系统等特点。

分布式数据库尽管有着天生的高贵血统，但它依赖高速网络，对事务的处理远没有传统数据库成熟，在很长一段时间内分布式数据存储将与传统数据存储共存。若想打造一个高可用、高性能、易扩展、可伸缩且安全的分布式应用系统，则分布式数据库将是必不可少的选择。

1.3　分布式数据库的实现原理

通过 1.2 节我们已经知道，分布式数据库具有逻辑整体性、物理分布性，正是因为其物理分布性才使得分布式数据库的实现变得更加复杂，因为数据划分后存储在不同的节点上，而为了保证可靠性，需要存储多个副本，所以产生了数据复制的问题。为了保证良好的性能，分布式数据库必须易于扩展。具体来讲分布式数据库应有 **4** 个优势：数据分片及复制管理、具有事务的可靠性存取、良好的性能、易于扩展，所以分布式数据库在设计上需要实现分布式数据库的目录管理、数据分片、分布式查询处理、分布式并发控制、分布式锁管理、分布式存储、分布式网络架构、分布式安全管理等。

1. 分布式数据库的目录管理

分布式数据库的目录存放着系统元数据及数据库的元数据的全部信息，这些数据的存在是为了正确、有效地访问数据。数据的增、删、改、查操作都需要用到目录，用户授权、安全管理及并发控制等也都需要用到目录，目录结构的合理性直接影响数据库的性能。目录一般包括各级模式的描述、访问方法的描述、关于数据库的统计数据和一致性信息等，系统根据这些信息将用户查询转换为物理数据库上的查询，选择一条最佳的存取路径进行事务管理及安全性、完整性检查等。

分布式数据库的目录可分为全局目录、分布式目录、全局与本地混合目录。

2. 数据分片

当数据库过于庞大，尤其是写入过于频繁且很难由一台主机支撑时，我们还是会面临扩展瓶颈。我们将存放在同一个数据库实例中的数据分散存放到多个数据库实例（主机）上，进行多台设备存取以提高性能，在切分数据的同时可以提高系统整体的可用性。

数据分片是指将数据全局地划分为相关的逻辑片段，有水平切分、垂直切分、混合切分三种类型。

- **水平切分**：按照某个字段的某种规则分散到多个节点库中，每个节点中包含一部分数据。可以将数据的水平切分简单理解为按照数据行进行切分，就是将表中的某些行切分到一个节点，将另外某些行切分到其他节点，从分布式的整体来看它们是一个整体的表。
- **垂直切分**：一个数据库由很多表构成，每个表对应不同的业务，垂直切分是指按照业务将表进行分类并分布到不同的节点上。垂直拆分简单明了，拆分规则明确，应用程序模块清晰、明确、容易整合，但是某个表的数据量达到一定程度后扩展起来比较困难。
- **混合切分**：为水平切分与垂直切分的结合。

3. 分布式查询处理

分布式查询处理的任务就是把一个分布式数据库上的高层次查询映射为在本地数据库上的操作，查询的解析必须拆分为代数查询的关系运算序列，将要查询的数据定位到各节点，使得查询在各节点进行，最后通过网络通信的操作汇聚查询结果。

4. 分布式并发控制

并发控制是分布式事务管理的基本任务之一，其目的是保证分布式数据库中的多个事务并发高效、正确地执行。并发控制用来保证事务的可串行性，也就是说事务的并发执行等价于它们按某种次序的串行执行，从而为用户提供并发的透明性。进行并发控制的方法主要有三种：加锁并发控制、时间戳控制、乐观并发控制。加锁并发控制应用广泛，但是容易发生死锁；时间戳控制消除了死锁，一旦发生冲突便会重启而不是等待，需要有全局的统一时钟；乐观并发控制对于冲突较少的系统较为适合，对于冲突多的系统则效率低下。

总体来看，实现分布式数据库不仅仅是如此简单的几个方面，在实际设计中还需要解决网络架构、分布式存储等问题。每个模块都可以作为一个课题去研究，本节仅是抛砖引玉。

1.4　Mycat 数据库中间件简介

Mycat 是一个彻底开源的面向企业应用开发的大数据库集群，支持事务、ACID，是可以替代 MySQL 的加强版数据库。Mycat 被视为 MySQL 集群的企业级数据库，用来替代昂贵的 Oracle 集群，它是融合了内存缓存技术、NoSQL 技术、HDFS 大数据的新型 SQL Server，是结合了传统数据库和新型分布式数据仓库的新一代的企业级数据库产品，也是一个优秀的数据库中间件。

1.4.1　Mycat 的历史与未来规划

Mycat 的诞生，要从其前身 Amoeba 和 Cobar 说起。Mycat 最初起源于 Amoeba（变形虫）项目，这个项目致力于 MySQL 的分布式数据库前端代理层，主要在应用层访问 MySQL 时充当 SQL 路由功能。阿里巴巴在 Amoeba 的基础上开发出了 Cobar，于 2012 年 6 月 19 日正式对外开源。Cobar 自诞生之日起，就受到广大程序员的追捧，但是 2013 年后几乎没有更新，在生产环境的使用中也出现了许多 Bug。

2013 年，Cobar 在某大型项目的使用过程中被发现存在一些比较严重的问题，于是广大开源爱好者将其改良，Mycat 1.0 由此而生。相比当时的 Cobar，Mycat 在 I/O 方面的重大改进就是后端由原来的 BIO 改成 NIO，在后来的版本中又增加了 AIO，并且通过配置可任选一种 I/O。

2014 年，Mycat 开源社区首次在上海举行的“中华架构师”大会上对外宣讲，引起了业界的关注，随后越来越多的项目采用了 Mycat。

2015 年是 Mycat 发展壮大的一年，发生了许多令人兴奋并激动的事情。在这一年里，Mycat 1.3 诞生，该版本是 Mycat 历史上非常重要的一个里程碑，需求、测试和功能开发等各项工作模式首次从以个人为主变为以开源团队为主，更多的人参与到需求、开发、测试及 Bug 修复活动中，已确定的 Bug 基本上会在 24 小时被内修复并有志愿者或用户确认修复。Mycat 1.3 的性能与 1.2 版本相比提升巨大，功能更完备。同年 10 月，Mycat 项目总共有 16 个 Committer。同年

11 月，超过 300 个项目采用了 Mycat，涵盖银行、电信、电子商务、物流、移动应用、O2O 等众多领域和公司。同年 12 月，超过 4000 名用户加群、研究讨论、测试或使用 Mycat。该年年底发布了 1.4 版本，其后完全脱离了基本的 Cobar 内核，结合 Mycat 集群管理、自动扩容、智能优化等成为高性能的中间件。

Mycat 支持 Oracle、PostgreSQL，从 1.3 版本开始支持 NoSQL（SequoiaDB 及 MongoDB）并引入了 Druid 解析器。Druid 完全开源，一方面可以降低 SQL 解析的入门门槛，使复杂的问题简单化；另一方面可以使解析性能得到很大的提高。Mycat 在 1.4 版本中去掉了 Fdbparser，只保留 Druidparser 方式。

2016 年，Mycat 发布了 1.5 版本，这是到目前为止比较稳定的版本。另外，在这一年里，Mycat 的 Committer 发起了众筹，成立了维护团队，这标志着 Mycat 形成了一个正式组织。

目前，Mycat 1.6 的 Beta 版本已经发布（正式版本预计 2017 年春节后发布），已有用户将其用于生产环境，暂时未发现重大 Bug，但是在生产环境中还是建议使用 Mycat 1.5 版本。

Mycat 1.6 正在开发的功能如下。

（1）对子查询的支持。

（2）数据一致性的增强：采用分布式事务或消息队列实现；全局表的一致性保证。

（3）基于 ZooKeeper 协调的主从切换。

（4）引入测试框架，可以配置用例和进行自动化测试。

（5）新增自动迁移功能，实现不停机迁移时不丢数据。

Mycat 1.6 对系统稳定性的开发任务如下。

（1）解决全局自增 ID 列无法作为分片字段的问题。

（2）解决分片表与对应分片规则的一致性检查问题。

（3）解决 DirectMemory 的管理问题。

（4）完善 reload config，支持不丢事务。

目前，Mycat 2.0 项目（https://github.com/MyCATApache/Mycat2.git）已经启动，核心代码已经提交。Mycat 是一个全新的项目，全部重构，目的是打造真正的分布式数据库中间件。其架构设计如图 1-1 所示。

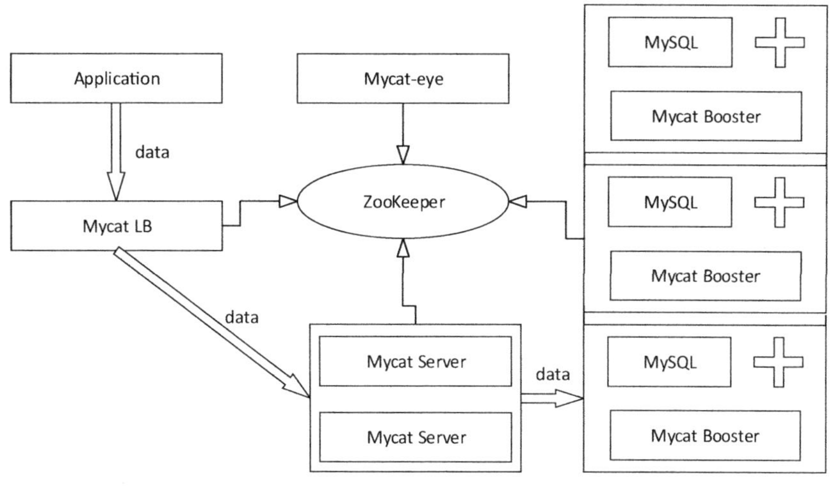

图 1-1

其核心功能如图 1-2 所示。

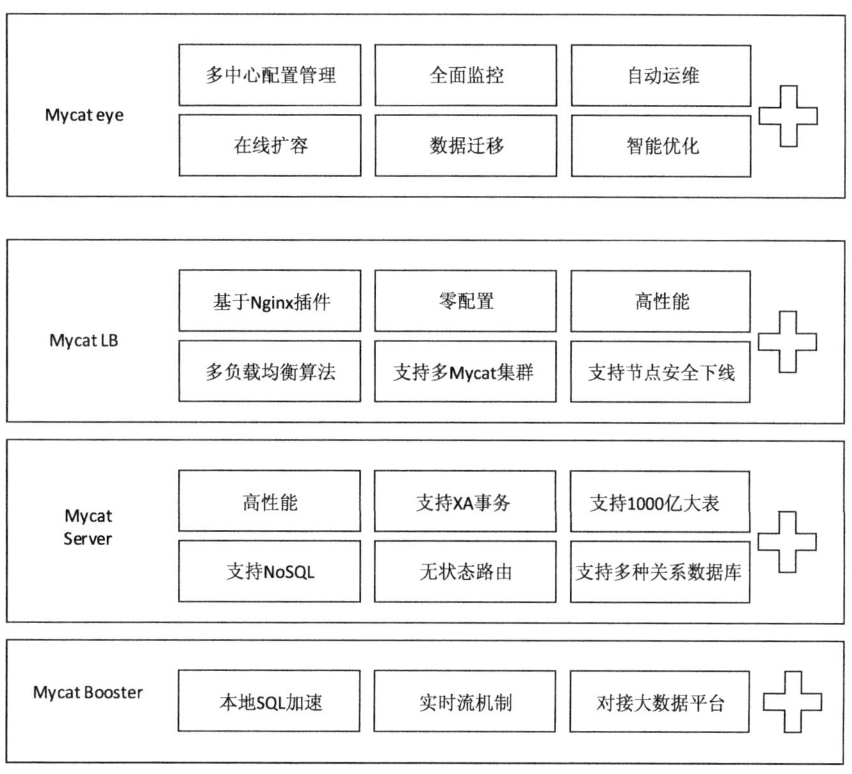

图 1-2

多中心支持如图 1-3 所示。

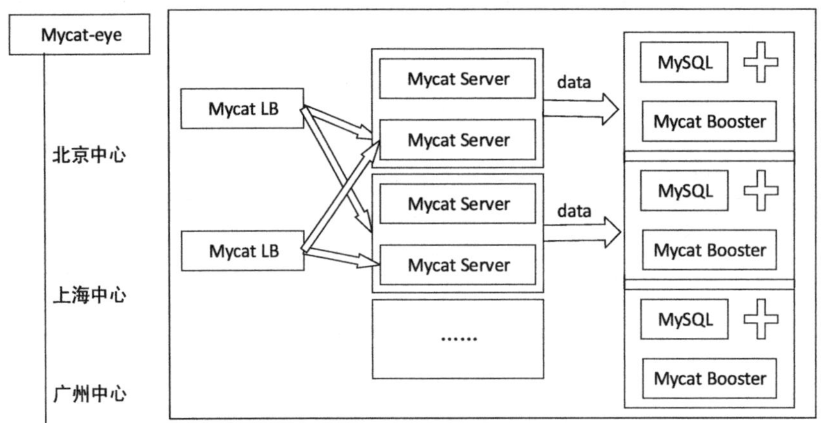

图 1-3

在 Mycat 2.0 的规划中将完全实现分布式事务，完全支持分布式；通过 Mycat web（eye）完成可视化配置及智能监控、自动化运维；通过 MySQL 的本地节点 Mycat Booster 全面解决数据扩容问题，实现自动扩容机制；支持基于 ZooKeeper 的主从切换及 Mycat 集群化管理；通过用 Mycat Balance 替代第三方的 HAProxy、LVS 等，全面兼容 Mycat 集群节点的动态上下线；接入 Spark 等第三方工具，解决数据分析及大数据聚合的业务场景；通过 Mycat 智能优化、分析分片热点，提供合理的分片、索引及数据切分实时业务建议；另外 Mycat 的堆外内存模块正在开发中，预计在 2.0 版本之前投入使用。

如今 Mycat 的发展已经超越了 Mycat-server 本身，希望将来我们可以看到一个以 Mycat 为中心的开源社区，并且这个开源社区将吸收更多的优秀开源项目，坚持开源不动摇，成为中国版的 Apache。

1.4.2 Mycat 与其他中间件的区别

目前的数据库中间件有很多，本节将介绍主流的中间件，并从各个维度将其与 Mycat 进行对比。

1. Mango

Mango 的中文名是“芒果”，它是一个轻量级的极速数据层访问框架，目前已有十多个大型线上项目在使用它。据称，某一支付系统利用 Mango 承载了每秒 12 万的支付下单请求，其超高的性能及超快的响应速度几乎相当于直接使用 JDBC。它采用接口与注解的形式定义 DAO，完美地结合了 db 与 cache 操作；支持动态 SQL，可以构造任意复杂的 SQL 语句；支持多数据源、分表、分库、事务；内嵌“函数式调用”功能，能将任意复杂的对象映射到数据库的表中。但是从整体上看 Mango 是一个 Java Dao Framework，是一个 jar 包，它的运行依赖于应用系统的项目代码和服务器，采用了 JDBC Shard 思想，与 TDDL 是同款产品。

2. Cobar

Cobar 是阿里巴巴研发的关系型数据的分布式处理系统（Amoeba 的升级版，该产品成功替代了原先基于 Oracle 的数据存储方案，目前已经接管了 3000 多个 MySQL 数据库的 schema，平均每天处理近 50 亿次的 SQL 执行请求。

3. Heisenberg

Heisenberg 源于 Cobar，结合了 Cobar 和 TDDL 的优势，让其分片策略变为分库分表策略，节约了大量的连接。优点是分库分表与应用脱离，分库分表如同使用单库表一样，减少了 db 连接数的压力，采用热重启配置，可水平扩容，并遵循 MySQL 原生协议，采用读写分离，无语言限制。MySQL Client、C、Java 等都可以使用 Heisenberg 服务器通过管理命令查看和调整连接数、线程池、节点等。Heisenberg 采用 velocity 的分库分表脚本进行自定义分库分表，相当灵活。

4. Atlas

Atlas 是由奇虎 360 的 Web 平台部基础架构团队开发维护的一个基于 MySQL 协议的数据库中间层项目。它在 MySQL-proxy 0.8.2 版本的基础上进行了优化，增加了一些新的功能和特性。奇虎 360 内部使用 Atlas 运行的 MySQL 业务，每天承载的读写请求数达几十亿条。Atlas 位于应用程序与 MySQL 之间，实现了 MySQL 的客户端与服务端协议，作为服务端与应用程序通信，同时作为客户端与 MySQL 通信。它对应用程序屏蔽了 db 的细节，同时为了降低 MySQL 的负担，维护了连接池。

5. Amoeba

Amoeba 是一个以 MySQL 为底层数据存储，并对应用提供 MySQL 协议接口的 Proxy。它集中响应应用的请求，依据用户事先设置的规则，将 SQL 请求发送到特定的数据库上执行，基于此可以实现负载均衡、读写分离、高可用性等需求。与 MySQL 官方的 MySQL Proxy 相比，Amoeba 的作者强调的是 Amoeba 配置的便捷性（基于 XML 的配置文件，用 SQLJEP 语法书写规则，比基于 Lua 脚本的 MySQL Proxy 简单）。

Mycat 与以上中间件的对比如表 1-1 所示。

表 1-1

对 比 项 目	Mycat	Mango	Cobar	Heisenberg	Altas	Amoeba
数据切片	支持	支持	支持	支持	支持	支持
读写分离	支持	支持	支持	支持	支持	支持
宕机自动切换	支持	不支持	支持	不支持	半支持，影响写	不支持
MySQL 协议	前后端支持	JDBC	前端支持	前后端支持	前后端支持	JDBC
支持的数据库	MySQL、Oracle、MongoDB、PostgreSQL	MySQL	MySQL	MySQL	MySQL	MySQL、MongoDB
社区活跃度	高	活跃	停滞	低	中等	停滞
文档资料	极丰富	较齐全	较齐全	缺少	中等	缺少
是否开源	开源	开源	开源	开源	开源	开源
是否支持事务	弱 XA	支持	单库强一致、分布式弱事务	单库强一致、多库弱事务	单库强一致、分布式弱事务	不支持

1.4.3 Mycat 的优势

Mycat 对 Cobar 代码进行了彻底重构，使用 NIO 重构了网络模块，并且优化了 Buffer 内核，增强了聚合、Join 等基本特性，同时兼容绝大多数数据库，成为通用的数据库中间件。Mycat 后端以 JDBC 方式支持 Oracle、DB2、SQL Server、MongoDB、SequoiaDB 等数据库，支持透明的读写分离机制，减轻了写库的压力，提高了数据库的并发查询能力，支持各种 MySQL 集群，包括标准的主从异步集群、MySQL Galera Cluster 多主同步集群等，提升了数据库的可用性与性能，还可以通过大表水平分片方式支持 100 亿级大表的分布式存储和秒级的并行查询能力，内建数据库集群故障切换机制，实现了自动切换，可满足大部分应用的高可用性要求。

Mycat 现在由一支强大的技术团队维护，吸引和聚集了一大批业内大数据和云计算方面的资深工程师，参与者都是有 5 年以上经验的资深软件工程师、架构师、DBA，优秀的技术团队保证了 Mycat 的产品质量。Mycat 不依托于任何一家商业公司，因此不像某些开源项目将一些重要的特性封闭在其商业产品中，Mycat 的发展和壮大主要基于开源社区志愿者的持续努力。

Mycat 还形成了一系列周边产品，比较有名的是 Mycat-web、Mycat-NIO、Mycat-Balance 等，已成为一个比较完整的数据处理解决方案，而不仅仅是中间件。

1.4.4 Mycat 的适用场合

想要用好 Mycat，就需要了解其适用场景，以下几个场景适合使用 Mycat。

1. 高可用性与 MySQL 读写分离

利用 Mycat 可以轻松实现热备份，当一台服务器停机时，可以由双机或集群中的另一台服务器自动接管其业务，从而在无须人工干预的情况下，保证系统持续提供服务。这个切换动作由 Mycat 自动完成。

Mycat 还可以轻松实现数据库的读写分离，实现主数据库处理事务的增、改、删（insert、update、delete）操作，而在数据库中处理查询（select）操作。需要强调的是，Mycat 的读写分离及自动切换都依赖于数据库产品的主从数据同步。

2. 业务数据分级存储保障

企业的数据量总是在无休止地增长，这些数据的格式不一样，访问频率不一样，重要性也不一样。如果将这些数据统一存储在高性能的数据库产品中，则不但价格昂贵，而且随着数据量的不断增长，读写的性能会受到影响，统一存储的性能也会逐渐下降，维护成本相当高，所以此时我们需要对数据进行分级存储。

数据分级存储是指数据客体存放在不同级别的存储设备（磁盘、磁盘阵列、光盘库、磁带库）中，通过分级存储管理软件实现数据客体在存储设备之间的自动迁移及自动访问切换。在企业的数据中，比较重要的交易类热数据通常存放在 Oracle 等关系型数据库中；历史数据通常存放在其他数据库中；日志数据通常存放在 NoSQL 或其他日志管理产品中；图片数据通常存放在图片分布式系统中。考虑到成本因素，对于不同的数据所采用的磁盘性能也不一样。如何统一管理这些不同的业务数据，并给业务系统提供一个统一的入口呢？此时，Mycat 可以派上用场。

Mycat 与生俱来的中间件特性决定了它能承担这一重任，并且只需几个配置就能实现。在 Mycat 中，一个逻辑库可以对应 Oracle、MySQL、MongoDB 等多个数据源，对于用户而言就像一个数据库，用户并不知道后端有哪些数据库，只需要关注 Mycat。

3. 100 亿大表水平分片，集群并行计算

数据切分是 Mycat 的核心功能，是指通过某种特定的条件，将存放在同一个数据库中的数据分散存放在多个数据库（主机）中，以达到分散单台设备负载的效果。数据切分有两种切分模式：一种是按照不同的表（或者 schema）将数据切分到不同的数据库（主机）中，这种切分

可以叫作数据的垂直（纵向）切分；另外一种则是根据表中数据的逻辑关系，将同一个表中的数据按照某种条件拆分到多个数据库（主机）中，这种切分叫作数据的水平（横向）切分。当数据量超过 800 万行且需要做分片时，可以利用 Mycat 实现数据切分。

4. 数据库路由器

Mycat 基于 MySQL 实例的连接池复用机制，可以让每个应用最大程度地共享一个 MySQL 实例的所有连接池，让数据库的并发访问能力大大提升。

5. 整合多种数据源

当一个项目需要用到多种数据源如 Oracle、MySQL、SQL Server、PostgreSQL 时，可以利用 Mycat 进行整合，只需访问 Mycat 这一个数据源就行。

第 2 章 Mycat 入门

本章主要讲解 Mycat 的安装、基本概念和基本原理，这些是学习 Mycat 的基础，也是深入理解第 3 章中各项配置属性的基础。

2.1 环境搭建

Mycat 是采用 Java 开发的开源分布式中间件，支持 Windows 和 Linux 运行环境，下面介绍 Mycat 的 Windows 和 Linux 环境搭建方法。

注意：在 Mycat 文件的目录中，bin 目录存放 Mycat 的执行文件；conf 目录存放 Mycat 的配置文件；lib 目录存放 Mycat 所依赖的 jar 包；logs 目录存放 Mycat 运行时所产生的日志文件；version.txt 记录当前的 Mycat 版本的信息。

2.1.1 Windows 环境搭建

1. 安装和配置 MySQL 数据库

首先需要下载 MySQL 安装包，推荐使用 MySQL 5.6，下载地址为 http://dev.MySQL.com/downloads/file/?id=461390。

双击安装包进行安装，安装过程比较简单，这里只介绍安装细节。使用 navicat 连接本地安装好的数据库，如图 2-1 所示。

图 2-1

2. 配置 Java 虚拟机的环境变量

通过 http://www.oracle.com/technetwork/java/javase/downloads/jdk8-downloads-2133151.html 下载合适的 JDK 版本并以默认方式进行安装，安装完成后将环境变量配置为 JAVA_HOME（值为 C:\Program Files\Java\jdk1.8.0_66）、CLASSPATH（值为%JAVA_HOME%\lib\dt.jar;%JAVA_HOME%\lib\tools.jar）和 Path（值为%JAVA_HOME%\bin;%JAVA_HOME%\jre\bin）。

注意：变量 JAVA_HOME 的值是 JDK 的安装路径，配置好环境变量后测试配置是否成功，同时按住 Win 和 R 键，桌面左下角会弹出“运行”窗口，输入 cmd 后按回车键。在弹出的 DOS 命令行窗口中依次输入 JAVAC、java -version，观察输出结果是否正确，如果不正确则环境变量配置失败。

3. 下载和启动 Mycat

可以在 GitHub 的 Mycat-Server 项目中找到 Mycat 安装包，将其解压到根目录。双击 startup_nowrap.bat 后将启动 Mycat（见图 2-2），如果未显示错误信息，则表示 Mycat 启动成功。

图 2-2

2.1.2　Linux 环境搭建

1. 安装和配置 MySQL 数据库

建议安装 MySQL 5.6 版本，下载地址为 http://dev.MySQL.com/get/MySQL-community-release-el6-5.noarch.rpm，把下载好的文件放到/opt 目录下。

安装 MySQL 仓库：

```
#sudo yum localinstall MySQL-community-release-el6-5.noarch.rpm
```

安装 MySQL 5.6：

```
# sudo yum install MySQL-community-server
```

2. 配置 Java 虚拟机的环境变量

查看当前仓库的 Java 版本（yum -y list java*），安装当前仓库的 JDK 1.8 版本，配置好环境变量后通过 java -version 查看当前安装完成的 Java 版本。

3. 下载和启动 Mycat

通过 https://github.com/MyCATApache/Mycat-download/tree/master/1.5-RELEASE 下载对应 Linux 系统的 Mycat 安装包并进行安装。使用浏览器下载 Linux 版本的压缩包，请勿采用 wget 方式下载，以此方式下载文件后在解压时会报错。下载完成后安装文件会存放到 opt 目录下，通过 tar 命令解压该安装文件。

```
tar －zxvf  Mycat-server-1.5.1-RELEASE-20160816173057-win.tar.gz。
```

通过/opt/mycat/bin/mycat 命令进行启动，/mycat console 如果没有报错，则说明 Mycat 启动成功。

2.2 Mycat 核心概念详解

2.2.1 逻辑库（schema）

业务开发人员通常在实际应用中并不需要知道中间件的存在，只需要关注数据库，所以数据库中间件可以被当作一个或多个数据库集群构成的逻辑库。

在云计算时代，数据库中间件可以以多租户的形式为一个或多个应用提供服务，每个应用访问的可能是一个独立或者共享的物理库，如图 2-3 所示为阿里云数据库服务器 RDS。

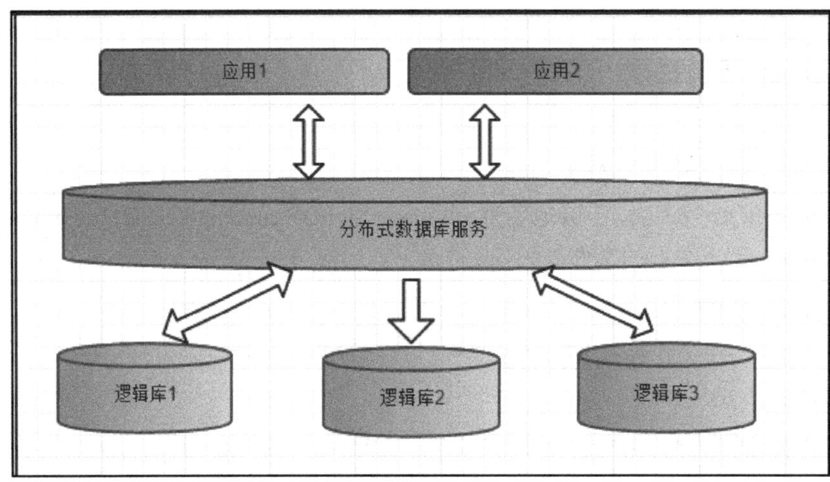

图 2-3

2.2.2 逻辑表（table）

既然有逻辑库，就会有逻辑表。在分布式数据库中，对于应用来说，读写数据的表就是逻辑表。逻辑表可以分布在一个或多个分片库中，也可以不分片。

1. 分片表

分片表是将指数据量很大的表切分到多个数据库实例中，所有分片组合起来构成了一张完整的表。例如在 Mycat 上配置 t_node 的分片表，数据按照规则被切分到 dn1、dn2 两个节点。

```
<table name="t_node" primaryKey="vid" autoIncrement="true" dataNode="dn1,dn2"
rule="rule1" />
```

2. 非分片表

并非所有的表在数据量很大时都需要进行分片，某些表可以不用分片。非分片表是相对于分片表而言的不需要进行数据切分的表。如下配置中的 t_node 只存在于节点 dn1 上。

```
<table name="t_node" primaryKey="vid" autoIncrement="true" dataNode="dn1" />
```

3. ER 表

关系型数据库是基于实体关系模型（Entity Relationship Model）的，Mycat 中的 ER 表便来源于此。基于此思想，Mycat 提出了基于 E-R 关系的数据分片策略，子表的记录与其所关联的父表的记录存放在同一个数据分片上，即子表依赖于父表，通过表分组（Table Group）保证数据关联查询不会跨库操作。

表分组（Table Group）是解决跨分片数据关联查询的一种很好的思路，也是数据切分的一条重要规则。

4. 全局表

在一个真实的业务场景中往往存在大量类似的字典表，这些字典表中的数据变动不频繁，而且数据规模不大，很少有超过数十万条的记录。

当业务表因为规模进行分片后，业务表与这些附属的字典表之间的关联查询就成了比较棘手的问题，所以在 Mycat 中通过数据冗余来解决这类表的关联查询，即所有分片都复制了一份数据，我们把这些冗余数据的表定义为全局表。

数据冗余是解决跨分片数据关联查询的一种很好的思路，也是数据切分规划的另一条重要规则。

2.2.3　分片节点（dataNode）

将数据切分后，一个大表被分到不同的分片数据库上，每个表分片所在的数据库就是分片节点。

2.2.4　节点主机（dataHost）

将数据切分后，每个分片节点不一定会独占一台机器，同一台机器上可以有多个分片数据库，这样一个或多个分片节点所在的机器就是节点主机。为了规避单节点主机并发数量的限制，尽量将读写压力高的分片节点均匀地放在不同的节点主机上。

2.3 Mycat 原理介绍

Mycat 原理并不复杂，复杂的是代码，如果代码也不复杂，那么 Mycat 早就成为一个传说了。Mycat 原理中最重要的一个动词是“拦截”，它拦截了用户发送过来的 SQL 语句，首先对 SQL 语句做了一些特定的分析，例如分片分析、路由分析、读写分离分析、缓存分析等，然后将此 SQL 语句发往后端的真实数据库，并将返回的结果做适当处理，最终再返回给用户，如图 2-4 所示。

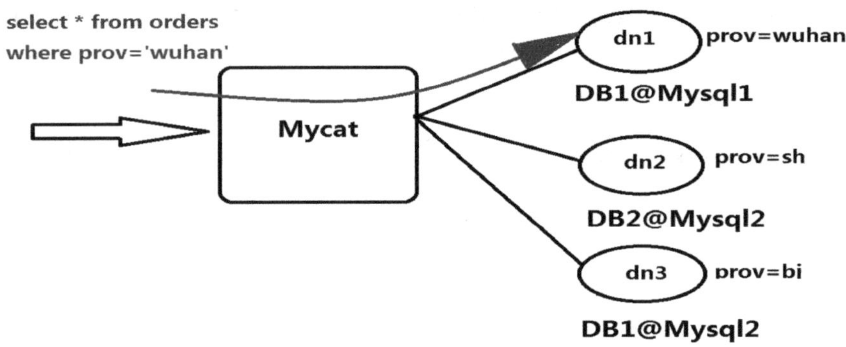

图 2-4

在图 2-4 中，Orders 表被分为三个分片节点 dn1、dn2、dn3，它们分布在两台 MySQL Server（dataHost）上，即 datanode=database@datahost，因此你可以用 1～*N* 台服务器来分片，分片规则（sharding rule）为典型的字符串枚举分片规则，一个规则的定义是分片字段（sharding column）+分片函数（rule function）。这里的分片字段为 prov，分片函数则为字符串枚举方式。

Mycat 收到一条 SQL 语句时，首先解析 SQL 语句涉及的表，接着查看此表的定义，如果该表存在分片规则，则获取 SQL 语句里分片字段的值，并匹配分片函数，得到该 SQL 语句对应的分片列表，然后将 SQL 语句发送到相应的分片去执行，最后处理所有分片返回的数据并返回给客户端。以 select * from Orders where prov=?语句为例，查找 prov=wuhan，按照分片函数，wuhan 值存放在 dn1 上，于是 SQL 语句被发送到 Mysql1，把 DB1 上的查询结果返回给用户。

如果将上述 SQL 语句改为 select * from Orders where prov in('wuhan', 'beijing')，那么 SQL 语句会被发送到 Mysql1 与 Mysql2 去执行，然后将结果集合并后输出给用户。但通常在业务中我们的 SQL 语句会有 Order By 及 Limit 翻页语法，此时就涉及结果集在 Mycat 端的二次处理，这部分代码也比较复杂，而最复杂的则是两个表的关联查询问题。为此，Mycat 提出了创新性的 ER 分片、全局表、HBT（Human Brain Tech，人工智能）的 catlet，以及结合了 Storm、

Spark 引擎等十八般武艺的解决方案，成为目前业界很强大的跨分片 Join 方案，这就是开源的力量！

2.4　参与 Mycat 源码开发

Mycat 是采用 Java 开发的，参与者需要有编程基础；熟悉 Java、Maven 及 MySQL、Oracle、PostgreSQL、MongoDB 等数据库；熟悉 SQL 优化与 NoSQL 技术；了解 JDBC 技术，可以完善 Mycat Server 中的 JDBC 驱动部分；了解 Java I/O、多线程、算法；了解连接池、线程池、缓冲池等概念；了解 Java 中的 BIO、NIO、AIO 等概念；了解 SQL 解析原理。如果有条件，则还可以深入学习 JVM 相关技术；如果想参与 Mycat Web 开发，则还需要掌握 Java Web 技术。

2.4.1　Mycat 源码环境搭建

首先，安装 GitHub 管理工具、JDK1.7+、Maven、Eclipse，如果操作系统是 x64 位的，则需要安装 64 位的 JDK 和 Eclipse。

可以利用 Eclipse 的 Git 插件直接获取源码（https://github.com/MycatApache/Mycat-Server.git）到 Eclipse 中，也可以利用 Git 工具先获取源码到本地，再导入 Eclipse。

因为 Maven 需要下载项目依赖的 jar 包，所以第一次导入项目时等待的时间较长。

2.4.2　Mycat 源码调试

Mycat 运行的 main class 是 MycatStartup，在调试前需要配置相关的启动参数，需要指定 Mycat_HOME 这个系统变量的值，可以为任意位置，不过一般指定为与源代码同级的目录。可以在 IDE 运行选项内配置 VM arguments，例如：

```
-DMycat_HOME=E:\Mycat\MycatGit\Mycat-Server\src\main
```

若启动报错，DirectBuffer 内存不够，则可以增加 JVM 系统参数：

```
-DMycat_HOME=E:\Mycat\MycatGit\Mycat-Server\src\main -XX:MaxDirectMemorySize=
2048M
```

设置好以上环境后，将后端数据库配置好，用鼠标右键单击 org.opencloudb.MycatStartup→run as，Mycat 就能启动了。下面是 Mycat 中各包的情况。

```
org
 │      └─opencloudb
 │           ├─backend #后端连接
 │           ├─buffer #缓冲池包
 │           ├─cache #缓存
 │           │  └─impl
 │           ├─classloader
 │           ├─config #配置管理
 │           │  ├─loader
 │           │  │  ├─xml
 │           │  │  └─zookeeper
 │           │  │      ├─entitiy
 │           │  │      └─loader
 │           │  ├─model
 │           │  │  └─rule
 │           │  └─util
 │           ├─exception #异常
 │           ├─handler #IO haddler
 │           ├─heartbeat 心跳包
 │           ├─interceptor
 │           │  └─impl
 │           ├─jdbc  后端实现 MongoDB 及 SequoiaDB
 │           │  ├─mongodb
 │           │  └─sequoiadb
 │           ├─manager
 │           ├─mpp
 │           │  ├─model
 │           │  └─tmp
 │           ├─MySQL#NIO
 │           │  └─nio
 │           │      └─handler
 │           ├─net #AIO NIO
 │           │  ├─factory
 │           │  ├─handler
 │           │  ├─MySQL
 │           │  └─postgres
 │           ├─parser #解析器
 │           │  ├─druid
 │           │  │  └─impl
 │           │  └─util
 │           ├─postgres #后端实现 PostgreSQL
 │           │  └─handler
 │           ├─response
 │           ├─route #路由
 │           │  ├─config
 │           │  ├─factory
 │           │  ├─function
```

```
|           │  ├─handler
|           │  ├─impl
|           │  └─util
|           ├─sequence
|           │  └─handler
|           ├─server #server connection
|           │  ├─handler
|           │  ├─parser
|           │  ├─response
|           │  └─util
|           ├─sharejoin
|           ├─sqlcmd
|           ├─sqlengine #SQL 作业等
|           ├─stat
|           ├─statistic
|           └─util #公共类
|               ├─cmd
|               └─rehasher
```

第 3 章 Mycat 进阶

本章主要讲解 Mycat 的配置、分片规则和管理命令等内容。

3.1 Mycat 配置详解

熟悉 Mycat 配置是使用 Mycat 的基础，也是熟练掌握 Mycat 的必经之路。schema.xml 是 Mycat 的重要配置文件，它管理着逻辑库、分片表、分片节点和分片主机等信息；server.xml 是系统参数的配置文件，要掌握 Mycat 的优化方法，则必须熟悉该文件的配置项；sequence 是全局序列的配置文件；log4j.xml 是 Mycat 的日志输出配置文件。

3.1.1 Mycat 支持的两种配置方式

Mycat 从 1.5 版本开始支持两种配置方式：ZooKeeper 及本地 XML 方式。Mycat 默认以本地加载 XML 的方式启动，如果需要配置成以 ZooKeeper 的方式启动，则应把 conf 目录下 zk.conf 文件中的 loadfromzk 参数设置成 true。ZooKeeper 配置方式解决了统一配置和管理的问题，同时解决了 Mycat 与周边组件的协调问题。

如果使用 ZooKeeper 配置方式，则所有的配置参数都在 zk-create.yaml 文件中，该文件涵盖了 conf 目录下的 schema、server 等文件中的所有配置项（见 3.1.5 节）。

ZooKeeper 的配置方式依赖于 ZooKeeper，所以需要先安装 ZooKeeper 并初始化其中的数据，然后执行 Mycat_HOME/bin 目录下初始化 ZooKeeper 的脚本。

3.1.2　server.xml 配置文件

server.xml 配置文件包含了 Mycat 的系统配置信息，对应的源码是 SystemConfig.java。它有两个重要的标签，分别是 user、system，掌握 system 标签的各项配置属性是 Mycat 调优的关键。

1. user 标签

```
<user name="test">
  <property name="password">test</property>
  <property name="schemas">TESTDB</property>
  <property name="readOnly">true</property>
  <property name="benchmark">11111</property>
  <property name="usingDecrypt">1</property>
</user>
```

server.xml 中的标签并不多，user 标签主要用于定义登录 Mycat 的用户和权限。在如上配置中定义了用户名和密码都为 test 的用户，该用户可以访问的 schema 只有 TESTDB。

若要在 schema.xml 中定义 TESTDB，则 TESTDB 必须先在 server.xml 中定义，否则该用户将无法访问该 TESTDB。如果使用了 use 命令，则 Mycat 会有如下错误提示：

```
ERROR 1044 (HY000): Access denied for user 'test' to database 'xxx'
```

我们可以修改 user 标签的 name 属性来指定用户名，修改 password 的值来修改密码，修改 readOnly 的值为 true 或 false 来限制用户的读写权限。如果需要同时访问多个 schema，则多个 schema 之间使用英文逗号隔开，例如：

```
<property name="schemas">TESTDB,db1,db2</property>
```

1）benchmark 属性

通过设置 benchmark 属性的值来限制前端的整体连接数量，如果其值为 0 或不对其进行设置，则表示不限制连接数量。例如：

```
<property name="benchmark">1000</property>
```

2）usingDecrypt 属性

通过设置 usingDecrypt 属性的值来开启密码加密功能。默认值为 0，表示不开启加密；值为 1 表示开启加密，同时使用加密程序对密码加密，加密命令如下：

```
java -cp Mycat-server-1.5.1-RELEASE.jar org.opencloudb.util.DecryptUtil
0:user:password
```

说明：采用 Mycat 1.5.1 执行 mycat jar 程序；Mycat-server-1.5.1-RELEASE.jar 为 mycat download 下载目录的 jar 文件。

2. system 标签

system 标签与系统配置有关，下面一一介绍该标签的所有属性。

1）charset 属性

```
<system><property name="charset">utf8</property></system>
```

如果需要配置特殊字符集如 utf8mb4，则可以在 index_to_charset.properties 中配置，配置格式为“ID=字符集”，例如“224=utf8mb4”。

配置字符集时一定要保证 Mycat 字符集与数据库字符集的一致性，可以通过变量来查询 MySQL 字符集：

```
show variables like 'collation_%';
show variables like 'character_set_%';
```

2）defaultSqlParser 属性

由于 Mycat 的最初版本使用了 Foundation DB 的 SQL 解析器，在 Mycat 1.3 后增加了 Druid 解析器，所以要使用 defaultSqlParser 属性来指定默认的解析器。目前有两个可用的选项：druidparser 和 fdbparser。Mycat 1.3 默认的解析器为 fdbparser；Mycat 1.4 默认的解析器为 druidparser；在 Mycat 1.4 之后的版本中 fdbparser 解析器作废。

3）processors 属性

processors 属性指定系统可用的线程数量，默认值为机器 CPU 核心×每个核心运行线程的数量，processors 值会影响 processorBufferPool、processorBufferLocalPercent、processorExecutor 属性。NIOProcessor 的个数也由 processors 属性决定，所以调优时可以适当地修改 processors 值。

4）processorBufferChunk 属性

processorBufferChunk 属性指定每次分配 Socket Direct Buffer 的默认值为 4096 个字节，也会影响 BufferPool 的长度，如果一次性获取的字节过多而导致 Buffer 不够用，则会经常出现警告，可以适当调大 processorBufferChunk 值。

5）processorBufferPool 属性

processorBufferPool 属性指定 BufferPool 的计算比例。由于每次执行 NIO 读、写操作都需要使用到 Buffer，所以 Mycat 初始化时会建立一定长度的 Buffer 池来加快 NIO 读、写效率，减少建立 Buffer 的时间。

Mycat 中有两个主要的 Buffer 池：BufferPool、ThreadLocalPool。BufferPool 使用 ThreadLocalPool 作为二级缓存，每次从 BufferPool 中获取时都会优先获取 ThreadLocalPool 中 Buffer 的值，如果 ThreadLocalPool 未命中，则会获取 BufferPool 中 Buffer 的值。ThreadLocalPool 中的 Buffer 在每个线程内部使用。然而，BufferPool 是每个 NIOProcessor 共享的。

processorBufferPool 的默认值为：bufferChunkSize(4096)×processors×1000。

如图 3-1 所示，BufferPool 的总长度为 BufferPool 与 BufferChunk 的比。如果 BufferPool 的长度不是 BufferChunk 的整数倍，则其总长度为前面计算得出的比值的整数部分加 1。

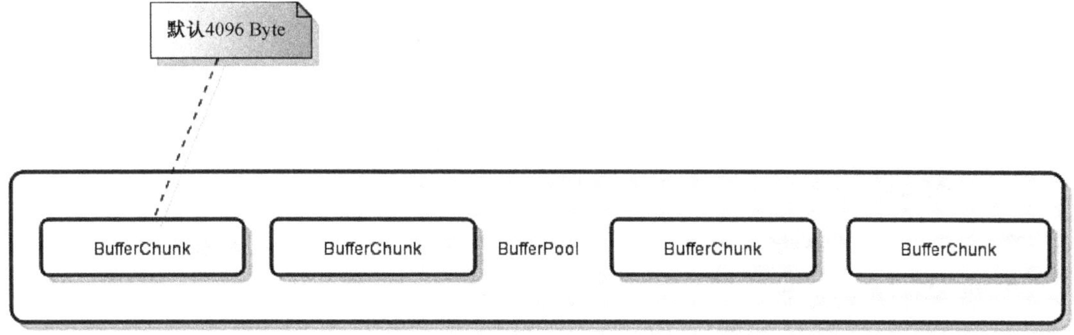

图 3-1

假设系统的线程数为 4，其他都为属性的默认值，则 BufferPool=4096×4×1000，BufferPool 的总长度为 4000，即 16384000/4096。

6）processorBufferLocalPercent 属性

processorBufferLocalPercent 属性用来控制 ThreadLocalPool 分配 Pool 的比例大小，这个属性的默认值为 100。

线程缓存百分比=bufferLocalPercent/processors。

例如，系统可以同时运行 4 个线程，使用默认值，根据上面的计算公式得出每个线程的百分比为 25，最后根据这个百分比可以计算出具体的 ThreadLocalPool 的长度公式如下：

ThreadLocalPool 长度=线程缓存百分比×BufferPool 长度/100

假设 BufferPool 的长度为 4000，其他保持默认值，则最后每个线程的 ThreadLocalPool 长度为 1000，即 25×4000/100。

7）processorExecutor 属性

processorExecutor 属性用于指定 NIOProcessor 上共享 businessExecutor 固定线程池的大小，Mycat 把异步处理任务提交到 businessExecutor 线程池中。在最新版本的 Mycat 中，这个连接池的使用频率不高，可以适当地把该值调小。

8）sequnceHandlerType 属性

sequnceHandlerType 属性指定 Mycat 全局序列的类型：0 为本地文件方式；1 为数据库方式；2 为时间戳序列方式。默认使用本地文件方式，文件方式主要用于测试。

9）TCP 连接的相关属性

- StandardSocketOptions.SO_RCVBUF
- StandardSocketOptions.SO_SNDBUF
- StandardSocketOptions.TCP_NODELAY

以上三个 TCP 的属性，衍生出针对前端和后端的 TCP 属性如下。

- frontSocketSoRcvbuf：默认值为 1024 * 1024。
- frontSocketSoSndbuf：默认值为 4 * 1024 * 1024。
- frontSocketNoDelay：默认值为 1。
- backSocketSoRcvbuf：默认值为 4 * 1024 * 1024。
- backSocketSoSndbuf：默认值为 1024 * 1024。
- backSocketNoDelay：默认值为 1。

Mycat 在每次建立前、后端连接时都会使用这些参数初始化 TCP 的属性。可以根据系统的实际情况适当地调整这些参数的大小，关于 TCP 连接参数的详细定义，请查看 Java API 文档（http://docs.oracle.com/javase/7/docs/api/java/net/StandardSocketOptions.html）。

10）MySQL 连接的相关属性

初始化 MySQL 前后端连接所涉及的一些属性如下。

- packetHeaderSize：指定 MySQL 协议中的报文头长度，默认值为 4 个字节。
- maxPacketSize：指定 MySQL 协议可以携带的数据的最大大小，默认值为 16MB。
- idleTimeout：指定连接的空闲时间的超时长度。如果某个连接的空闲时间超过 idleTimeout 的值，则该连接将关闭资源并回收，单位为毫秒，默认为 30 分钟。
- charset：初始化连接字符集，默认为 utf8。
- txIsolation：初始化前端连接事务的隔离级别，后续的 txIsolation 值为客户端的配置值。默认值为 REPEATED_READ，对应的数字为 3。

```
READ_UNCOMMITTED = 1;
READ_COMMITTED = 2;
REPEATED_READ = 3;
SERIALIZABLE = 4;
```

- sqlExecuteTimeout：执行 SQL 语句的超时时间，若 SQL 语句的执行时间超过这个值，则会直接关闭连接，单位为秒，默认值为 300 秒。

11）心跳属性

- processorCheckPeriod：清理 NIOProcessor 前后端空闲、超时、关闭连接的时间间隔，单位为毫秒，默认为 1 秒。
- dataNodeIdleCheckPeriod：对后端连接进行空闲、超时检查的时间间隔，单位为毫秒，默认为 300 秒。
- dataNodeHeartbeatPeriod：对后端的所有读、写库发起心跳的间隔时间，单位为毫秒，默认为 10 秒。

12）服务相关属性

这里介绍与 Mycat 服务相关的属性，主要影响外部系统对 Mycat 的感知。

- bindIp:Mycat：服务监听的 IP 地址，默认值为 0.0.0.0。
- serverPort：定义 Mycat 的使用端口，默认值为 8066。
- managerPort：定义 Mycat 的管理端口，默认值为 9066。

13）fakeMySQLVersion 属性

Mycat 使用 MySQL 的通信协议模拟了一个 MySQL 服务器，默认为 5.6 版本，一般不要修改其值，目前支持设置 5.5、5.6 版本，其他版本可能会有问题。Mycat 从 1.6 版本开始支持此属性。

14）分布式事务开关属性

handleDistributedTransactions 是分布式事务开关：0 为不过滤分布式事务；1 为过滤分布式事务（如果分布式事务内只涉及全局表，则不过滤）；2 为不过滤分布式事务，但是记录分布式事务日志。主要用于控制是否允许跨库事务，配置如下：

```
<property name="handleDistributedTransactions">0</property>
```

Mycat 从 1.6 版本开始支持此属性。

15）useOffHeapForMerge 属性

该属性用于配置是否启用非堆内存处理跨分片结果集，1 为开启，0 为关闭，Mycat 从 1.6 版本开始支持此属性。配置如下：

```
<property name="useOffHeapForMerge">1</property>
```

16）全局表一致性检测

其原理是通过在全局表中增加_MYCAT_OP_TIME 字段来进行一致性检测，为 BIGINT 类型。create 语句通过 Mycat 执行时会自动加上这个字段，其他情况下需要手工添加。1 为开启、0 为关闭，Mycat 从 1.6 版本开始支持此属性。配置如下：

```
<property name="useGlobleTableCheck">0</property>
```

全局表一致性检测功能的使用说明及步骤如下。

（1）在所有全局表中增加一个 BIGINT 类型的内部列，列名为_mycat_op_time（alter table t add column_mycat_op_time bigint [not null default 0]），同时建议在该列建立索引（alter table t add index _op_idx(_mycat_op_time)）。

（2）在对全局表进行 crud 时，可以将内部列当作不存在，建议不要对内部列进行 update、insert 等操作，否则会在 Log 日志中出现警告语句“不用操作内部列”。

（3）因为全局表中多了一个内部列，所以在对全局表进行 insert 时必须带有列名，意味着 SQL 插入的语句必须是 insert into t(id,name) values(xx,xx)，而不能使用 insert into t values(xx,xx)，否则会报列数不对的异常。这样的操作可能给开发工程师带来不便，将来会改善这个问题。

17）useSqlStat 属性

开启 SQL 实时统计，1 为开启、0 为关闭。配置如下：

```
<property name="useSqlStat">0</property>
```

3.1.3 schema.xml 配置文件

schema.xml 作为 Mycat 中重要的配置文件之一，涵盖了 Mycat 的逻辑库、表、分片规则、分片节点及数据源。

1. schema 标签

```
<schema name="TESTDB" checkSQLschema="false" sqlMaxLimit="100">
</schema>
```

schema 标签用于定义 Mycat 实例中的逻辑库。Mycat 可以有多个逻辑库，每个逻辑库都有自己的相关配置。可以使用 schema 标签来划分不同的逻辑库，如果不配置 schema 标签，则所有的表配置都会属于同一个默认的逻辑库。

```
    <schema name="TESTDB" checkSQLschema="false" sqlMaxLimit="100">
<table name="travelrecord" dataNode="dn1,dn2,dn3" rule="auto-sharding-long"
></table>
    </schema>
    <schema name="USERDB" checkSQLschema="false" sqlMaxLimit="100">
    <table  name="company" dataNode="dn10,dn11,dn12" rule="auto-sharding-long">
</table>
    </schema>
```

如上所示配置了两个不同的逻辑库，逻辑库的概念等同于 MySQL 数据库中的 Database 概念，我们在查询逻辑库中的表时，需要切换到该逻辑库下才可以查询其中的表。

```
MySQL > use schema1;
```

schema 标签的相关属性如表 3-1 所示。

表 3-1

属　性　名	值	数 量 限 制
dataNode	任意 String	(0..1)
checkSQLschema	Boolean	(1)
sqlMaxLimit	Integer	(1)

1）dataNode 属性

该属性用于绑定逻辑库到具体的 Database 上，Mycat 1.3 如果配置了 dataNode，则不可以配置分片表；Mycat 的 1.4 和 1.5 版本可以配置默认分片，配置需要分片的表即可。

Mycat 1.3 版本的配置如下：

```
<schema name="USERDB" checkSQLschema="false" sqlMaxLimit="100" dataNode="dn1">
    <!--里面不能配置任何表-->
</schema>
```

Mycat 的 1.4 和 1.5 版本的配置如下：

```
<schema name="USERDB" checkSQLschema="false" sqlMaxLimit="100" dataNode="dn2">
    <!--配置需要分片的表-->
<table name="tuser" dataNode= "dn1"/>
</schema>
```

如 Mycat 1.5 的配置所示，tuser 表绑定到 dn1 所配置的具体的 database 上，可以直接访问这个 database，没有配置的表则会走默认的节点 dn2。需要注意，用 navicat 工具不能查询到没有配置在分片规则里的表（如默认的表），但是程序可以正常使用。

2）checkSQLschema 属性

当该值设置为 true 时，如果我们执行语句**select * from TESTDB.travelrecord;**，则 Mycat 会把 schema 字符去掉，把 SQL 语句修改为**select * from travelrecord;**可避免发送到后端数据库执行时报“**(ERROR 1146 (42S02): Table 'testdb.travelrecord' doesn't exist)”错误。

不过，即使设置该值为 true，如果语句所带的 schema 名字不是 schema 指定的名字，例如**select * from db1.travelrecord;**，那么 Mycat 并不会删除 db1 这个字符串。如果没有定义该库，则会报错，在 SQL 语句中最好不带这个字段。

3）sqlMaxLimit 属性

当该属性设置为某个数值时，每次执行的 SQL 语句如果没有加上 limit 语句，MyCat 也会自动在 limit 语句后面加上对应的数值。例如设置值为 100，则执行**select*from TESTDB.travelrecord;**的效果和执行**select * from TESTDB.travelrecord limit 100;**的效果一样。

如果不设置该值，则 Mycat 默认会把查询到的信息全部返回，所以在正常使用的过程中还是建议设置该值，避免过多的数据返回。

当然，如果在 SQL 语句中也显式地指定了 limit 的大小则不受该属性的约束。需要注意的是，如果运行 SQL 语句的 schema 为非拆分库的，那么该属性不会生效，需要手动在 SQL 语句后面添加 limit。

2. table 标签

```
<table name="travelrecord" dataNode="dn1,dn2,dn3" rule="auto-sharding-long">
</table>
```

table 标签定义了 Mycat 中的逻辑表，所有需要拆分的表都需要在 table 标签中定义。

table 标签的相关属性如表 3-2 所示。

表 3-2

属 性 名	值	数 量 限 制
name	String	(1)
dataNode	String	(1..*)
rule	String	(0..1)
ruleRequired	boolean	(0..1)
primaryKey	String	(1)
type	String	(0..1)
autoIncrement	boolean	(0..1)
subTables	String	(1)
needAddLimit	boolean	(0..1)

1）name 属性

定义逻辑表的名称，如同我们在数据库中执行 create table 语句的表名一样，同一个 schema 标签中定义的 table 的名字必须唯一。

2）dataNode 属性

定义逻辑表所属的 dataNode，该属性的值需要与 dataNode 标签中 name 属性的值相互对应。如果需要定义过多的 dn，则可以使用如下方法减少配置：

```
<table name="travelrecord" dataNode="multipleDn$0-99,multipleDn2$100-199" rule
="auto-sharding-long" >
```

```
</table>
<dataNode name="multipleDn" dataHost="localhost1" database="db$0-99" >
</dataNode>
<dataNode name="multipleDn2" dataHost="localhost1" database=" db$0-99" >
</dataNode>
```

这里需要注意的是 database 属性所指定的真实数据库的名称。在上面的例子中，需要在真实的 MySQL 上建立名为 db0～db99 的数据库。

3）rule 属性

该属性用于指定逻辑表要使用的规则的名字，规则的名字在 rule.xml 中定义，必须与 tableRule 标签中 name 属性的值一一对应。

4）ruleRequired 属性

该属性用于指定表是否绑定分片规则，如果配置为 true，但没有配置具体的 rule，则程序会报错。

5）primaryKey 属性

逻辑表对应真实表的主键，例如：分片的规则是使用非主键进行分片，那么在使用主键查询时，就会发送查询语句到所有配置的 dn 上；如果使用该属性配置真实表的主键，那么 Mycat 会缓存主键与具体 dn 的信息，再次使用主键进行查询时就不会进行广播式的查询了，而是直接把 SQL 语句发送到具体的 dn。但是尽管配置了该属性，如果缓存并没有命中，则还是会把该 SQL 语句发送给所有的 dn 执行来获得数据。

6）type 属性

该属性定义了逻辑表的类型，目前逻辑表只有“全局表”和“普通表”两种类型。

- 全局表：type 的值是 global，代表全局表。
- 普通表：不指定该值为 global 的所有表。

7）autoIncrement 属性

MySQL 对于非自增长主键使用 last_insert_id()是不会返回结果的，只会返回 0。所以，只有对定义了自增长主键的表使用 last_insert_id()才可以返回主键的值。Mycat 目前提供了自增长主键功能，但是如果对应的 MySQL 节点上的表没有定义 auto_increment，那么在 Mycat 层调用 last_insert_id()也是不会返回结果的。

由于 insert 操作时没有带入分片键，所以 Mycat 会先取下这个表对应的全局序列，然后赋值给分片键。

如果要使用这个功能，则最好配合数据库模式的全局序列。使用 autoIncrement="true"指定这个表使用自增长主键，这样 Mycat 才不会抛出“分片键找不到”的异常。使用 autoIncrement="false"来禁用这个功能，autoIncrement 的值默认为 false。

8）subTables 属性

使用 subTables="t_order$1-2,t_order3"方式添加，目前 Mycat 在 1.6 版本以后才开始支持分表，并且 dataNode 在分表条件下只能配置一个，不支持各种条件的 Join 关联查询语句。

9）needAddLimit 属性

指定表是否需要自动在每个语句的后面加上 limit 限制。由于使用了分库分表，所以数据量有时会特别大。如果恰巧忘记加上数量限制，那么查询所有的数据需要一定的时间。

所以，添加该属性后 Mycat 将会自动为我们在查询语句后面加上 LIMIT 100。如果语句中有 limit 限制，则不会重复添加。该属性默认为 true，你也可以把该值设置为 false 来禁用默认的行为。

3. childTable 标签

childTable 标签用于定义 E-R 分片的子表，通过标签上的属性与父表进行关联。

childTable 标签的相关属性如表 3-3 所示。

表 3-3

属 性 名	值	数 量 限 制
name	String	(1)
joinKey	String	(1)
parentKey	String	(1)
primaryKey	String	(0..1)
needAddLimit	boolean	(0..1)

1）name 属性

定义子表的名称。

2）joinKey 属性

插入子表时会使用这个值查找父表存储的数据节点。

3）parentKey 属性

parentKey 为与父表建立关联关系的列名。程序首先获取 joinKey 的值，再通过 parentKey 属性指定的列名产生查询语句，通过执行该语句得知父表存储在哪个分片上，从而确定子表存储的位置。

4）primaryKey 属性

同 table 标签所描述的。

5）needAddLimit 属性

同 table 标签所描述的。

4. dataNode 标签

dataNode 标签定义了 Mycat 中的数据节点，也就是我们通常所说的数据分片。一个 dataNode 标签就是一个独立的数据分片。

如下所示为使用名为 lch3307 的数据库实例上的 db1 物理数据库组成一个数据分片，我们通过名字 dn1 标识这个分片。

```
<dataNode name="dn1" dataHost="lch3307" database="db1" ></dataNode>
```

dataNode 标签的相关属性如表 3-4 所示。

表 3-4

属 性 名	值	数 量 限 制
name	String	(1)
dataHost	String	(1)
database	String	(1)

1）name 属性

定义数据节点的唯一名字，我们需要在 table 标签上应用这个名字，来建立表与分片的对应关系。

2）dataHost 属性

该属性用于定义该分片所属的数据库实例，属性值引用自 dataHost 标签上定义的 name 属性。

3）database 属性

该属性用于定义该分片所属数据库实例上的具体的库，这里使用两个维度来定义分片：实例+具体的库。因为每个库上的表结构是一样的，所以这样就可以轻松地对表进行水平拆分。

5. dataHost 标签

作为 schema.xml 中的最后一个标签，该标签在 Mycat 逻辑库中作为底层标签存在，直接定义了具体的数据库实例、读写分离和心跳语句。

```
<dataHost name="localhost1" maxCon="1000" minCon="10" balance="0"
    writeType="0" dbType="MySQL" dbDriver="native">
    <heartbeat>select user()</heartbeat>
    <!-- can have multi write hosts -->
    <writeHost host="hostM1" url="localhost:3306" user="root"
      password="123456">
      <!-- can have multi read hosts -->
      <!-- <readHost host="hostS1" url="localhost:3306" user="root"
password="123456"
        /> -->
    </writeHost>
    <!-- <writeHost host="hostM2" url="localhost:3316" user="root"
password="123456"/> -->
</dataHost>
```

dataHost 标签的相关属性如表 3-5 所示。

表 3-5

属 性 名	值	数 量 限 制
name	String	(1)
maxCon	Integer	(1)
minCon	Integer	(1)
balance	Integer	(1)
writeType	Integer	(1)
dbType	String	(1)
dbDriver	String	(1)

1）name 属性

唯一标识 dataHost 标签，供上层标签使用。

2）maxCon 属性

指定每个读写实例连接池的最大连接数。内嵌标签 writeHost、readHost 都会使用这个属性的值来实例化连接池的最大连接数。

3）minCon 属性

指定每个读写实例连接池的最小连接数，初始化连接池的大小。

4）balance 属性

负载均衡类型，目前的取值有如下 4 种。

- balance="0"：不开启读写分离机制，所有读操作都发送到当前可用的 writeHost 上。
- balance="1"：全部的 readHost 与 stand by writeHost 都参与 select 语句的负载均衡，简

而言之，当为双主双从模式（M1→S1，M2→S2，并且 M1 与 M2 互为主备）时，在正常情况下，M2、S1 和 S2 都参与 select 语句的负载均衡。

- balance="2"：所有的读操作都随机地在 writeHost、readHost 上分发。
- balance="3"：所有的读请求都随机分发到 writeHost 对应的 readHost 上执行，writeHost 不负担读压力，注意 balance=3 只在 Mycat 1.4 及之后的版本中有，在 Mycat 1.3 中没有。

5）writeType 属性

负载均衡类型目前的取值有两种。

- writeType="0"：所有的写操作都发送到配置的第 1 个 writeHost 上，writeHost1 挂了则切到 writeHost2 上，重新恢复 writeHost1 节点后，不会再切回来，还是以 writeHost2 为准，切换记录在配置文件 dnindex.properties 中。
- writeType="1"：所有的写操作都随机地发送到配置的 writeHost 上，Mycat 1.5 版本以后不再推荐使用该值。

6）dbType 属性

指定后端连接的数据库类型，目前除了支持二进制的 MySQL 协议，还支持使用 JDBC 连接的数据库，例如 MongoDB、Oracle 等。

7）dbDriver 属性

指定连接后端数据库使用的 Driver，目前可选的值有 native 和 JDBC。因为 native 执行的是二进制的 MySQL 协议，所以可以使用 MySQL 和 MariaDB。其他类型的数据库则需要使用 JDBC 驱动来支持，如果使用 JDBC 方式，则需要将符合 JDBC4 标准的驱动 jar 包放到 MYCAT\lib 目录下。

8）switchType 属性

- -1 表示不自动切换。
- 1 为默认值，表示自动切换。
- 2 表示基于 MySQL 主从同步的状态决定是否切换，心跳语句如下：

```
Show slave status
```

- 3 表示基于 MySQL Galary Cluster 的切换机制（适合集群，Mycat 1.4.1 及以上版本支持），心跳语句如下：

```
show status like 'wsrep%'
```

9）tempReadHostAvailable 属性

如果配置了 writeHost 属性，下面的 readHost 依旧可用，则默认值为 0。

6. heartbeat 标签

这个标签内指明了用于后端数据库进行心跳检查的语句。例如，MySQL 可以使用 select user()，Oracle 可以使用 select 1 from dual 等。

这个标签还有 connectionInitSql 属性，当使用 Oracle 时，需要执行的初始化 SQL 语句会放到这里。例如 alter session set nls_date_format='yyyy-mm-dd hh24:mi:ss'。

Mycat 1.4 中主从切换的语句必须是 show slave status。

7. writeHost 标签、readHost 标签

这两个标签都指定 Mycat 后端数据库的相关配置，用于实例化后端连接池。唯一的不同是，writeHost 指定写实例，readHost 指定读实例，组成这些读写实例来满足系统的要求。

在一个 dataHost 内可以定义多个 writeHost 和 readHost。但是，如果 writeHost 指定的后端数据库宕机，那么这个 writeHost 绑定的所有 readHost 也将不可用；另一方面，Mycat 会自动检测到 writeHost 宕机，并切换到备用的 writeHost 上。

这两个标签的属性相同，如表 3-6 所示。

表 3-6

属 性 名	值	数 量 限 制
host	String	(1)
url	String	(1)
password	String	(1)
user	String	(1)
weight	String	(1)
usingDecrypt	String	(1)

1）host 属性

用于标识不同的实例，对于 writeHost，我们一般使用*M1；对于 readHost，我们一般使用*S1。

2）url 属性

后端实例的连接地址，如果使用 native 的 dbDriver，则一般为 address:port 形式；如果使用 JDBC 或其他 dbDriver，则需要特殊指定。在使用 JDBC 时，则可以写为 jdbc:MySQL://localhost:3306/。

3）user 属性

后端存储实例的用户名。

4）password 属性

后端存储实例的密码。

5）weight 属性

在 readHost 中作为读节点的权重（Mycat 在 1.4 版本以后才有）。

6）usingDecrypt 属性

同 server.xml 中 usingDecrypt 的配置。

3.1.4　sequence 配置文件

在实现分库分表的情况下，数据库自增主键已经无法保证在集群中是全局唯一的主键，因此，Mycat 提供了全局 sequence，并且提供了本地配置、数据库配置等多种实现方式。

1. 本地文件方式

采用该方式，Mycat 将 sequence 配置到 classpath 目录的 sequence_conf.properties 文件中。

在 sequence_conf.properties 文件中做如下配置：

```
GLOBAL_SEQ.HISIDS=
GLOBAL_SEQ.MINID=1001
GLOBAL_SEQ.MAXID=1000000000
GLOBAL_SEQ.CURID=1000
```

其中 HISIDS 表示使用过的历史分段（一般无特殊需要则可不配置），MINID 表示最小的 ID 值，MAXID 表示最大的 ID 值，CURID 表示当前的 ID 值。

要启用这种方式，则首先需要在 server.xml 中配置如下参数：

```
<system><property name="sequnceHandlerType">0</property></system>
```

注意：sequnceHandlerType 配置为 0，表示使用本地文件方式。

```
insert into table1(id,name) values(next value for MYCATSEQ_GLOBAL,'test');
```

采用这种方式的缺点是 Mycat 重新发布后，配置文件中的 sequence 会恢复到初始值；优点是本地加载且读取速度较快。

2. 数据库方式

在数据库中创建一张名为 sequence 的表，有 sequence 的当前值（current_value）、步长（increment int 类型，指每次读取多少个 sequence，假设为 *K*）等信息。

sequence 的获取步骤如下。

（1）初次使用 sequence 时，根据传入的 sequence 名称，从数据库表中读取 current_value、increment 到 Mycat 中，并将数据库中的 current_value 修改为 current_value+increment 的值。

（2）Mycat 将读取到的 current_value+increment 作为本次使用的 sequence 值，在下次使用时，sequence 自动加 1，当使用 increment 次后，执行与步骤 1 相同的操作。

（3）Mycat 负责维护这张表，用到那些 sequence 时，只需要在这张表中插入一条记录即可。若某次读取的 sequence 没有用完系统就宕机了，则本次已经读取 sequence 且未使用的值将会被丢弃。

要启用这种方式，则需要在 server.xml 中配置如下参数：

```
<system><property name="sequnceHandlerType">1</property></system>
```

注意：sequnceHandlerType 需要配置为 1，表示使用数据库方式生成 sequence。

数据库配置如下。

（1）创建存放 MYCAT_SEQUENCE 的表：

```
CREATE TABLE MYCAT_SEQUENCE (name VARCHAR(50) NOT NULL,current_value
INT NOT NULL,increment INT NOT NULL DEFAULT 100, PRIMARY KEY(name))
ENGINE=InnoDB;
```

name、current_value 和 increment 分别是 sequence 的名称、当前 value 和增长步长。increment 可理解为 Mycat 从数据库中批量读取 100 个（默认值）sequence 来使用，用完这些值后，再从数据库中读取。

插入一条 sequence 语句：

```
INSERT INTO MYCAT_SEQUENCE(name,current_value,increment) VALUES
('GLOBAL', 100000, 100);
```

（2）创建相关的 function。

```
-获取当前的 sequence 值
DROP FUNCTION IF EXISTS mycat_seq_currval;
DELIMITER
CREATE FUNCTION mycat_seq_currval(seq_name VARCHAR(50)) RETURNS
varchar(64) CHARSET utf-8
DETERMINISTIC
BEGIN
DECLARE retval VARCHAR(64);
SET retval="-999999999,null";
SELECT concat(CAST(current_value AS CHAR),",",CAST(increment AS CHAR))
INTO retval FROM MYCAT_SEQUENCE WHERE name = seq_name;
RETURN retval;
END
```

```
DELIMITER;
 - 设置 sequence 值
DROP FUNCTION IF EXISTS mycat_seq_setval;
DELIMITER
CREATE FUNCTION mycat_seq_setval(seq_name VARCHAR(50),value INTEGER)
RETURNS varchar(64) CHARSET utf-8
DETERMINISTIC
BEGIN
UPDATE MYCAT_SEQUENCE
SET current_value = value
WHERE name = seq_name;
RETURN mycat_seq_currval(seq_name);
END
DELIMITER;
 -获取下一个 sequence 值
DROP FUNCTION IF EXISTS mycat_seq_nextval;
DELIMITER
CREATE FUNCTION mycat_seq_nextval(seq_name VARCHAR(50)) RETURNS
varchar(64) CHARSET utf-8
DETERMINISTIC
BEGIN
UPDATE MYCAT_SEQUENCE
SET current_value = current_value + increment WHERE name = seq_name;
RETURN mycat_seq_currval(seq_name);
END
DELIMITER;
```

（3）进行与 sequence_db_conf.properties 相关的配置，指定与 sequence 相关的配置所在的节点，例如 USER_SEQ=test_dn1。

注意：MYCAT_SEQUENCE 表和以上 3 个 function 需要放在同一个节点上。function 直接在具体节点的数据库上执行，如果执行时报“you might want to use the less safe log_bin_trust_function_creators variable”，则需要对数据库做如下设置。

- 在 Windows 环境下：`my.ini[MySQLd]+log_bin_trust_function_creators=1`。
- 在 Linux 环境的/etc/my.cnf 目录下：`my.ini[MySQLd]+log_bin_trust_function_creators=1`。

修改完后，即可在 MySQL 中执行上面的函数。使用示例如下：

```
insert into table1(id,name) values(next value for MYCATSEQ_GLOBAL 'test');
```

3. 本地时间戳方式

ID=64 位二进制［42 位（毫秒）+5 位（机器 ID）+5 位（业务编码）+12 位（重复累加）］，换算成十进制为 18 位数的 long 类型，每毫秒可以并发 12 位二进制的累加。

使用方式如下。

（1）配置 server.xml：

```
<property name="sequnceHandlerType">2</property>
```

（2）配置 sequence_time_conf.properties 文件。

- WORKID=0～31，可取 0～31 中的任意整数。
- DATACENTERID=0～31，可取 0～31 中的任意整数。

多个 Mycat 节点下每个 Mycat 配置的 WORKID、DATACENTERID 都不一样，组成唯一标识，总共支持 32×32=1024 种组合。

ID 示例：56763083475511

4. 其他方式

（1）使用 catlet 注解方式：

```
/*!mycat:catlet=demo.catlets.BatchGetSequence */
SELECT mycat_get_seq('GLOBAL',100);
```

注意：此方法表示获取 GLOBAL 的 100 个 sequence 值，例如当前 GLOBAL 的最大 sequence 值为 5000，则通过此方式返回的是 5001，同时更新数据库中的 GLOBAL 的最大 sequence 值为 5100。

（2）也可以使用 ZooKeeper 方式实现。

5. 自增长主键

Mycat 自增长主键和返回生成主键 ID 的实现如下。

（1）MySQL 本身对非自增长主键使用 last_insert_id()只会返回 0。

（2）MySQL 对定义自增长的主键才可以用 last_insert_id()返回主键的值。

Mycat 目前提供了自增长主键功能，但是如果对应的 MySQL 节点上的数据表没有定义 auto_increment，那么在 Mycat 层调用 last_insert_id()也是不会返回结果的。

正确的配置方式如下。

（1）MySQL 定义自增长主键。

```
CREATE TABLE table1('id_' INT(10) UNSIGNED NOT NULL AUTO_INCREMENT,
'name_' INT(10) UNSIGNED NOT NULL,PRIMARY KEY ('id_'))
ENGINE=MYISAM AUTO_INCREMENT=6 DEFAULT CHARSET=utf8;
```

（2）Mycat 定义主键自增。

在 table 标签中增加 autoIncrement="true"：

```
<table name="A" primaryKey="id" autoIncrement="true"
dataNode="dn1,dn2" rule="mod-long">
<table>
```

（3）Mycat 对应 sequence_db_conf.properties 增加相应的设置。

```
TABLE1=dn1
```

（4）在数据库的 mycat_sequence 表中增加 TABLE1 表的 sequence 记录。

3.1.5　zk-create.yaml 配置文件

在介绍配置之前，先介绍几个概念。Mycat Zone 指的是分布于不同地域（Zone）的 Mycat Cluster，Zone 的命名建议用地理位置来标识，比如北京联通机房 1。Cluster 是 Mycat 集群，一个 Cluster 包含一个或多个 Mycat Server。一般来讲，一个 Zone 都有一组主备 Mycat 负载均衡器 LB，LB 与同一中心内的 Mycat Cluster 组成一对多关系，即一个 LB 可以服务一个中心内的所有 Cluster 的负载均衡请求，也可以是多个 LB，每个负担不同的 Mycat Cluster 的流量。此外，建议每个 LB 都有一个 Backup，Backup 平时并不连接 Mycat Cluster，但监测到 LB Master 下线以后，就立即开始连接 Mycat Cluster 并开始工作。它们的关系大概可以用一组箭头来表示：Zone→Mycat Cluster→Mycat Server→MySQL，如图 3-2 所示。

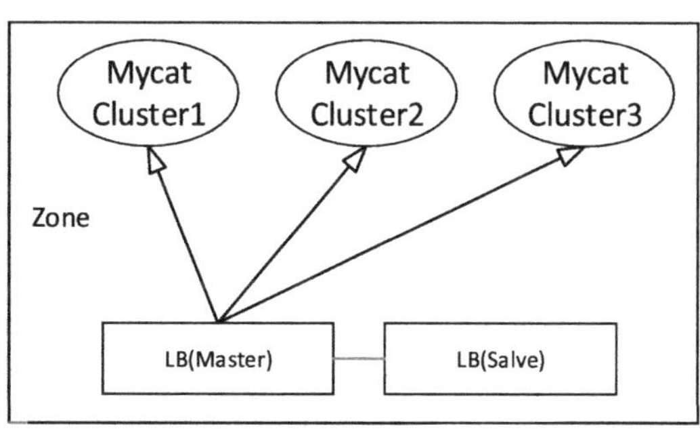

图 3-2

zk-create 的总体结构如下：

```
zkURL : 127.0.0.1:2181 ZooKeeper 的连接地址
+mycat-cluster: 集群中所有的主机信息
+mycat-hosts: 主机信息
+mycat-zones: Mycat 中心
+mycat-nodes: Mycat 节点
+mycat-MySQLs: MySQL 信息
```

```
+mycat-MySQLgroup: 复制组的信息
+mycat-lbs: mycat lb
```

下面分别介绍各部分的配置。

（1）mycat-zones 的配置如下：

```
mycat-zones:
  wh:
    name: 武汉中心
  fz:
    name: 福州中心
```

注意：wh、fz 为 zone 识别的名称，即 name 的备注名字。

（2）mycat-cluster 的配置如图 3-3 所示。

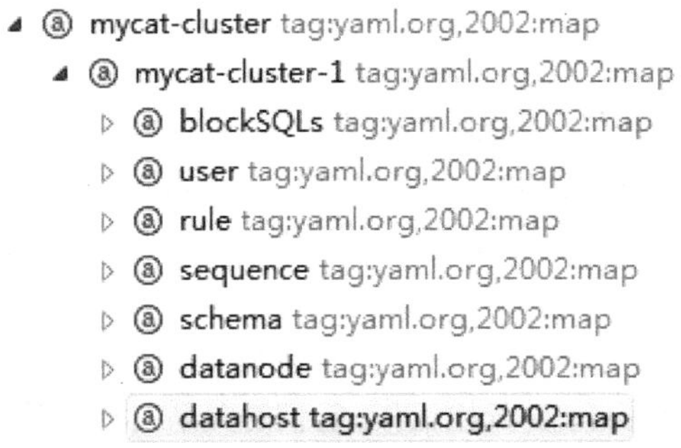

mycat-cluster tag:yaml.org,2002:map
mycat-cluster-1 tag:yaml.org,2002:map
blockSQLs tag:yaml.org,2002:map
user tag:yaml.org,2002:map
rule tag:yaml.org,2002:map
sequence tag:yaml.org,2002:map
schema tag:yaml.org,2002:map
datanode tag:yaml.org,2002:map
datahost tag:yaml.org,2002:map

图 3-3

```
mycat-cluster:
  mycat-cluster-1:
    +blockSQLs:
    +user:
    +rule:
    +sequence:
    +schema:
    +datanode:
    +datahost:
```

如上面的代码片段所示，blockSQLs 表示 SQL 名单，填写格式如下：

```
blockSQLs:
      sql1:
        name:sql1
      sql2:
        name:sql2
      sql3:
        name:sql3
```

- user 相当于 server.xml 的 user 标签。

- rule 是 rule.xml 的 function、tableRule 标签的合并。
- sequence 是前面介绍过的全局序列配置。
- schema 相当于 schema.xml 里的 schema 标签，与 datanode 类似。

（3）mycat-hosts 配置如下：

```
#集群中所有的主机信息
mycat-hosts:
  fz_vm1:
    hostname:fz_vm1
    ip:192.168.10.2
    root:root
    password:admin
```

注意：这里指的是操作系统级别的主机信息，分别为主机名、IP 地址、用户及密码。

（4）mycat-nodes 配置如下。Mycat 的节点配置记录了这个 Mycat 实例所属集群的 zone、主机名等信息，systemParams 是 server.xml 中的系统属性。

```
#zone 内的 Mycat 实例配置,名字为 mycat 实例的 myid.
mycat-nodes:
  mycat_fz_01:
    name: mycat_fz_01
    hostname: fz_vm1
    zone: fz
    cluster: mycat-cluster-1
    weigth: 1
    leader: 1
    state: red
    systemParams:
      defaultsqlparser : druidparser
      serverport : 8066
  sequncehandlertype : 1
```

（5）mycat-MySQLs 配置如下。这里配置 MySQL 的连接信息及所属主机、机房。

```
mycat-MySQLs:
  MySQL_1:
    ip: 127.0.0.1
    port: 3306
    user: root
    password: admin
    hostId: host
zone: bj
mycat-MySQLgroup :
  MySQL_rep_1:
    name: MySQL_rep_1
    repType: 0
```

```
    zone: bj
    servers:
      - MySQL_1
    cur-write-server: MySQL_1
    auto-write-switch: true
heartbeatSQL : select user()
```

（6）MySQL 组的配置如下：

```
mycat-MySQLgroup :
  MySQL_rep_1:
    name: MySQL_rep_1
    repType: 0
    zone: bj
    servers:
      - MySQL_1  #MySQL服务器
    cur-write-server: MySQL_1   #当前写节点
    auto-write-switch: true   #是否自动切换
heartbeatSQL : select user() #心跳语句
```

（7）Mycat 的负载均衡信息如下：

```
mycat-lbs:
    lb_v1:
      hostId: fz_vm1
      name: yyyyy
      ip: 192.168.10.2
      port: 9999
      zone: fz
      cluster: mycat-cluster-1
```

3.1.6 其他配置文件

1. 缓存文件配置

Mycat 支持 encache、mapdb、leveldb 缓存，可以通过配置文件 cacheservice.properties 决定使用哪种缓存框架。

```
#used for mycat cache service conf
factory.encache=org.opencloudb.cache.impl.EnchachePooFactory
#key is pool name ,value is type,max size, expire seconds
pool.SQLRouteCache=encache,10000,1800
pool.ER_SQL2PARENTID=encache,1000,1800
layedpool.TableID2DataNodeCache=encache,10000,18000
layedpool.TableID2DataNodeCache.TESTDB_ORDERS=50000,18000
```

第一项 factory.encache 指定缓存的实现类，不同的实现类对应不同的缓存框架，有如下两个选项：

- org.opencloudb.cache.impl.LevelDBCachePooFactory；
- org.opencloudb.cache.impl.MapDBCachePooFactory。

pool.SQLRouteCache=encache,10000,1800 指定了缓存框架、缓存大小、过期时间，后面三项的意义同 pool.SQLRouteCache，具体内容参见 6.4 节。

2. 日志配置

Mycat 使用 log4j 作为日志管理工具，log4j 为开发人员最熟悉不过的配置文件，这里对其进行简要介绍。

${MYCAT_HOME}/logs/mycat.log 为 Mycat 日志的输出路径，<level value="warn" />配置日志的输出级别。

```
<log4j:configuration xmlns:log4j="http://jakarta.apache.org/log4j/">
<appender name="ConsoleAppender" class="org.apache.log4j.ConsoleAppender">
  <layout class="org.apache.log4j.PatternLayout">
   <param name="ConversionPattern" value="%d{MM-dd HH:mm:ss.SSS}  %5p [%t]
(%F:%L) -%m%n" />
  </layout>
 </appender>
  <appender name="FILE" class="org.apache.log4j.RollingFileAppender">
    <param name="file" value="${MYCAT_HOME}/logs/mycat.log" />
    <param name="Append" value="false"/>
    <param name="MaxFileSize" value="10000KB"/>
    <param name="MaxBackupIndex" value="10"/>
    <param name="encoding" value="UTF-8" />
    <layout class="org.apache.log4j.PatternLayout">
      <param name="ConversionPattern" value="%d{MM/dd HH:mm:ss.SSS}  %5p [%t]
(%F:%L) -%m%n" />
    </layout>
  </appender>
  <root>
    <level value="warn" />
    <appender-ref ref="ConsoleAppender" />
     <!--<appender-ref ref="FILE" />-->
  </root>
```

3. AIO、NIO 的配置

在 server.xml 中 usingAIO 配置 I/O 类型：0 为 NIO、1 为 AIO。

```
<property name="usingAIO">0</property>
```

3.2 Mycat 分片规则详解

3.2.1 分片表与非分片表

Mycat 位于应用与数据库的中间层，可以灵活解耦应用与数据库，后端数据库可以位于不同的主机上。在 Mycat 中将表分为两种大的概念：对于数据量小且不需要做数据切分的表，称之为非分片表；对于数据量大到单库性能、容量不足以支撑，数据需要通过水平切分均匀分布到不同的数据库中的表，称之为分片表。而中间件最终需要处理的事情是对数据切分、聚合。下面介绍如何通过不同的规则对数据进行切分。

3.2.2 ER 关系分片表

ER 模型是实体关系模型，广泛采用概念模型设计方法，基本元素是实体、关系和属性。Mycat 创新性地将它引入数据切分规则中，使得有互相依赖的表能够按照某一规则切分到相同的节点上，避免跨库 Join 关联查询。

下面介绍一组模型。订单（order）、订单明细（order_detail）、明细表会依赖于订单，也就是说会存在表的主从关系，这种业务的切分可以设计出合适的切分规则，比如根据用户 ID 切分，其他相关的表都依赖于用户 ID 或者根据订单 ID 切分，总之部分业务可以抽象出父子关系的表。这类表适用于 ER 分片表，子表的记录与所关联的父表记录存放在同一个数据分片上，避免数据跨库 Join 操作。

以 order 与 order_detail 为例，在 schema.xml 中定义如下分片配置：order、order_detail。根据 order_id 进行数据切分，保证相同 order_id 的数据分到同一个分片上，在进行数据插入操作时，Mycat 会获取 order 所在的分片，然后将 order_detail 也插入 order 所在的分片。

```
    <table name="order" dataNode="dn$1-32" rule="mod-long">
    <childTable name="order_detail" primaryKey="id" joinKey="order_id"
parentKey="order_id" />
    </table>
```

3.2.3 分片规则 rule.xml 文件详解

rule.xml 位于$MYCAT_HOME/conf 目录，它定义了所有拆分表的规则。在使用的过程中可以灵活使用不同的分片算法，或者对同一个分片算法使用不同的参数，它让分片过程可配置化，只需要简单的几步就可以让运维人员及数据库管理员轻松将数据拆分到不同的物理库中。该文件包含两个重要标签，分别是 Function 和 tableRule，下面对它们进行讲解。

1. Function 标签

```
<function name="rang-mod" class=
"org.opencloudb.route.function.PartitionByRangeMod">
     <property name="mapFile">partition-range-mod.txt</property>
</function>
```

- name 属性指定算法的名称，在该文件中唯一。
- class 属性对应具体的分片算法，需要指定算法的具体类。
- property 属性根据算法的要求指定，后面详细说明。

2. tableRule 标签

```
<tableRule name="auto-sharding-rang-mod">
       <rule>
           <columns>id</columns>
           <algorithm>rang-mod</algorithm>
       </rule>
</tableRule>
```

- name 属性指定分片唯一算法的名称。
- rule 属性指定分片算法的具体内容，包含 columns 和 algorithm 两个属性。
- columns 属性指定对应的表中用于分片的列名。
- algorithm 属性对应 function 中指定的算法的名称。

3.2.4　取模分片

配置如下：

```
<tableRule name="mod-long">
       <rule>
           <columns>id</columns>
           <algorithm>mod-long</algorithm>
       </rule>
</tableRule>
<function name="mod-long" class=
"org.opencloudb.route.function.PartitionByMod">
       <!-- how many data nodes -->
       <property name="count">3</property>
</function>
```

配置说明如下。

- columns 用来标识将要分片的表字段。
- algorithm 指定分片函数与 function 对应。

此分片算法根据 id 进行十进制求模计算，相比固定的分片 hash，这种分片算法在批量插入时会增加事务一致性的难度。

3.2.5 枚举分片

通过在配置文件中配置可能的枚举 id，指定数据分布到不同物理节点上，本规则适用于按照省份或区县来拆分数据类业务。配置如下：

```
<tableRule name="sharding-by-intfile">
    <rule>
        <columns>sharding_id</columns>
        <algorithm>hash-int</algorithm>
    </rule>
</tableRule>
<function name="hash-int"
    class="org.opencloudb.route.function.PartitionByFileMap">
    <property name="mapFile">partition-hash-int.txt</property>
    <property name="type">0</property>
    <property name="defaultNode">0</property>
</function>
```

其中的 partition-hash-int.txt 内容如下：

```
10000=0
10010=1
DEFAULT_NODE=1
```

配置说明如下。

- columns 指定分片的表列名。
- algorithm 指定分片函数的名称，与 function 中 mapFile 配置文件标识的名称一致。
- type 的默认值为 0，0 表示 Integer，非零表示 String。

注意：所有的节点配置都是从 0 开始的，0 代表节点 1。

```
/**
* defaultNode 默认节点:小于 0 表示不设置默认节点，大于等于 0 表示设置默认节点
* 默认节点的作用：枚举分片时，如果碰到不识别的枚举值，就让它路由到默认节点
* 如果不配置默认节点（defaultNode 值小于 0 表示不配置默认节点），则碰到识别的枚举值就会报错
* like this: can't find datanode for sharding column:column_name val:ffffffff
*/
```

3.2.6　范围分片

适用于想明确知道分片字段的某个范围属于哪个分片时，配置如下：

```
<tableRule name="auto-sharding-long">
       <rule>
          <columns>id</columns>
          <algorithm>rang-long</algorithm>
       </rule>
</tableRule>
<function name="rang-long"
       class="org.opencloudb.route.function.AutoPartitionByLong">
       <property name="mapFile">autopartition-long.txt</property>
       <property name="defaultNode">0</property>
</function>
```

autopartition-long.txt 的配置如下：

```
# range start-end ,data node index
# K=1000,M=10000.
0-3000000=0
3000001-4000000=1
4000001-6000000=2
```

配置说明如下。

- columns 指定分片的表列名。
- algorithm 指定分片函数与 function 对应。
- rang-long 函数中的 mapFile 代表配置文件的路径。
- defaultNode 为超过范围后的默认节点。

所有的节点配置都是从 0 开始的，0 代表节点 1，此配置非常简单，即预先设计好某个分片的 id 范围。

3.2.7　范围求模算法

该算法为先进行范围分片，计算出分片组，组内再求模，综合了范围分片和求模分片的优点。分片组内使用求模可以保证组内的数据分布比较均匀，分片组之间采用范围分片可以兼顾范围分片的特点。事先规定好分片的数量，数据扩容时按分片组扩容，则原有分片组的数据不需要迁移。由于分片组内的数据分布比较均匀，所以分片组内可以避免热点数据问题。

配置如下：

```
<tableRule name="auto-sharding-rang-mod">
```

```
<rule>
<columns>id</columns>
<algorithm>rang-mod</algorithm>
</rule>
</tableRule>
<function name="rang-mod" class=
"org.opencloudb.route.function.PartitionByRangeMod">
<property name="mapFile">partition-range-mod.txt</property>
<property name="defaultNode">21</property>
</function>
```

配置说明如下。

- columns 指定分片的表列名。
- algorithm 指定分片函数与 function 对应。
- mapFile 指定分片的配置文件。
- 未包含以上规则的数据存储在 defaultNode 节点中，节点从 0 开始。

partition-range-mod.txt 配置格式如下：

```
range start-end ,data node group size
```

具体 partition-range-mod.txt 配置如下，等号前面的范围代表一个分片组，等号后面的数字代表该分片组所拥有的分片数量。

```
0-200M=5  //代表有 5 个分片节点
200M1-400M=1
400M1-600M=4
600M1-800M=4
800M1-1000M=6
```

3.2.8 固定分片 hash 算法

类似于十进制的求模运算，但是为二进制的操作，取 id 的二进制低 10 位，即 id 二进制 &1111111111。

此算法的优点在于如果按照十进制取模运算，则在连续插入 1～10 时，1～10 会被分到 1～10 个分片，增大了插入事务的控制难度。而此算法根据二进制则可能会分到连续的分片，降低了插入事务的控制难度。

配置如下：

```
<tableRule name="rule1">
<rule>
<columns>user_id</columns>
```

```
<algorithm>func1</algorithm>
</rule>
</tableRule>
<function name="func1" class="org.opencloudb.route.function.PartitionByLong">
<property name="partitionCount">2,1</property>
<property name="partitionLength">256,512</property>
</function>
```

配置说明如下。

- columns 标识将要分片的表字段。
- algorithm 为分片函数。
- partitionCount 为分片个数列表。
- partitionLength 为分片范围列表，分区长度默认最大为 2^n=1024，即最大支持 1024 个分区。

约束如下。

- count、length 两个数组的长度必须一致。
- 1024 = sum((count[i]*length[i]))，count 和 length 两个向量的点积恒等于 1024。

用法如下所示，分区策略希望将数据水平分成 3 份，前两份各占 25%，第 3 份占 50%（故本例为非均匀分区）。

```
// |<————————1024——————————————>|
// |<—-256—>|<—-256—>|<————-512—————->|
// | partition0 | partition1 | partition2 |
// | 共 2 份,故 count[0]=2 | 共 1 份，故 count[1]=1 |
   int[] count = new int[] { 2, 1 };
   int[] length = new int[] { 256, 512 };
   PartitionUtil pu = new PartitionUtil(count, length);
```

下面的代码演示了分别以 offerId 字段或 memberId 字段根据上述分区策略拆分的结果：

```
int DEFAULT_STR_HEAD_LEN = 8; // cobar 默认会配置为此值
long offerId = 12345;
String memberId = "qiushuo"; // 若根据 offerId 分配，partNo1 将等于 0，即
按照上述分区策略，offerId 为 12345 时将会被分配到 partition0 中
int partNo1 = pu.partition(offerId); // 若根据 memberId 分配，partNo2 将
等于 2，即按照上述分区策略，memberId 为 qiushuo 时将会被分到 partition2 中
int partNo2 = pu.partition(memberId, 0, DEFAULT_STR_HEAD_LEN);
```

如果需要设置为平均分片，比如平均分为 4 个分片，则 partitionCount×partitionLength=1024。

```
<function name="func1" class="org.opencloudb.route.function.PartitionByLong">
<property name="partitionCount">4</property>
<property name="partitionLength">256</property>
</function>
```

3.2.9 取模范围算法

取模运算与范围约束的结合主要是为后续的数据迁移做准备，即可以自主决定取模后数据的节点分布，配置如下：

```
<tableRule name="sharding-by-pattern">
<rule>
<columns>user_id</columns>
<algorithm>sharding-by-pattern</algorithm>
</rule>
</tableRule>
<function name="sharding-by-pattern" class=
"org.opencloudb.route.function.PartitionByPattern">
<property name="patternValue">256</property>
<property name="defaultNode">2</property>
<property name="mapFile">partition-pattern.txt</property>
</function>
```

partition-pattern.txt 配置如下：

```
# id partition range start-end ,data node index
###### first host configuration
1-32=0
33-64=1
65-96=2
97-128=3
######## second host configuration
129-160=4
161-192=5
193-224=6
225-256=7
0-0=7
```

配置说明如下。

- columns 标识将要分片的表字段。
- algorithm 为分片函数。
- patternValue 为求模基数。
- defaultNode 为默认节点，如果采用默认配置，则不会进行求模运算。

在 mapFile 配置文件中，1～32 即代表 id%256 后分布的范围。如果在 1～32，则在分区 1，以此类推，如果 id 不是数据，则会分配在 defaultNode（默认节点）上。

```
String idVal = "0";
Assert.assertEquals(true, 7 == autoPartition.calculate(idVal));
idVal = "45a";
Assert.assertEquals(true, 2 == autoPartition.calculate(idVal));
```

3.2.10　字符串 hash 求模范围算法

与取模范围算法类似，该算法支持数值、符号、字母取模，配置如下：

```
<tableRule name="sharding-by-prefixpattern">
<rule>
<columns>user_id</columns>
<algorithm>sharding-by-prefixpattern</algorithm>
</rule>
</tableRule>
<function name="sharding-by-pattern" class=
"org.opencloudb.route.function.PartitionByPrefixPattern">
<property name="patternValue">256</property>
<property name="prefixLength">5</property>
<property name="mapFile">partition-pattern.txt</property>
</function>
```

partition-pattern.txt 配置如下：

```
# range start-end ,data node index
# ASCII
# 8-57=0-9 阿拉伯数字
# 64、65-90=@、A-Z # 97-122=a-z
###### first host configuration
1-4=0
5-8=1
9-12=2
13-16=3
###### second host configuration
17-20=4
21-24=5
25-28=6
29-32=7
0-0=7
```

配置说明如下。

- columns 标识将要分片的表字段。
- algorithm 为分片函数。
- patternValue 为求模基数。
- prefixLength 为截取的位数。
- mapFile 为配置文件。
- 1～32 代表 id%256 后分布的范围，如果在 1～32，则在分区 1，其他类推。

该算法与取模范围算法类似，截取长度为 prefixLength 的子串，再对子串中每个字符的

ASCII 码进行求和得出 sum，然后对 sum 值进行求模运算（sum%patternValue），可以计算出 prefixLength 长度的子串分片数。

```
String idVal="gf89f9a";
   Assert.assertEquals(true, 0==autoPartition.calculate(idVal));
idVal="8df99a";
   Assert.assertEquals(true, 4==autoPartition.calculate(idVal));
idVal="8dhdf99a";
   Assert.assertEquals(true, 3==autoPartition.calculate(idVal));
```

3.2.11 应用指定的算法

在运行阶段由应用自主决定路由到哪个分片，配置如下：

```
<tableRule name="sharding-by-substring">
<rule>
<columns>user_id</columns>
<algorithm>sharding-by-substring</algorithm>
</rule>
</tableRule>
<function name="sharding-by-substring" class=
"org.opencloudb.route.function.PartitionDirectBySubString">
<property name="startIndex">0</property>
<!-- zero-based -->
<property name="size">2</property>
<property name="partitionCount">8</property>
<property name="defaultPartition">0</property>
</function>
```

配置说明如下。

- columns 标识将要分片的表字段。
- algorithm 为分片函数。

直接根据字符子串（必须是数字）计算分区号（由应用传递参数，显式指定分区号）。例如 id=05-100000002，其中 id 是从 startIndex=0 开始的，截取长度为两位数字，即 05，05 就是获取的分区，默认分配到 defaultPartition。

3.2.12 字符串 hash 解析算法

截取字符串中的 int 数值 hash 分片，配置如下：

```
<tableRule name="sharding-by-stringhash">
<rule>
```

```
<columns>user_id</columns>
<algorithm>sharding-by-stringhash</algorithm>
</rule>
</tableRule>
<function name="sharding-by-stringhash" class=
"org.opencloudb.route.function.PartitionByString">
<property name=length>512</property>
<!-- zero-based -->
<property name="count">2</property>
<property name="hashSlice">0:2</property>
</function>
```

配置说明如下。

- columns 标识将要分片的表字段。
- algorithm 为分片函数。
- length 为字符串 hash 的求模基数。
- count 为分区数。
- hashSlice 为预算位，即根据子字符串中的 int 值进行 hash 运算。

```
hashSlice :  0 means str.length(), -1 means str.length()-1
/**
* "2" -> (0,2)
* "1:2" -> (1,2)
* "1:" -> (1,0)
* "-1:" -> (-1,0)
* ":-1" -> (0,-1)
* ":" -> (0,0)
*/
```

3.2.13 一致性 hash 算法

一致性 hash 算法有效解决了分布式数据的扩容问题，配置如下：

```
<tableRule name="sharding-by-murmur">
<rule>
<columns>user_id</columns>
<algorithm>murmur</algorithm>
</rule>
</tableRule>
<function name="murmur" class=
"org.opencloudb.route.function.PartitionByMurmurHash">
<property name="seed">0</property><!-- 默认是 0-->
<property name="count">2</property><!-- 要分片的数据库节点数量，必须指定，
```

```
    否则没法分片-->
    <property name="virtualBucketTimes">160</property>
    <!-- 一个实际的数据库节点被映射为这么多虚拟节点，默认是 160 倍，也就是虚拟节点数是物理节点数
的 160 倍-->
    <!-- <property name="weightMapFile">weightMapFile</property>
    节点的权重，没有指定权重的节点默认是 1。以 properties 文件的格式填写，以从 0 开始
    到 count-1 的整数值也就是节点索引为 key，以节点权重值为值。所有权重值必须是正整数，
    否则以 1 代替 -->
    <!-- <property
    name="bucketMapPath">/etc/mycat/bucketMapPath</property>
    用于测试时观察各物理节点与虚拟节点的分布情况，如果指定了这个属性，则会把虚拟节点的 murmur
hash 值与物理节点的映射按行输出到这个文件，没有默认值，如果不指定，就不会输出任何东西 -->
    </function>
```

3.2.14 按日期（天）分片算法

配置如下：

```
<tableRule name="sharding-by-date">
<rule>
<columns>create_time</columns>
<algorithm>sharding-by-date</algorithm>
</rule>
</tableRule>
<function name="sharding-by-date" class=
"org.opencloudb.route.function.PartitionByDate">
<property name="dateFormat">yyyy-MM-dd</property>
<property name="sBeginDate">2014-01-01</property>
<property name="sEndDate">2014-01-02</property>
<property name="sPartionDay">10</property>
</function>
```

配置说明如下。

- columns 标识将要分片的表字段。
- algorithm 为分片函数。
- dateFormat 为日期格式。
- sBeginDate 为开始日期。
- sEndDate 为结束日期。
- sPartionDay 为分区天数，默认从开始日期算起，每隔 10 天一个分区。

如果配置了 sEndDate，则代表数据达到了这个日期的分片后会重复从开始分片插入。

```
Assert.assertEquals(true, 0 == partition.calculate("2014-01-01"));
```

```
Assert.assertEquals(true, 0 == partition.calculate("2014-01-10"));
Assert.assertEquals(true, 1 == partition.calculate("2014-01-11"));
Assert.assertEquals(true, 12 == partition.calculate("2014-05-01"));
```

3.2.15　按单月小时算法

单月内按照小时拆分，最小粒度是小时，一天最多可以有 24 个分片，最少 1 个分片，下个月从头开始循环，每个月末需要手工清理数据。

配置如下：

```
<tableRule name="sharding-by-hour">
<rule>
<columns>create_time</columns>
<algorithm>sharding-by-hour</algorithm>
</rule>
</tableRule>
<function name="sharding-by-hour"
class="org.opencloudb.route.function.LatestMonthPartion">
<property name="splitOneDay">24</property>
</function>
```

配置说明如下。

- columns 为拆分字段，字符串类型（yyyymmddHH），格式需要符合 Java 标准。
- splitOneDay 为一天切分的分片数。

部分代码示例如下：

```
LatestMonthPartion partion = new LatestMonthPartion();
partion.setSplitOneDay(24);
Integer val = partion.calculate("2015020100");
assertTrue(val == 0);
val = partion.calculate("2015020216");
assertTrue(val == 40);
val = partion.calculate("2015022823");
assertTrue(val == 27 * 24 + 23);
Integer[] span = partion.calculateRange("2015020100", "2015022823");
assertTrue(span.length == 27 * 24 + 23 + 1);
assertTrue(span[0] == 0 && span[span.length - 1] == 27 * 24 + 23);
span = partion.calculateRange("2015020100", "2015020123");
assertTrue(span.length == 24);
assertTrue(span[0] == 0 && span[span.length - 1] == 23);
```

3.2.16 自然月分片算法

使用场景为按月份列分区，每个自然月一个分片，查询条例时使用 between and，配置如下：

```
<tableRule name="sharding-by-month">
<rule>
<columns>create_time</columns>
<algorithm>sharding-by-month</algorithm>
</rule>
</tableRule>
<function name="sharding-by-month" class=
"org.opencloudb.route.function.PartitionByMonth">
<property name="dateFormat">yyyy-MM-dd</property>
<property name="sBeginDate">2014-01-01</property>
</function>
```

配置说明如下。

- columns 为分片字段，字符串类型，与 dateFormat 格式一致。
- algorithm 为分片函数。
- dateFormat 为日期字符串格式。
- sBeginDate 为开始日期。

部分源码示例如下：

```
PartitionByMonth partition = new PartitionByMonth();
partition.setDateFormat("yyyy-MM-dd");
partition.setsBeginDate("2014-01-01");
partition.init();
Assert.assertEquals(true, 0 == partition.calculate("2014-01-01"));
Assert.assertEquals(true, 0 == partition.calculate("2014-01-10"));
Assert.assertEquals(true, 0 == partition.calculate("2014-01-31"));
Assert.assertEquals(true, 1 == partition.calculate("2014-02-01"));
Assert.assertEquals(true, 1 == partition.calculate("2014-02-28"));
Assert.assertEquals(true, 2 == partition.calculate("2014-03-1"));
Assert.assertEquals(true, 11 == partition.calculate("2014-12-31"));
Assert.assertEquals(true, 12 == partition.calculate("2015-01-31"));
Assert.assertEquals(true, 23 == partition.calculate("2015-12-31"));
```

3.2.17 日期范围 hash 算法

其思想与范围求模一致，由于日期取模方法会出现数据热点问题，所以先根据日期分组，再根据时间 hash 使得短期内数据分布得更均匀。其优点是可以避免扩容时的数据迁移，又可以在一定程度上避免范围分片的热点问题，要求日期格式尽量精确，不然达不到局部均匀的目的。

配置如下：

```
<tableRule name="rangeDateHash">
<rule>
<columns>col_date</columns>
<algorithm>range-date-hash</algorithm>
</rule>
</tableRule>
<function name="range-date-hash" class=
"org.opencloudb.route.function.PartitionByRangeDateHash">
<property name="sBeginDate">2014-01-01 00:00:00</property>
<property name="sPartionDay">3</property>
<property name="dateFormat">yyyy-MM-dd HH:mm:ss</property>
<property name="groupPartionSize">6</property>
</function>
```

配置说明如下。

- columns 标识将要分片的表字段。
- algorithm 为分片函数。
- sBeginDate 指定开始的日期，与 dateFormat 格式一致。
- sPartionDay 代表多少天一组。
- dateFormat 为指定的日期格式，符合 Java 标准。
- groupPartionSize 为每组的分片数量。

3.3 Mycat 管理命令详解

Mycat 提供类似数据库的管理监控方式，可以通过 MySQL 命令行登录管理端口（9066）执行相应的 SQL 语句进行管理，也可以通过 JDBC 方式进行远程连接管理，本节主要讲解命令行的操作。

目前 Mycat 有两个端口：8066 数据端口和 9066 管理端口，登录方式类似于 MySQL 的服务端登录。

```
MySQL -h127.0.0.1 -utest -ptest -P9066 [-dmycat]
```

- -h 后面是主机，即当前 Mycat 安装的主机 IP 地址。
- -u Mycat server.xml 中配置逻辑库的用户。
- -p Mycat server.xml 中配置逻辑库的密码。

- -P 后面是管理端口号，注意 P 是大写。
- -d Mycat server.xml 中配置逻辑库。

数据端口默认为 8066，管理端口默认为 9066，如果需要修改，则需要配置 serve.xml：

```
<system>
    <property name="serverPort">8067</property>
    <property name="managerPort">9066</property>
</system>
```

通过 show @@help;可以查看所有命令：

```
MySQL> show @@help;
+---------------------------------------+--------------------------
--------+
| STATEMENT                             | DESCRIPTION                    |
+---------------------------------------+--------------------------
--------+
| clear @@slow where datanode = ?      | Clear slow sql by datanode      |
| clear @@slow where schema = ?        | Clear slow sql by schema        |
| kill @@connection id1,id2,...     | Kill the specified connections   |
| offline                              | Change MyCat status to OFF      |
| online                               | Change MyCat status to ON       |
| reload @@config                      | Reload all config from file     |
| reload @@route                       | Reload route config from file   |
| reload @@user                        | Reload user config from file    |
| rollback @@config                    | Rollback all config from memory  |
| rollback @@route                    | Rollback route config from memory |
| rollback @@user                     | Rollback user config from memory  |
| show @@backend                      | Report backend connection status  |
| show @@cache                         | Report system cache usage        |
| show @@command                       | Report commands status           |
| show @@connection                    | Report connection status         |
| show @@connection.sql                | Report connection sql            |
| show @@database                      | Report databases                 |
| show @@datanode                      | Report dataNodes                 |
| show @@datanode where schema = ?     | Report dataNodes                 |
| show @@datasource                    | Report dataSources               |
| show @@datasource where dataNode = ? | Report dataSources               |
| show @@heartbeat                     | Report heartbeat status          |
| show @@parser                        | Report parser status             |
| show @@processor                     | Report processor status          |
| show @@router                        | Report router status             |
| show @@server                        | Report server status             |
| show @@session                       | Report front session details     |
| show @@slow where datanode = ?       | Report datanode slow sql         |
| show @@slow where schema = ?         | Report schema slow sql           |
| show @@sql where id = ?              | Report specify SQL               |
```

```
| show @@sql.detail where id = ?     | Report execute detail status     |
| show @@sql.execute                 | Report execute status            |
| show @@sql.slow                    | Report slow SQL                  |
| show @@threadpool                  | Report threadPool status         |
| show @@time.current                | Report current timestamp         |
| show @@time.startup                | Report startup timestamp         |
| show @@version                     | Report Mycat Server version      |
| stop @@heartbeat name:time         | Pause dataNode heartbeat         |
| switch @@datasource name:index     | Switch dataSource                |
+--------------------------------------+--------------------------
--------+
39 rows in set (0.00 sec)
```

3.3.1　Reload 命令

1. reload @@config

该命令用于更新配置文件，例如更新 schema.xml 文件后在命令行窗口中输入该命令，不用重启即可进行配置文件更新，运行结果参考如下：

```
MySQL> reload @@config;
Query OK, 1 row affected (0.29 sec)
Reload config success
```

对应的 reload 配置有：

```
reload @@config              Reload all config from file
reload @@config_all          Reload all config from file
reload @@route               Reload route config from file （未实现）
reload @@user                Reload user config from file （未实现）
rollback @@config            Rollback all config from memory
rollback @@route             Rollback route config from memory （未实现）
rollback @@user              Rollback user config from memory  （未实现）
```

2. reload @@sqlstat

Mycat 1.5 版本新增开启和关闭 SQL 监控分析的指令，需要在 QPS 测试时关闭 SQL 监控分析功能，否则测试结果比较差。

开启 SQL 监控分析功能：

```
reload @@sqlstat=open             Open real-time sql stat analyzer
```

关闭 SQL 监控分析功能：

```
reload @@sqlstat=close             Close real-time sql stat analyzer
```

设置慢 SQL 时间阈值：

```
reload @@sqlslow=                    Set Slow SQL Time(ms)
```

重置 SQL 监控分析的数据：

```
reload @@user_stat              Reset show @@sql  @@sql.sum @@sql.slow
```

3.3.2 Show 命令

1. show @@database

该命令用于显示 Mycat 数据库列表，运行结果对应 schema.xml 配置文件的 schema 子节点。

```
MySQL> show @@database;
+----------+
| DATABASE |
+----------+
| mycat    |
+----------+
1 row in set (0.00 sec)
```

2. show @@datanode

该命令用于显示 Mycat 数据节点列表，运行结果对应 schema.xml 配置文件的 dataNode 节点。

```
MySQL>  show @@datanode;
+------+-----------+-------+-------+--------+------+------+--------
-+------------+----------+---------+---------------+
| NAME | DATHOST   | INDEX | TYPE  | ACTIVE | IDLE | SIZE | EXECUTE |
TOTAL_TIME | MAX_TIME | MAX_SQL | RECOVERY_TIME |
+------+-----------+-------+-------+--------+------+------+--------
-+------------+----------+---------+---------------+
| blog | blog/blog |     0 | MySQL |      0 |   13 |  100 |  329521 |
0 |        0 |       0 |            -1 |
+------+-----------+-------+-------+--------+------+------+--------
-+------------+----------+---------+---------------+
1 row in set (0.00 sec)
```

- NAME 表示 dataNode 的名称。
- DATAHOST 表示对应的 dataHost 属性的值，即数据主机。
- ACTIVE 表示活跃连接数，IDLE 表示闲置连接数，SIZE 对应总连接数量。

运行如下命令，可查找对应的 schema 的 dataNode 列表：

```
MySQL> show @@datanode where schema = mycat;
+------+-----------+-------+-------+--------+------+------+--------
-+-----------+----------+---------+---------------+
| NAME | DATHOST   | INDEX | TYPE  | ACTIVE | IDLE | SIZE | EXECUTE |
TOTAL_TIME | MAX_TIME | MAX_SQL | RECOVERY_TIME |
+------+-----------+-------+-------+--------+------+------+--------
-+-----------+----------+---------+---------------+
| blog | blog/blog |     0 | MySQL |      0 |   13 |  100 |  329541 |
0 |        0 |       0 |             -1 |
+------+-----------+-------+-------+--------+------+------+--------
-+-----------+----------+---------+---------------+
1 row in set (0.00 sec)
```

3. show @@heartbeat

该命令用于报告心跳状态。

RS_CODE 状态如下。

- OK_STATUS=1 代表正常状态。
- ERROR_STATUS=-1 代表连接出错。
- TIMEOUT_STATUS = -2 代表连接超时。
- INIT_STATUS = 0 代表初始化状态。

若节点发生故障，则会连续进行默认的 5 个周期检测，心跳连续失败后就会变成-1，节点故障确认，然后可能发生切换，运行结果参考如下：

```
MySQL> show @@heartbeat;
+--------+-------+----------------+------+---------+-------+-------
-+---------+----------------+---------------------+-------+
| NAME   | TYPE  | HOST           | PORT | RS_CODE | RETRY | STATUS | TIMEOUT
| EXECUTE_TIME   | LAST_ACTIVE_TIME    | STOP  |
+--------+-------+----------------+------+---------+-------+-------
-+---------+----------------+---------------------+-------+
| master | MySQL | 121.40.121.133 | 3306 |       1 |     0 | idle   |
30000 | 8334,7833,5722 | 2015-04-08 21:34:33 | false |
+--------+-------+----------------+------+---------+-------+-------
-+---------+----------------+---------------------+-------+
1 row in set (0.00 sec)
```

4. show @@version

该命令用于获取 Mycat 的版本，参考运行结果如下：

```
MySQL>  show @@version ;
```

```
+-----------------+
| VERSION         |
+-----------------+
| 5.5.8-mycat-1.3 |
+-----------------+
1 row in set (0.00 sec)
```

5. show @@connection

该命令用于获取 Mycat 的前端连接状态，即应用于 Mycat 的连接。

6. kill @@connection id,id,id

用于关闭连接。运行结果参考如下：

```
MySQL> show @@connection;
  +------------+------+---------------+------+------------+--------+--------
  -+--------+---------+---------------+------------+------------+---------+
  ------------+ | PROCESSOR   | ID   | HOST          | PORT | LOCAL_PORT | SCHEMA
  | CHARSET | NET_IN | NET_OUT | ALIVE_TIME(S) | RECV_BUFFER | SEND_QUEUE | txlevel
  | autocommit |
  +------------+------+---------------+------+------------+--------+--------
  -+--------+---------+---------------+------------+------------+---------+
  ------------+ | Processor0 |    7 | 101.44.170.64 | 8066 |      13694 | mycat
  | utf8    |    233 |     968 |           105 |       4096 |          0 | 3       |
  true       | | Processor0 |    2 | 127.0.0.1     | 9066 |      34774 | NULL    |
  utf8    |  2014 |  33646 |           720 |        4096 |          0 | NULL    |
  NULL       | | Processor0 |    1 | 127.0.0.1     | 8066 |      44751 | mycat   |
  utf8    |  2502 |  85432 |           727 |        4096 |          0 | 3       |
  true       | | Processor0 |    4 | 101.44.170.64 | 8066 |      13626 | mycat   |
  utf8    |  1244 |   3462 |           209 |        4096 |          0 | 3       | true
  |
  +------------+------+---------------+------+------------+--------+--------
  -+--------+---------+---------------+------------+------------+---------+
  ------------+ 4 rows in set (0.00 sec)
```

强制关闭连接如下：

```
MySQL> kill @@connection 7;
Query OK, 1 row affected (0.01 sec)
```

7. show @@backend

用于查看后端的连接状态：

```
MySQL> show @@backend;
+------------+------+---------+---------------+------+--------+-----------+
```

```
   ---------+---------+--------+----------+------------+--------+---------+--
   ----------+
| processor  | id   | MySQLId | host           | port | l_port | net_in    | net_out
   | life     | closed | borrowed | SEND_QUEUE | schema | txlevel | autocommit |
+------------+------+---------+---------------+------+--------+-----------+
   ---------+---------+--------+----------+------------+--------+---------+--
   ----------+
| Processor0 |   12 |    4768 | 121.40.121.133 | 3306 |   37141 | 236533254
...
...
| Processor0 |    3 |    4635 | 121.40.121.133 | 3306 |   59893 | 305185063 | 3618816
   | 1296826 | false   | false     |          0 | blog   | 3       | true       |
| Processor0 |   11 |       0 | 121.40.121.133 | 3306 |   59896 |   7261962 | 1685851
   | 1296825 | false   | false     |          0 | NULL   | NULL    | NULL       |
| Processor0 |    4 |    4629 | 121.40.121.133 | 3306 |   59887 | 296327067 | 3631921
   | 1296826 | false   | false     |          0 | blog   | 3       | true       |
+------------+------+---------+---------------+------+--------+-----------+
   ---------+---------+--------+----------+------------+--------+---------+--
   ----------+
14 rows in set (0.00 sec)
```

8. show @@cache

用于查看 Mycat 缓存。

- SQLRouteCache：SQL 语句路由缓存。
- TableID2DataNodeCache：缓存表主键与分片的对应关系。
- ER_SQL2PARENTID：缓存 ER 分片中子表与父表的关系。

```
MySQL> show @@cache;
+-------------------------------------+-------+------+--------+------+------
   +---------------+----------+
| CACHE                               | MAX   | CUR  | ACCESS | HIT  | PUT  |
   LAST_ACCESS     | LAST_PUT |
+-------------------------------------+-------+------+--------+------+------
   +---------------+----------+
| SQLRouteCache                       | 10000 |    0 | 298175 |    0 |    0 |
   1428815230596 |        0 |
| TableID2DataNodeCache.TESTDB_ORDERS | 50000 |    0 |      0 |    0 |    0 |
   0 |        0 |
| ER_SQL2PARENTID                     |  1000 |    0 |      0 |    0 |    0 |
   0 |        0 |
+-------------------------------------+-------+------+--------+------+------
   +---------------+----------+
3 rows in set (0.00 sec)
```

9. show @@datasource

查看数据源的状态，如果配置了主从或者多主，则可以切换。

```
MySQL> show @@datasource;
+----------+--------+-------+---------------+------+------+--------+------
   +------+---------+
| DATANODE | NAME    | TYPE  | HOST           | PORT | W/R  | ACTIVE | IDLE | SIZE
   | EXECUTE |
+----------+--------+-------+---------------+------+------+--------+------
   +------+---------+
| blog     | master  | MySQL | 121.40.121.133 | 3306 | W    |      0 |   10 |  100
   |      16 | | blog     | master2 | MySQL | 127.0.0.1      | 3306 | W    |      0
   |    0 |  100 |       0 |
+----------+--------+-------+---------------+------+------+--------+------
   +------+---------+
2 rows in set (0.00 sec)
```

10. switch @@datasource name:index

用于切换数据源。

- name：schema 中配置的 dataHost 中的 name。
- index：schema 中配置的 dataHost 的 writeHost index 位标，即按照从上到下的配置顺序，从 0 开始。

切换数据源时会将原数据源所有的连接池中的连接关闭，并且从新数据源创建新的连接，此时 Mycat 服务不可用。

dnindex.properties 文件记录了当前活跃的 writer。

```
MySQL> switch @@datasource blog:1;
Query OK, 1 row affected (1 min 0.05 sec)
```

注意：在 reload @@config、switch @@datasource name:index 这两个命令执行的过程中 Mycat 服务不可用，应谨慎处理，防止正在提交的事务出错。

11. show @@syslog limit

用于显示系统日志。

- 端口号：该命令工作在 9066 端口，用来在客户端命令窗口中显示系统的日志信息，通常用于远程查看 Mycat Server 的日志信息。
- 参数：limit=后接正整数，该数值用来限定每次显示的日志条数的最大数量。

12. reload@@user_stat

用于清除缓存。该命令工作在 9066 端口，用来将客户端执行 show @@sql ;show@@sql.sum; show@@slow.success ;命令之后所缓存的信息清空。

13. SQL 统计命令

- show@@sql：显示在 Mycat 中执行过的 SQL 语句。
- show@@sql.slow：显示慢 SQL 语句。
- show@@sql.sum：显示 SQL 语句的整体执行情况、读写比例等。

```
MySQL> show @@sql;
+------+------+---------------+-------------+-------------------------------
-------------------------+
| ID   | USER | START_TIME    | EXECUTE_TIME | SQL                              |
+------+------+---------------+-------------+-------------------------------
-------------------------+
|   49 | cat  | 1468901663738 |           3 | SELECT * FROM `c` LIMIT 0, 1000 |
|   48 | cat  | 1468901661900 |           7 | SELECT * FROM `b` LIMIT 0, 1000 |
|   47 | cat  | 1468901660031 |           4 | SELECT * FROM `a` LIMIT 0, 1000 |
|   46 | cat  | 1468901656677 |           8 | SELECT @@character_set_database,
@@collation_database |
+------+------+---------------+-------------+-------------------------------
-------------------------+
4 rows in set (0.00 sec)

MySQL> show @@sql.slow;
Empty set (0.01 sec)
MySQL> show @@sql.sum;
+------+------+------+------+------+------+-------------+-------------+---
------------+
| ID   | USER | R    | W    | R%   | MAX  | TIME_COUNT   | TTL_COUNT   | LAST_TIME |
+------+------+------+------+------+------+-------------+-------------+---
------------+
|   1 | cat  |   4 |   0 | 1.00 | 1    | [0, 4, 0, 0] | [4, 0, 0, 0] | 1468901663741 |
+------+------+------+------+------+------+-------------+-------------+---
------------+
1 row in set (0.00 sec)
```

第 4 章 Mycat 高级技术实战

互联网的高速发展使分布式技术兴起，当系统的压力越来越大时，人们首先想到的解决方案就是向上扩展（SCALE UP），简单来说就是不断增加硬件性能来解决这个问题，采用这种方案的硬件成本比较高。还有另外一种方案是水平扩展，通过新增服务器的节点采用负载均衡来解决问题，这样在应用服务器层面我们可以不停地增加服务器，来满足高并发和高访问量。这里我们只需要考虑 session 的处理，所有的压力都会指向数据库，数据库却不能新增节点来完成扩容，必须采用其他方式来实现。这时各种分布式数据库中间件应运而生，出现了很多解决方案，比如 Amoeba、Atlas、Cobar、Mycat、MySQL Proxy 等，而 Mycat 是目前开源的数据库中间件中非常成熟的解决方案，社区也非常活跃。

4.1 用 Mycat 搭建读写分离

读写分离，简单地说是把对数据库的读和写操作分开，以对应不同的数据库服务器。主数据库提供写操作，从数据库提供读操作，这样能有效地减轻单台数据库的压力。主数据库进行写操作后，数据及时同步到所读的数据库，尽可能保证读、写数据库的数据一致，比如 MySQL 的主从复制、Oracle 的 data guard、SQL Server 的复制订阅等。在很多系统中，读操作的比例远远高于写操作，所以对应读操作的数据库可以有多台，通过负载均衡技术进一步分摊读操作的压力，让整个数据库系统高效、平稳地运行。而这些读写分离所需的复杂的数据库架构，要对开发人员和应用程序透明。我们通过 Mycat 即可轻易实现上述功能，不仅可以支持 MySQL，也可以支持 Oracle 和 SQL Server。

Mycat 控制后台数据库的读写分离和负载均衡由 schema.xml 文件 datahost 标签的 balance

属性控制，关于 balance 属性的描述详见 2.2.3 节。

下面将通过实验验证并结合 log 来说明各种情况下的读写分离。为了节省写作篇幅，log 只截取了关键的部分。读者可以根据书中的介绍，自己完成实验及观察完整的 log。为了准确区分各个场景的实验结果，每次修改完配置文件都做 mycat restart 处理（在生产环境中建议使用 reload @@config 或者 reload @@config_all，详见第 3 章）。

4.1.1　MySQL 读写分离

对于 MySQL，主流的读写分离是 master-slave 和 galera cluster，下面分别介绍如何通过 Mycat 来实现这两种模式的读写分离。

MySQL 主从读写分离的准备环境见表 4-1。

表 4-1

项　目	Mycat	MySQL-master	MySQL-master-standby	MySQL-slave
IP	10.230.4.131	10.230.3.194	10.230.3.195	10.230.2.132
Port	8077/9077	3306	3306	3307

MySQL 为二主一从，3.195 为 3.194 的 standby，2.132 为 3.194 的 slave。本书重在介绍 Mycat 的用法，关于 MySQL 主从及双主配置 MHA 的内容请见 4.4 节。

实验用例表如下：

```
CREATE TABLE `yf.travelrecord` (
  `id` int(11) NOT NULL AUTO_INCREMENT,
  `org_code` varchar(20) NOT NULL,
  `test_name` varchar(20) DEFAULT NULL,
  PRIMARY KEY (`id`)
) ENGINE=InnoDB DEFAULT CHARSET=utf8;
```

当 balance 为 0 时，不开启读写分离机制，所有读操作都发送到当前的 writeHost 上。Mycat 的基本配置如下：

```
<?xml version="1.0"?>
<!DOCTYPE mycat:schema SYSTEM "schema.dtd">
<mycat:schema xmlns:mycat="http://org.opencloudb/">
        <schema name="mycat01" checkSQLschema="false" sqlMaxLimit="100">
            <table name="travelrecord" primaryKey="id" dataNode="MySQL" />
        </schema>
        <dataNode name="MySQL" dataHost="MySQL_Host" database="yf" />
        <dataHost name="MySQL_Host" maxCon="1000" minCon="3" balance="0"
writeType="0" dbType="MySQL" dbDriver="native" switchType="1">
              <heartbeat>select user()</heartbeat>
```

```
                <writeHost host="MySQL_M1" url="10.230.3.194:3306" user="xjw_dba"
password="123">
                    <readHost host="MySQL_S1" url="10.230.2.132:3307" user=
"xjw_dba" password="123" />
                </writeHost>
                <writeHost host="MySQL_M2" url="10.230.3.195:3306" user="xjw_dba"
password="123">
                </writeHost>
            </dataHost>
    </mycat:schema>
```

通过 show @@datasource 查看后端数据库的分布情况：

```
MySQL>show @@datasource;
+----------+----------+-------+--------------+------+------+--------+------+
------+---------+
| DATANODE | NAME     | TYPE  | HOST         | PORT | W/R  | ACTIVE | IDLE | SIZE
| EXECUTE |
+----------+----------+-------+--------------+------+------+--------+------+
------+---------+
| MySQL    | MySQL_M1 | MySQL | 10.230.3.194 | 3306 | W    |      0 |    3 | 1000
|       4 |
| MySQL    | MySQL_M2 | MySQL | 10.230.3.195 | 3306 | W    |      0 |    1 | 1000
|       1 |
| MySQL    | MySQL_S1 | MySQL | 10.230.2.132 | 3307 | R    |      0 |    2 | 1000
|       2 |
+----------+----------+-------+--------------+------+------+--------+------+
------+---------+
```

多次执行查询语句 MySQL> select * from travelrecord。通过分析 Mycat 中的日志文件，可以看出所有读操作都分发到了 MySQL_M1。

```
02/22 16:00:19.603  DEBUG [$_NIOREACTOR-7-RW] (NonBlockingSession.java:113)
-ServerConnection [id=2, schema=mycat01, host=10.230.3.191, user=xjw_dba,
txIsolation=3, autocommit=true, schema=mycat01]select * from travelrecord, route={
   1 -> MySQL{SELECT *
FROM travelrecord
LIMIT 100}
} rrs
02/22 16:00:19.603  DEBUG [$_NIOREACTOR-7-RW] (PhysicalDBPool.java:452) -select
read source MySQL_M1 for dataHost:MySQL_Host
……
```

当 balance 为 1 时，所有读操作都发送到了当前的 writeHost 对应的 readhost 和备用的 writehost，这样就可以减轻主库的压力，高效地提供写操作，而由其他服务器承担比较耗费资源的读操作。

```
MySQL> show @@datasource;
+----------+----------+-------+--------------+------+------+--------+------+
```

```
------+---------+
    | DATANODE | NAME      | TYPE  | HOST          | PORT | W/R  | ACTIVE | IDLE | SIZE
| EXECUTE |
    +----------+----------+-------+--------------+------+------+--------+------+
------+---------+
    | MySQL    | MySQL_M1 | MySQL | 10.230.3.194 | 3306 | W    |      0 |    3 | 1000
|       4 |
    | MySQL    | MySQL_M2 | MySQL | 10.230.3.195 | 3306 | W    |      0 |    1 | 1000
|       1 |
    | MySQL    | MySQL_S1 | MySQL | 10.230.2.132 | 3307 | R    |      0 |    2 | 1000
|       2 |
    +----------+----------+-------+--------------+------+------+--------+------+
------+---------+
    MySQL> /*多次执行此条 SQL*/ select * from travelrecord;
```

分析 Mycat 中的日志，可以看出所有读操作都随机分发到了 MySQL_S1 和 MySQL_M2：

```
    03/03 17:21:17.594  DEBUG [$_NIOREACTOR-5-RW] (NonBlockingSession.java:113)
-ServerConnection [id=7, schema=mycat01, host=10.230.3.191, user=xjw_dba,
txIsolation=3, autocommit=true, schema=mycat01]select * from travelrecord, route={
       1 -> MySQL{SELECT *
    FROM travelrecord
    LIMIT 100}
    } rrs
    03/03 17:21:17.594  DEBUG [$_NIOREACTOR-5-RW] (PhysicalDBPool.java:452) -select
read source MySQL_S1 for dataHost:MySQL_Host
    ……
    03/03 17:21:18.875  DEBUG [$_NIOREACTOR-5-RW] (PhysicalDBPool.java:452) -select
read source MySQL_M2 for dataHost:MySQL_Host
    ……
```

当 balance 为 2 时，所有读操作都发送到所有的 writeHost 和 readhost 上，即当前 datahost 内的服务器都参与分担读操作。这个场景比较适合主库压力不是很大时，也可以分担读操作，更合理地利用资源。

```
    MySQL> show @@datasource;
    +----------+----------+-------+--------------+------+------+--------+------+
------+---------+
    | DATANODE | NAME      | TYPE  | HOST          | PORT | W/R  | ACTIVE | IDLE | SIZE
| EXECUTE |
    +----------+----------+-------+--------------+------+------+--------+------+
------+---------+
    | MySQL    | MySQL_M1 | MySQL | 10.230.3.194 | 3306 | W    |      0 |    3 | 1000
|       8 |
    | MySQL    | MySQL_M2 | MySQL | 10.230.3.195 | 3306 | W    |      0 |    1 | 1000
|       5 |
    | MySQL    | MySQL_S1 | MySQL | 10.230.2.132 | 3307 | R    |      0 |    2 | 1000
|       6 |
```

```
+----------+----------+-------+-------------+------+------+--------+------+
------+---------+
```

分析 Mycat 中的日志，可以看出所有读操作都随机分发到了 MySQL_S1、MySQL_M1 和 MySQL_M2：

```
03/03 17:53:34.617  DEBUG [$_NIOREACTOR-0-RW] (NonBlockingSession.java:113)
-ServerConnection [id=2, schema=mycat01, host=10.230.3.191, user=xjw_dba,
txIsolation=3, autocommit=true, schema=mycat01]select * from travelrecord, route={
   1 -> MySQL{SELECT *
FROM travelrecord
LIMIT 100}
} rrs
03/03 17:53:34.618  DEBUG [$_NIOREACTOR-0-RW] (PhysicalDBPool.java:452) -select
read source MySQL_S1 for dataHost:MySQL_Host
......
03/03 17:53:35.161  DEBUG [$_NIOREACTOR-0-RW] (PhysicalDBPool.java:452) -select
read source MySQL_M1 for dataHost:MySQL_Host
......
03/03 17:53:35.653  DEBUG [$_NIOREACTOR-0-RW] (PhysicalDBPool.java:452) -select
read source MySQL_M2 for dataHost:MySQL_Host
```

当 balance 为 3 时，所有读操作只发送到当前 writeHost 对应的 readhost 上。在实际应用时，比较适合双主热备、多个 slave 的情况。

```
MySQL> show @@datasource;
+----------+----------+-------+-------------+------+------+--------+------+
------+---------+
| DATANODE | NAME     | TYPE  | HOST        | PORT | W/R  | ACTIVE | IDLE | SIZE
| EXECUTE |
+----------+----------+-------+-------------+------+------+--------+------+
------+---------+
| MySQL    | MySQL_M1 | MySQL | 10.230.3.194 | 3306 | W    |      0 |    3 | 1000
|       6 |
| MySQL    | MySQL_M2 | MySQL | 10.230.3.195 | 3306 | W    |      0 |    1 | 1000
|       3 |
| MySQL    | MySQL_S1 | MySQL | 10.230.2.132 | 3307 | R    |      0 |    2 | 1000
|       4 |
+----------+----------+-------+-------------+------+------+--------+------+
------+---------+
```

分析 Mycat 中的日志，可以看出所有读操作都分发到了 MySQL_S1：

```
03/03 17:58:57.645  DEBUG [$_NIOREACTOR-0-RW] (NonBlockingSession.java:113)
-ServerConnection [id=2, schema=mycat01, host=10.230.3.191, user=xjw_dba,
txIsolation=3, autocommit=true, schema=mycat01]select * from travelrecord, route={
   1 -> MySQL{SELECT *
FROM travelrecord
LIMIT 100}
```

```
} rrs
03/03 17:58:57.645  DEBUG [$_NIOREACTOR-0-RW] (PhysicalDBPool.java:452) -select
read source MySQL_S1 for dataHost:MySQL_Host
......
```

通过上面的验证可以看出，Mycat 很容易实现 MySQL master-slave 架构的读写分离。对于应用程序，只有一个数据库地址和端口，即 Mycat 的地址和端口，至于后台数据库是 master-slave 还是多级 slave 或者 MHA，这些对前端应用都是透明的。到 MySQL 5.6 时，slave 的延时问题还是没能有效地解决，这就导致在某些情况下读节点滞后写节点，特别是在某些场景中，更改数据后马上又要查询，此时简单地进行读写分离已经不能满足需求。Mycat 针对这类问题也有相应的解决方法，就是使用/*balance*/注解，强制某些读操作走写节点。关于 Mycat 注解的详细介绍请见 4.5 节。在 4.1.2 节介绍的 Galera Cluster 用作读写分离时，也能很好地解决读延迟的问题。

4.1.2　MySQL Galera Cluster 读写分离

MySQL Galera Cluster 是一套基于同步复制的多主 MySQL 集群解决方案，使用简单，没有单点故障，可用性高，能很好地保证业务量不断增长时数据的安全和随时扩展。Galera 本质上是一个 wsrep 提供者（provider），其运行依赖于 wsrep 的 API 接口。wsrep API 定义了一系列应用回调和复制调用库，来实现事务数据库同步写集（writeset）复制及相似的应用。目前的主流分支有 MariaDB Galera Cluster 和 Percona Xtradb Cluster。

MySQL Galera Cluster 的主要特点有：多主服务器的拓扑结构；真正的多主架构，任何节点都可以进行读写及同步复制，各节点之间无延迟且节点宕机不会导致数据丢失；紧密耦合，所有节点均保持相同的状态，节点之间没有不同的数据，无须主从切换操作或使用 vip；在 Failover 的过程中无停机时间（其实不需要 Failover），无须手工备份当前数据库并复制至新节点；支持 InnoDB 存储引擎；对应用透明，无须更改应用或进行极小的更改便可以在任意节点上进行读写；自动剔除故障节点，自动加入新节点；真正行级别的并发复制，客户端连接与操作单台 MySQL 数据库的体验一致。如图 4-1 所示为 MySQL Galera Cluster 结构图。

Galera 官方建议一套 Cluster 至少有 3 个节点，理论上可以多写，但是在多个实验环境和生产案例中，多写会引起较为严重的锁等待，所以使用 Galera 时推荐用一写多读。这样，Galera Cluster 结合 Mycat，很容易配置出一套读写分离的数据库架构。

在以下实例中，我们以 Percona Xtradb Cluster 为例，演示 Mycat 如何实现读写分离。

准备环境如表 4-2 所示。

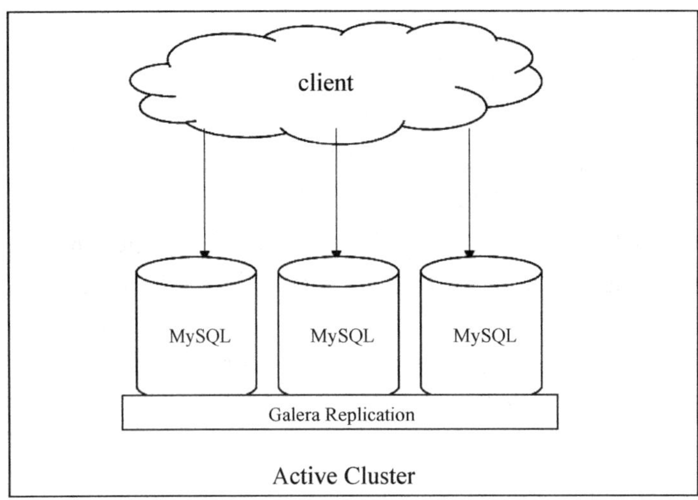

图 4-1

表 4-2

项　目	Mycat	Pxc1	Pxc2	Pxc3	Pxc1_slave
IP	10.230.4.131	10.230.3.194	10.230.3.195	10.230.3.196	10.230.2.132
Hostname	xjw-cc-03	xjw-dba-01	xjw-dba-02	xjw-dba-03	rfd-dba-01
port	8077/9077	3306	3306	3306	3307

3.194、3.195、3.196 为一套 Percona Xtradb Cluster，2.132 为 3.194 节点的 slave。关于 Percona Xtradb Cluster 配置的内容，详见 4.3 节。

实验用例表如下：

```
CREATE TABLE `yf.travelrecord` (
  `id` int(11) NOT NULL AUTO_INCREMENT,
  `org_code` varchar(20) NOT NULL,
  `test_name` varchar(20) DEFAULT NULL,
  PRIMARY KEY (`id`)
) ENGINE=InnoDB DEFAULT CHARSET=utf8;
```

Mycat 基本配置如下：

```
schema.xml
<?xml version="1.0"?>
<!DOCTYPE mycat:schema SYSTEM "schema.dtd">
<mycat:schema xmlns:mycat="http://org.opencloudb/">
        <schema name="mycat01" checkSQLschema="false" sqlMaxLimit="100">
            <table name="travelrecord" primaryKey="id" dataNode="pxc" />
        </schema>
        <dataNode name="pxc" dataHost="pxc_Host" database="yf" />
```

```
        <dataHost name="pxc_Host" maxCon="1000" minCon="3" balance="0"
writeType="0" dbType="MySQL" dbDriver="native" switchType="3">
                <heartbeat>show status like 'wsrep%'</heartbeat>
                <writeHost host="pxc_M1" url="10.230.3.194:3306" user="xjw_dba"
password="123">
                </writeHost>
                <writeHost host="pxc_M2" url="10.230.3.195:3306" user="xjw_dba"
password="123">
                </writeHost>
                <writeHost host="pxc_M3" url="10.230.3.196:3306" user="xjw_dba"
password="123">
                </writeHost>
        </dataHost>
    </mycat:schema>
```

下面通过 balance 为 0、1、2 的情况，分析 Mycat 是如何“路由”读写 SQL 的。

当 balance 为 0 时，是不进行读写分离的，所有读写都走我们所配置的 dataHost 的第 1 个 writeHost（故障切换后的情况会有所不同，会在下面的章节中介绍）。

```
MySQL> show @@datasource;
+----------+--------+-------+--------------+------+------+--------+------+--
----+---------+
| DATANODE | NAME   | TYPE  | HOST         | PORT | W/R  | ACTIVE | IDLE | SIZE | EXECUTE |
+----------+--------+-------+--------------+------+------+--------+------+--
----+---------+
| pxc      | pxc_M1 | MySQL | 10.230.3.194 | 3306 | W    |      0 |    3 | 1000 |       8 |
| pxc      | pxc_M2 | MySQL | 10.230.3.195 | 3306 | W    |      0 |    1 | 1000 |       5 |
| pxc      | pxc_M3 | MySQL | 10.230.3.196 | 3306 | W    |      0 |    1 | 1000 |       5 |
+----------+--------+-------+--------------+------+------+--------+------+--
----+---------+
```

对表 travelrecord 插入一条数据：

```
MySQL> insert into travelrecord(org_code,test_name) values ('mycat',
@@HOSTNAME);
Query OK, 1 row affected (0.11 sec)
```

分析 Mycat 中的日志，这条 insert 语句进入了 10.230.3.194，是按照我们的 dataHost 配置执行的。

```
02/19 11:13:27.620  DEBUG [$_NIOREACTOR-7-RW] (NonBlockingSession.java:113)
-ServerConnection [id=2, schema=mycat01, host=10.230.3.191, user=xjw_dba,
txIsolation=3, autocommit=true, schema=mycat01]insert into travelrecord
(org_code,test_name) values ('mycat',@@HOSTNAME), route={
    1 -> pxc{insert into travelrecord(org_code,test_name) values ('mycat',
@@HOSTNAME)}
} rrs
02/19 11:13:27.623  DEBUG [$_NIOREACTOR-7-RW] (MySQLConnection.java:445) -con
```

```
need syn ,total syn cmd 1 commands SET SESSION TRANSACTION ISOLATION LEVEL REPEATABLE
READ;schema change:false con:MySQLConnection [id=3, lastTime=1455851607623,
user=xjw_dba, schema=yf, old shema=yf, borrowed=true, fromslaveDB=false,
threadId=594, charset=utf8, txIsolation=0, autocommit=true, attachment=pxc{insert
into travelrecord(org_code,test_name) values ('mycat',@@HOSTNAME)}, respHandler=
SingleNodeHandler [node=pxc{insert into travelrecord(org_code,test_name) values
('mycat',@@HOSTNAME)}, packetId=0], host=10.230.3.194, port=3306, statusSync=null,
writeQueue=0, modifiedSQLExecuted=true]
```

执行查询语句如下：

```
MySQL> select * from travelrecord;
+----+----------+------------+
| id | org_code | test_name  |
+----+----------+------------+
|  1 | mycat    | xjw-dba-01 |
+----+----------+------------+
1 rows in set (0.07 sec)
```

分析 Mycat 中的日志，此时的查询仍然选择 pxc_M1（10.230.3.194），符合我们预期的 dataHost 配置。

```
02/19 11:15:38.138  DEBUG [$_NIOREACTOR-1-RW] (NonBlockingSession.java:229)
-release connection MySQLConnection [id=1, lastTime=1455851738130, user=xjw_dba,
schema=yf, old shema=yf, borrowed=true, fromslaveDB=false, threadId=593,
charset=utf8, txIsolation=3, autocommit=true, attachment=pxc{SELECT *
FROM travelrecord
LIMIT 100}, respHandler=SingleNodeHandler [node=pxc{SELECT *
FROM travelrecord
LIMIT 100}, packetId=7], host=10.230.3.194, port=3306, statusSync=
org.opencloudb.MySQL.nio.MySQLConnection$StatusSync@2ff2ab91, writeQueue=0,
modifiedSQLExecuted=false]
```

当 balance 为 1 时，全部的 readHost 与 stand by writeHost 都参与 select 语句的负载均衡，如果 Pxc 的节点本身没带 slave，则三个 Pxc 节点中的一个作为写，另外两个作为读。按照我们的配置，pxc_M1 将作为写节点，pxc_M2、pxc_M3 参与读。

```
MySQL> show @@datasource;
+----------+--------+-------+--------------+------+------+--------+------+------+---------+
| DATANODE | NAME   | TYPE  | HOST         | PORT | W/R  | ACTIVE | IDLE | SIZE | EXECUTE |
+----------+--------+-------+--------------+------+------+--------+------+------+---------+
| pxc    | pxc_M1 | MySQL | 10.230.3.194 | 3306 | W |  0 |  3 | 1000 |  4 |
| pxc    | pxc_M2 | MySQL | 10.230.3.195 | 3306 | W |  0 |  1 | 1000 |  1 |
| pxc    | pxc_M3 | MySQL | 10.230.3.196 | 3306 | W |  0 |  1 | 1000 |  1 |
+----------+--------+-------+--------------+------+------+--------+------+------+---------+
```

```
MySQL> insert into travelrecord(org_code,test_name) values ('mycat',@@HOSTNAME);
Query OK, 1 row affected (0.01 sec)
```

执行上面 insert 语句后，Mycat 的日志输出如下：

```
02/19 11:28:13.899  DEBUG [$_NIOREACTOR-7-RW] (NonBlockingSession.java:113)
-ServerConnection [id=2, schema=mycat01, host=10.230.3.191, user=xjw_dba,
txIsolation=3, autocommit=true, schema=mycat01]insert into travelrecord(org_code,
test_name) values ('mycat',@@HOSTNAME), route={
   1 -> pxc{insert into travelrecord(org_code,test_name) values
('mycat',@@HOSTNAME)}
} rrs
...
MySQL> select * from travelrecord;
+----+----------+------------+
| id | org_code | test_name  |
+----+----------+------------+
|  1 | mycat    | xjw-dba-01 |
+----+----------+------------+
```

执行上面 select 语句后，Mycat 的日志输出如下：

```
02/19 12:59:35.637  DEBUG [$_NIOREACTOR-0-RW] (NonBlockingSession.java:113)
-ServerConnection [id=3, schema=mycat01, host=10.230.3.191, user=xjw_dba,
txIsolation=3, autocommit=true, schema=mycat01]select * from travelrecord, route={
   1 -> pxc{SELECT *
FROM travelrecord
LIMIT 100}
} rrs
```

在第 1 个节点增加一台 slave，Mycat 配置如下：

```
<?xml version="1.0"?>
<!DOCTYPE mycat:schema SYSTEM "schema.dtd">
<mycat:schema xmlns:mycat="http://org.opencloudb/">
      <schema name="mycat01" checkSQLschema="false" sqlMaxLimit="100">
        <table name="travelrecord" primaryKey="id" dataNode="pxc" />
      </schema>
      <dataNode name="pxc" dataHost="pxc_Host" database="yf" />
      <dataHost name="pxc_Host" maxCon="1000" minCon="3" balance="1"
writeType="0" dbType="MySQL" dbDriver="native" switchType="3">
            <heartbeat>show status like 'wsrep%'</heartbeat>
            <writeHost host="pxc_M1" url="10.230.3.194:3306" user="xjw_dba"
password="123">
                <readHost host="pxc_S1" url="10.230.2.132:3307" user=
"xjw_dba" password="123" />
            </writeHost>
            <writeHost host="pxc_M2" url="10.230.3.195:3306" user="xjw_dba"
password="123">
```

```
                </writeHost>
                <writeHost host="pxc_M3" url="10.230.3.196:3306" user="xjw_dba"
password="123">
                </writeHost>
        </dataHost>
</mycat:schema>
```

这时 pxc_S1 也作为读节点加入读负载均衡，多次执行查询：

```
MySQL> show @@datasource;
+----------+--------+-------+--------------+------+------+--------+------+--
----+---------+
| DATANODE | NAME   | TYPE  | HOST         | PORT | W/R  | ACTIVE | IDLE | SIZE |
EXECUTE |
+----------+--------+-------+--------------+------+------+--------+------+--
----+---------+
| pxc    | pxc_M1 | MySQL | 10.230.3.194 | 3306 | W    |    0 |    3 | 1000 |    3 |
| pxc    | pxc_M2 | MySQL | 10.230.3.195 | 3306 | W    |    0 |    1 | 1000 |    1 |
| pxc    | pxc_M3 | MySQL | 10.230.3.196 | 3306 | W    |    0 |    1 | 1000 |    1 |
| pxc    | pxc_S1 | MySQL | 10.230.2.132 | 3307 | R    |    0 |    1 | 1000 |    2 |
+----------+--------+-------+--------------+------+------+--------+------+--
----+---------+

MySQL> select * from travelrecord;
+----+----------+------------+
| id | org_code | test_name  |
+----+----------+------------+
|  1 | mycat    | xjw-dba-01 |
+----+----------+------------+
1 rows in set (0.01 sec)
```

分析 Mycat 中的日志，查询被随机分配到 pxc_M2、pxc_M3、pxc_S3 上：

```
02/22 14:19:06.682  DEBUG [$_NIOREACTOR-0-RW] (NonBlockingSession.java:113)
-ServerConnection [id=1, schema=mycat01, host=10.230.3.191, user=xjw_dba,
txIsolation=3, autocommit=true, schema=mycat01]select * from travelrecord, route={
   1 -> pxc{SELECT *
FROM travelrecord
LIMIT 100}
} rrs
02/22 14:19:06.682  DEBUG [$_NIOREACTOR-0-RW] (PhysicalDBPool.java:452) -select
read source pxc_M3 for dataHost:pxc_Host
...
```

当 balance 为 2 时，所有读操作都随机地在 writeHost、readHost 上分发。此时作为写节点的 pxc_M1 也参与读操作。

```
MySQL> show @@datasource;
+----------+--------+-------+--------------+------+------+--------+------+--
```

```
----+---------+
    | DATANODE | NAME   | TYPE  | HOST         | PORT | W/R  | ACTIVE | IDLE | SIZE |
EXECUTE |
    +----------+--------+-------+--------------+------+------+--------+------+--
----+---------+
    | pxc    | pxc_M1 | MySQL | 10.230.3.194 | 3306 | W    |    0 |    3 | 1000 |    4 |
    | pxc    | pxc_M2 | MySQL | 10.230.3.195 | 3306 | W    |    0 |    1 | 1000 |    2 |
    | pxc    | pxc_M3 | MySQL | 10.230.3.196 | 3306 | W    |    0 |    1 | 1000 |    2 |
    +----------+--------+-------+--------------+------+------+--------+------+--
----+---------+

    MySQL> insert into travelrecord(org_code,test_name) values ('mycat',@@HOSTNAME);
    Query OK, 1 row affected (0.01 sec)
```

执行上面的 insert 语句后，Mycat 日志输出：

```
    02/19 13:29:32.950  DEBUG [$_NIOREACTOR-7-RW] (ServerQueryHandler.java:56)
-ServerConnection [id=2, schema=mycat01, host=10.230.3.191, user=xjw_dba,
txIsolation=3, autocommit=true, schema=mycat01]insert into travelrecord
(org_code,test_name) values ('mycat',@@HOSTNAME)
    02/19 13:29:33.061  DEBUG [$_NIOREACTOR-7-RW] (NonBlockingSession.java:113)
-ServerConnection [id=2, schema=mycat01, host=10.230.3.191, user=xjw_dba,
txIsolation=3, autocommit=true, schema=mycat01]insert into travelrecord
(org_code,test_name) values ('mycat',@@HOSTNAME), route={
       1 -> pxc{insert into travelrecord(org_code,test_name) values ('mycat',
@@HOSTNAME)}
    } rrs
    ...

    MySQL> select * from travelrecord;
    +----+----------+------------+
    | id | org_code | test_name  |
    +----+----------+------------+
    |  1 | mycat    | xjw-dba-01 |
    +----+----------+------------+
    1 rows in set (0.07 sec)
```

多次执行查询操作后，Mycat 日志输出：

```
    02/19 13:30:40.048  DEBUG [$_NIOREACTOR-7-RW] (NonBlockingSession.java:113)
-ServerConnection [id=2, schema=mycat01, host=10.230.3.191, user=xjw_dba,
txIsolation=3, autocommit=true, schema=mycat01]select * from travelrecord, route={
       1 -> pxc{SELECT *
    FROM travelrecord
    LIMIT 100}
    } rrs
    02/19 13:30:40.049  DEBUG [$_NIOREACTOR-7-RW] (PhysicalDBPool.java:452) -select
read source pxc_M2 for dataHost:pxc_Host
    ...
```

在第 1 个节点再增加一台 slave：

```
<?xml version="1.0"?>
<!DOCTYPE mycat:schema SYSTEM "schema.dtd">
<mycat:schema xmlns:mycat="http://org.opencloudb/">
        <schema name="mycat01" checkSQLschema="false" sqlMaxLimit="100">
            <table name="travelrecord" primaryKey="id" dataNode="pxc" />
        </schema>
        <dataNode name="pxc" dataHost="pxc_Host" database="yf" />
        <dataHost name="pxc_Host" maxCon="1000" minCon="3" balance="2"
writeType="0" dbType="MySQL" dbDriver="native" switchType="3">
                <heartbeat>show status like 'wsrep%'</heartbeat>
                <writeHost host="pxc_M1" url="10.230.3.194:3306" user="xjw_dba"
password="123">
                    <readHost host="pxc_S1" url="10.230.2.132:3307" user=
"xjw_dba" password="123" />
                </writeHost>
                <writeHost host="pxc_M2" url="10.230.3.195:3306" user="xjw_dba"
password="123">
                </writeHost>
                <writeHost host="pxc_M3" url="10.230.3.196:3306" user="xjw_dba"
password="123">
                </writeHost>
        </dataHost>
</mycat:schema>
```

```
MySQL> show @@datasource;
+----------+--------+-------+--------------+------+------+--------+------+--
----+---------+
| DATANODE | NAME   | TYPE  | HOST         | PORT | W/R  | ACTIVE | IDLE | SIZE |
EXECUTE |
+----------+--------+-------+--------------+------+------+--------+------+--
----+---------+
| pxc    | pxc_M1 | MySQL | 10.230.3.194 | 3306 | W    |    0 |    3 | 1000 |     9 |
| pxc    | pxc_M2 | MySQL | 10.230.3.195 | 3306 | W    |    0 |    1 | 1000 |     6 |
| pxc    | pxc_M3 | MySQL | 10.230.3.196 | 3306 | W    |    0 |    1 | 1000 |     6 |
| pxc    | pxc_S1 | MySQL | 10.230.2.132 | 3307 | R    |    0 |    1 | 1000 |     7 |
+----------+--------+-------+--------------+------+------+--------+------+--
----+---------+
```

多次执行查询后，Mycat 日志输出：

```
MySQL> select * from travelrecord;
02/22 14:27:43.160  DEBUG [$_NIOREACTOR-0-RW] (NonBlockingSession.java:113)
-ServerConnection [id=1, schema=mycat01, host=10.230.3.191, user=xjw_dba,
txIsolation=3, autocommit=true, schema=mycat01]select * from travelrecord, route={
    1 -> pxc{SELECT *
FROM travelrecord
```

```
LIMIT 100}
} rrs
02/22 14:27:43.160  DEBUG [$_NIOREACTOR-0-RW] (PhysicalDBPool.java:452) -select
read source pxc_M1 for dataHost:pxc_Host
......
```

当 balance 为 3 时，所有读请求都随机地分发至 writeHost 对应的 readHost 执行，writerHost 不负担读压力。如果 writeHost 没有对应的 readHost，则查询分发至 writeHost。Mycat 配置如下：

```
<?xml version="1.0"?>
<!DOCTYPE mycat:schema SYSTEM "schema.dtd">
<mycat:schema xmlns:mycat="http://org.opencloudb/">
      <schema name="mycat01" checkSQLschema="false" sqlMaxLimit="100">
         <table name="travelrecord" primaryKey="id" dataNode="pxc" />
      </schema>
      <dataNode name="pxc" dataHost="pxc_Host" database="yf" />
      <dataHost name="pxc_Host" maxCon="1000" minCon="3" balance="2"
writeType="0" dbType="MySQL" dbDriver="native" switchType="3">
            <heartbeat>show status like 'wsrep%'</heartbeat>
            <writeHost host="pxc_M1" url="10.230.3.194:3306" user="xjw_dba"
password="123">
               <readHost host="pxc_S1" url="10.230.2.132:3307" user=
"xjw_dba" password="123" />
            </writeHost>
            <writeHost host="pxc_M2" url="10.230.3.195:3306" user="xjw_dba"
password="123">
            </writeHost>
            <writeHost host="pxc_M3" url="10.230.3.196:3306" user="xjw_dba"
password="123">
            </writeHost>
      </dataHost>
</mycat:schema>
```

多次执行查询语句：

```
MySQL> select * from travelrecord;
```

分析 Mycat 日志可以看出，所有的查询都分发至 pxc_S1。

```
02/22 14:45:35.729  DEBUG [$_NIOREACTOR-0-RW] (NonBlockingSession.java:113)
-ServerConnection [id=1, schema=mycat01, host=10.230.3.191, user=xjw_dba,
txIsolation=3, autocommit=true, schema=mycat01]select * from travelrecord, route={
   1 -> pxc{SELECT *
FROM travelrecord
LIMIT 100}
} rrs
...
```

如果去掉 pxc_M1 对应的 pxc_S1，则配置如下：

```
<?xml version="1.0"?>
<!DOCTYPE mycat:schema SYSTEM "schema.dtd">
<mycat:schema xmlns:mycat="http://org.opencloudb/">
        <schema name="mycat01" checkSQLschema="false" sqlMaxLimit="100">
                <table name="travelrecord" primaryKey="id" dataNode="pxc" />
        </schema>
        <dataNode name="pxc" dataHost="pxc_Host" database="yf" />
        <dataHost name="pxc_Host" maxCon="1000" minCon="3" balance="3"
writeType="0" dbType="MySQL" dbDriver="native" switchType="3">
                <heartbeat>show status like 'wsrep%'</heartbeat>
                <writeHost host="pxc_M1" url="10.230.3.194:3306" user="xjw_dba"
password="123">
                </writeHost>
                <writeHost host="pxc_M2" url="10.230.3.195:3306" user="xjw_dba"
password="123">
                </writeHost>
                <writeHost host="pxc_M3" url="10.230.3.196:3306" user="xjw_dba"
password="123">
                </writeHost>
        </dataHost>
</mycat:schema>
```

多次执行查询：

```
MySQL> select * from travelrecord;
```

分析 Mycat 日志可以看出，所有查询都分发至 pxc_M1。

```
02/22 14:50:52.314  DEBUG [$_NIOREACTOR-6-RW] (NonBlockingSession.java:113)
-ServerConnection [id=1, schema=mycat01, host=10.230.3.191, user=xjw_dba,
txIsolation=3, autocommit=true, schema=mycat01]select * from travelrecord, route={
   1 -> pxc{SELECT *
FROM travelrecord
LIMIT 100}
} rrs
02/22 14:50:52.314  DEBUG [$_NIOREACTOR-6-RW] (PhysicalDBPool.java:452) -select
read source pxc_M1 for dataHost:pxc_Host
...
```

从上面的各种验证中可以看出，Galera Cluster 有多个节点时，就能通过 Mycat 实现读写分离，而且从 Galera 的原理可以看出，节点间的同步是“实时”的，这就为那些对读写延迟要求严格的系统提供了很好的选择。如果某个读节点查询的资源消耗很大，影响了整个集群的性能，则这时可以在节点上加一个 slave，把大查询再分离出去。增加 slave 或者调整读节点的操作对前端都是透明的，只需修改 Mycat 配置文件，重启或者重载 Mycat 即可。Galera Cluster 的读写分离在某种意义上与 Oracle ADG 非常相似。

4.1.3 SQL Server 读写分离

SQL Server 也是目前流行的关系型数据库之一，有其特有的读写分离结构，比如复制订阅、allwaysone 等，Mycat 同样支持 SQL Server。本节将通过 SQL Server 的复制订阅架构，来说明 Mycat 是如何支持对它的读写分离的。首先简单介绍一下 SQL Server 复制订阅。比起 MySQL 的复制，SQL Server 复制相对强大，配置也较为复杂。

复制的架构组成如下。

- 发布服务器：生产维护数据源，审阅所有出版数据的更改并发送给分发服务器。
- 分发服务器：分发服务器包括分发数据库，并且存储元数据、历史数据和事务。
- 订阅服务器：保持数据的副本，并接收对所修改出版的更改。取决于所实现的复制选项，可能还允许更新者更新数据，并将其复制给服务器或者其他订阅者。

按照复制的类型，可将复制分为快照复制、事务复制、合并复制。按照复制模式又分为推模式（Push）和拉模式（Pull）

准备环境如表 4-3 所示。

表 4-3

项　目	Mycat	Sql1-master	Sql2-disp	Sql3-r
IP	10.230.4.131	10.230.3.200	10.230.3.201	10.230.3.201
Port	8077/9077	1433	1433	1433

注意：Mycat 连接 SQL Server 的驱动 sqljdbc4.jar 需放入 MYCAT_HOME/lib。

Mycat 的基本配置如下：

```
<?xml version="1.0"?>
<!DOCTYPE mycat:schema SYSTEM "schema.dtd">
<mycat:schema xmlns:mycat="http://org.opencloudb/">
        <schema name="mycat01" checkSQLschema="false" sqlMaxLimit="100" >
            <table name="yf" primaryKey="ID" dataNode="sql_dn" />
        </schema>
        <dataNode name="sql_dn" dataHost="sql01" database="xjw_test" />
        <dataHost name="sql01" maxCon="1000" minCon="1" balance="0" writeType=
"0" dbType="sqlserver" dbDriver="jdbc">
        <heartbeat >select 1</heartbeat>
        <writeHost host="sql01_m" url="jdbc:sqlserver://10.230.3.200:1433"
user="xjw_dba" password="xjw123" >
                <readHost host="sql01_s" url="jdbc:sqlserver:
//10.230.3.202:1433" user="xjw_dba" password="xjw123" />
         </writeHost>
        </dataHost>
```

```
</mycat:schema>
```

当 balance 为 0 时，不开启读写分离机制，所有读操作都发送到当前的 writeHost 上。

```
MySQL> show @@datasource;
+----------+---------+-----------+-------------+------+------+--------+------
+------+---------+
| DATANODE | NAME    | TYPE      | HOST        | PORT | W/R  | ACTIVE | IDLE | SIZE
| EXECUTE |
+----------+---------+-----------+-------------+------+------+--------+------
+------+---------+
| sql_dn   | sql01_m | sqlserver | 10.230.3.200 | 1433 | W    |      0 |    1 | 1000
|      11 |
| sql_dn   | sql01_s | sqlserver | 10.230.3.202 | 1433 | R    |      0 |    0 | 1000
|       0 |
+----------+---------+-----------+-------------+------+------+--------+------
+------+---------+
```

多次执行查询：

```
MySQL> select * from yf;
```

分析 Mycat 日志可以看出，balance 为 0 时实际上读写分离没有开启，读操作仍然是访问 writeHost。

```
03/08 15:30:53.402  DEBUG [$_NIOREACTOR-2-RW] (NonBlockingSession.java:113)
-ServerConnection [id=2, schema=mycat01, host=10.230.3.191, user=xjw_dba,
txIsolation=3, autocommit=true, schema=mycat01]select * from yf, route={
   1 -> sql_dn{SELECT TOP 100 *
FROM yf
ORDER BY (select 0)}
} rrs
03/08 15:30:53.402  DEBUG [$_NIOREACTOR-2-RW] (PhysicalDBPool.java:452) -select
read source sql01_m for dataHost:sql01
......
```

balance=1 时，执行 show @@datasource 命令的结果如下：

```
MySQL> show @@datasource;
+----------+---------+-----------+-------------+------+------+--------+------
+------+---------+
| DATANODE | NAME    | TYPE      | HOST        | PORT | W/R  | ACTIVE | IDLE | SIZE
| EXECUTE |
+----------+---------+-----------+-------------+------+------+--------+------
+------+---------+
| sql_dn   | sql01_m | sqlserver | 10.230.3.200 | 1433 | W    |      0 |    1 | 1000
|       1 |
| sql_dn   | sql01_s | sqlserver | 10.230.3.202 | 1433 | R    |      0 |    0 | 1000
|       0 |
+----------+---------+-----------+-------------+------+------+--------+------
```

```
--+------+---------+
```

多次执行查询：

```
MySQL> select * from yf;
```

分析 Mycat 日志可以看出，因为没有其他 writeHost，所以读操作只能分发至 readHost：

```
03/08 15:55:04.631  DEBUG [$_NIOREACTOR-2-RW] (NonBlockingSession.java:113)
-ServerConnection [id=2, schema=mycat01, host=10.230.3.191, user=xjw_dba,
txIsolation=3, autocommit=true, schema=mycat01]select * from yf, route={
    1 -> sql_dn{SELECT TOP 100 *
FROM yf
ORDER BY (select 0)}
} rrs
03/08 15:55:04.631  DEBUG [$_NIOREACTOR-2-RW] (PhysicalDBPool.java:452) -select
read source sql01_s for dataHost:sql01
```

balance=2 时，执行 show @@datasource 命令的结果如下：

```
MySQL> show @@datasource;
+----------+---------+-----------+--------------+------+------+--------+----
--+------+---------+
| DATANODE | NAME    | TYPE      | HOST         | PORT | W/R  | ACTIVE | IDLE | SIZE
| EXECUTE |
+----------+---------+-----------+--------------+------+------+--------+----
--+------+---------+
| sql_dn   | sql01_m | sqlserver | 10.230.3.200 | 1433 | W    |      0 |    1 | 1000
|       1 |
| sql_dn   | sql01_s | sqlserver | 10.230.3.202 | 1433 | R    |      0 |    0 | 1000
|       0 |
+----------+---------+-----------+--------------+------+------+--------+----
--+------+---------+
```

多次执行查询：

```
MySQL> select * from yf;
```

分析 Mycat 日志可以看出，读操作随机分发至 writeHost 和 readHost：

```
03/08 16:55:59.286  DEBUG [$_NIOREACTOR-4-RW] (NonBlockingSession.java:113)
-ServerConnection [id=4, schema=mycat01, host=10.230.3.191, user=xjw_dba,
txIsolation=3, autocommit=true, schema=mycat01]select * from yf, route={
    1 -> sql_dn{SELECT TOP 100 *
FROM yf
ORDER BY (select 0)}
} rrs
03/08 16:55:59.286  DEBUG [$_NIOREACTOR-4-RW] (PhysicalDBPool.java:452) -select
read source sql01_s for dataHost:sql01
……
```

balance=2 时，执行 show @@datasource 命令的结果如下：

```
MySQL> show @@datasource;
+----------+---------+-----------+-------------+------+------+--------+------+----------+
| DATANODE | NAME    | TYPE      | HOST        | PORT | W/R  | ACTIVE | IDLE | SIZE | EXECUTE |
+----------+---------+-----------+-------------+------+------+--------+------+----------+
| sql_dn   | sql01_m | sqlserver | 10.230.3.200 | 1433 | W    |      0 |    1 | 1000 |       1 |
| sql_dn   | sql01_s | sqlserver | 10.230.3.202 | 1433 | R    |      0 |    0 | 1000 |       0 |
+----------+---------+-----------+-------------+------+------+--------+------+----------+
```

Mycat 输出日志如下：

```
03/08 17:13:22.937  DEBUG [$_NIOREACTOR-3-RW] (NonBlockingSession.java:113)
-ServerConnection [id=3, schema=mycat01, host=10.230.3.191, user=xjw_dba,
txIsolation=3, autocommit=true, schema=mycat01]select * from yf, route={
   1 -> sql_dn{SELECT TOP 100 *
FROM yf
ORDER BY (select 0)}
} rrs
03/08 17:13:22.937  DEBUG [$_NIOREACTOR-3-RW] (PhysicalDBPool.java:452) -select
read source sql01_s for dataHost:sql01
……
```

通过上面的几种验证可以看出，SQL Server 如果有多个订阅库（本文中的 sql01_s）存在，则可以选择 balance 为 1，这样主库就不参加读操作了。当然，如果主库（本文中的 sql01_m）的写压力较小，则也可以选择 balance 为 2，将一部读操作分配到主库以更好地利用资源。

4.2 Mycat 故障切换

4.2.1 Mycat 主从切换

如图 4-2 所示，MySQL 节点开启主从复制的配置方案，并将主节点配置为 Mycat 的 dataHost 里的 writeNode，将从节点配置为 readNode，同时 Mycat 内部定期对一个 dataHost 里的所有 writeHost 与 readHost 节点发起心跳检测。在正常情况下，Mycat 会将第 1 个 writeHost 作为写节点，所有的 DML SQL 都会发送到此节点，若 Mycat 开启了读写分离，则查询节点会根据读写分离的策略发往 readHost（和 writeHost）执行。在一个 dataHost 里面配置了两个或多个 writeHost 的情况下，如果第 1 个 writeHost 宕机，则 Mycat 会在默认的 3 次心跳检查失败后，自动切换到下一个可用的 writeHost 执行 DML SQL 语句，并在 conf/dnindex.properties 文件里记录当前所用的 writeHost 的 index（第 1 个为 0，第 2 个为 1，以此类推）。注意，此文件不能删除和擅自改变，除非你深刻理解了它的作用及你的目的。

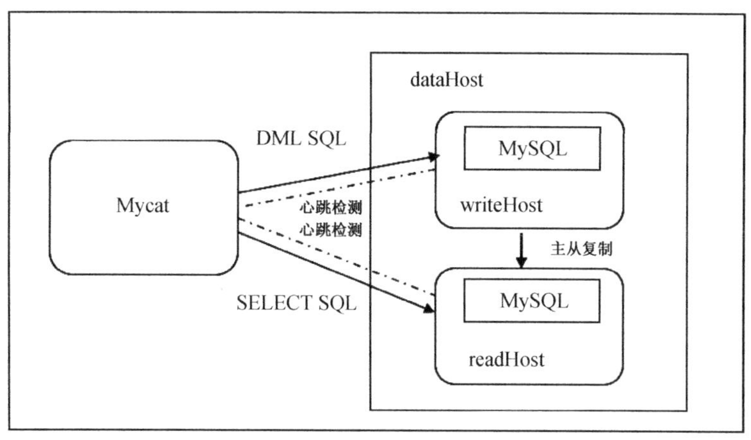

图 4-2

当原来配置的 MySQL 写节点宕机恢复以后，保持现有状态的不变，将恢复后的 MySQL 节点作为从节点，跟随新的主节点，重新配置主从同步，原先跟随该节点做同步的其他节点同样重新配置同步源，这些节点的数据手工完成同步以后，再加入 Mycat 里。

下面通过一个实例展示 Mycat 是如何控制 MySQL 主从故障切换的（实例来源于 leader-us 培训课程优秀作业，作者是郝坚剑）。

dataHost 配置如图 4-3 所示。

```
<dataHost name="localhost1" maxCon="1000" minCon="10" balance="1"
    writeType="0" dbType="mysql" dbDriver="native" switchType="2"  slaveThreshold="20">
    <heartbeat>show slave status</heartbeat>
    <!-- can have multi write hosts -->
    <writeHost host="hostM1" url="192.168.100.57:3306" user="mycat" password="mycat" >

    </writeHost>

    <writeHost host="hostM2" url="192.168.100.70:3306" user="mycat" password="mycat" >

    </writeHost>
    <!-- <writeHost host="hostM2" url="localhost:3316" user="root" password="123456"/> -->
</dataHost>
```

图 4-3

测试用例设计如表 4-4 所示。

表 4-4

测试用例	参数设置				测试内容
	balance	writeType	switchType	slaveThreshold	
Mycat-MS-01	1	0	2	20	主从同步状态下，主节点宕机后节点的切换情况
Mycat-MS-02	1	0	2	20	主从同步状态下，备节点宕机后节点切换的情况
Mycat-MS-03	1	0	2	20	主从同步延时超过 20 秒的状态下，当前节点宕机后节点的切换情况
Mycat-MS-05	1	0	2	20	测试停止 Mycat 服务，修改 schemma.xml 是否会对选择写节点产生影响

下面测试服务器的 Mycat-MS-01 节点，测试预期如下。

- 在 hostM1、hostM2 节点均正常的情况下，Mycat 会选择 hostM1 节点插入数据。
- hostM1 节点宕机，根据配置自动切换至 hostM2 节点。
- hostM1 恢复服务，根据配置继续使用 hostM2 节点作为写节点。
- 重启 Mycat Server，根据配置选择 hostM2 节点作为写节点。

测试过程如下，当前写节点为 hostM1，备用写节点为 hostM2。

注意：要先清除 travelrecord 表。

（1）执行如下语句：

```
insert into travelrecord (org_code,test_name) values
('01',@@hostname);
```

预期结果为：hostM1 为当前写节点，根据 hostM1 自增长键步长配置，插入的记录 ID 值为奇数，结果如图 4-4、图 4-5 所示。

图 4-4

图 4-5

（2）查看 hostM2 节点 slave status。

（3）停止 hostM1 节点的 MySQL 服务./mysql.server stop。

预期结果为：停止 hostM1 节点的 MySQL 服务后，根据配置及主从状态，当前写节点切换为 hostM2。

（4）执行如下语句：

```
insert into travelrecord (org_code,test_name) values
('01',@@hostname);
```

预期结果为：hostM2 为当前写节点，根据 hostM2 自增长键步长配置，插入的记录 ID 值为偶数，结果如图 4-6、图 4-7 所示。

图 4-6

图 4-7

（5）启动 hostM1 节点的 MySQL 服务./mysql.server start。

预期结果为：根据配置，hostM2 为写节点，hostM1 为备用写节点。

（6）执行如下语句：

```
insert into travelrecord (org_code,test_name) values
('01',@@hostname);
```

预期结果为：hostM2 为当前写节点，根据 hostM2 自增长键步长配置，插入的记录 ID 值为偶数，结果如图 4-8 所示。

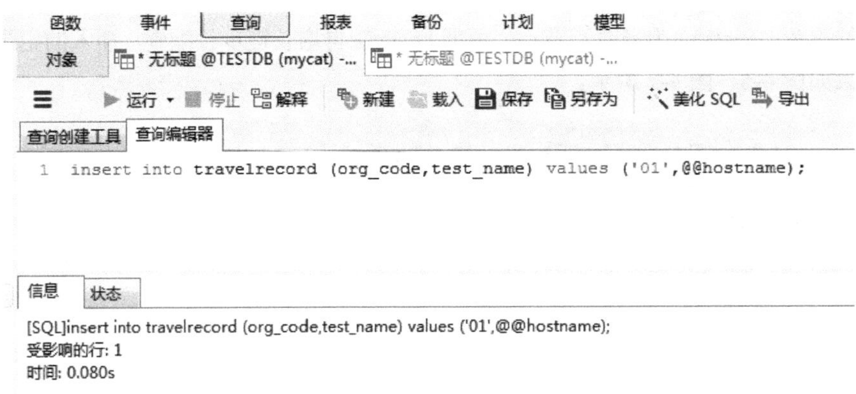

图 4-8

根据 balance 配置，应从 hostM1 节点读取数据，故查询到结果集 test_name 的值应为 oracle，如图 4-9 所示。

图 4-9

（7）重新启动 Mycat 服务/mycat start。

预期结果为：根据配置及上述测试过程中的节点切换，hostM2 为写节点，hostM1 为备用写节点。

（8）执行如下语句：

```
insert into travelrecord (org_code,test_name) values
('01',@@hostname);
```

预期结果为：hostM2 为当前写节点，根据 hostM2 自增长键步长配置，插入的记录 ID 值为偶数，结果如图 4-10、图 4-11 所示。

图 4-10

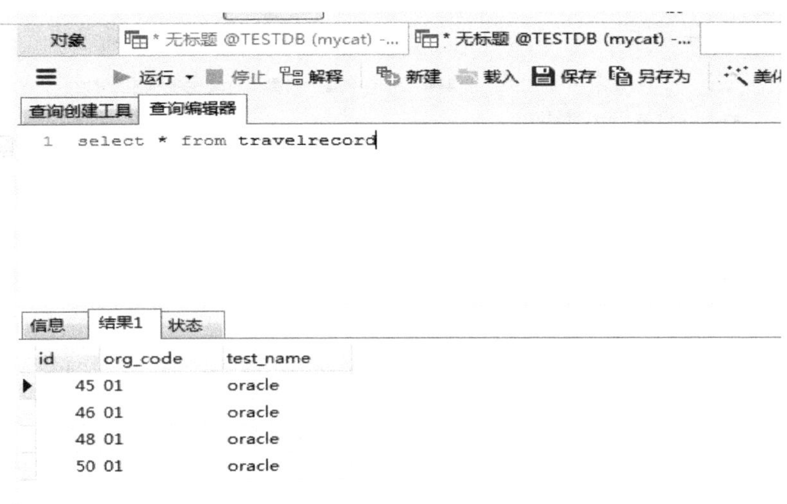

图 4-11

测试结论为：符合预期。

下面测试服务器的 Mycat-MS-02 节点。基于 Mycat 4.3 的测试流程，Mycat 的当前写节点为 hostM2，备用节点为 hostM1。

测试预期为：服务正常，通过 hostM2 写入数据，通过 hostM1 查询数据。停止 hostM1 节点的 MySQL 服务，读、写均通过 hostM2 完成。

测试过程如下。

（1）执行如下语句：

```
insert into travelrecord (org_code,test_name) values
('01',@@hostname);
```

预期结果为：hostM2 为当前写节点，根据 hostM2 自增长键步长配置，插入的记录 ID 值为偶数，结果如图 4-12、图 4-13 所示。

（2）停止 hostM1 节点的 MySQL 服务./mysql.server stop。

（3）执行如下语句：

```
insert into travelrecord (org_code,test_name) values
('01',@@hostname);
```

预期结果为：hostM2 为当前写节点，根据 hostM2 自增长键步长配置，插入的记录 ID 值为偶数，结果如图 4-14、图 4-15 所示。

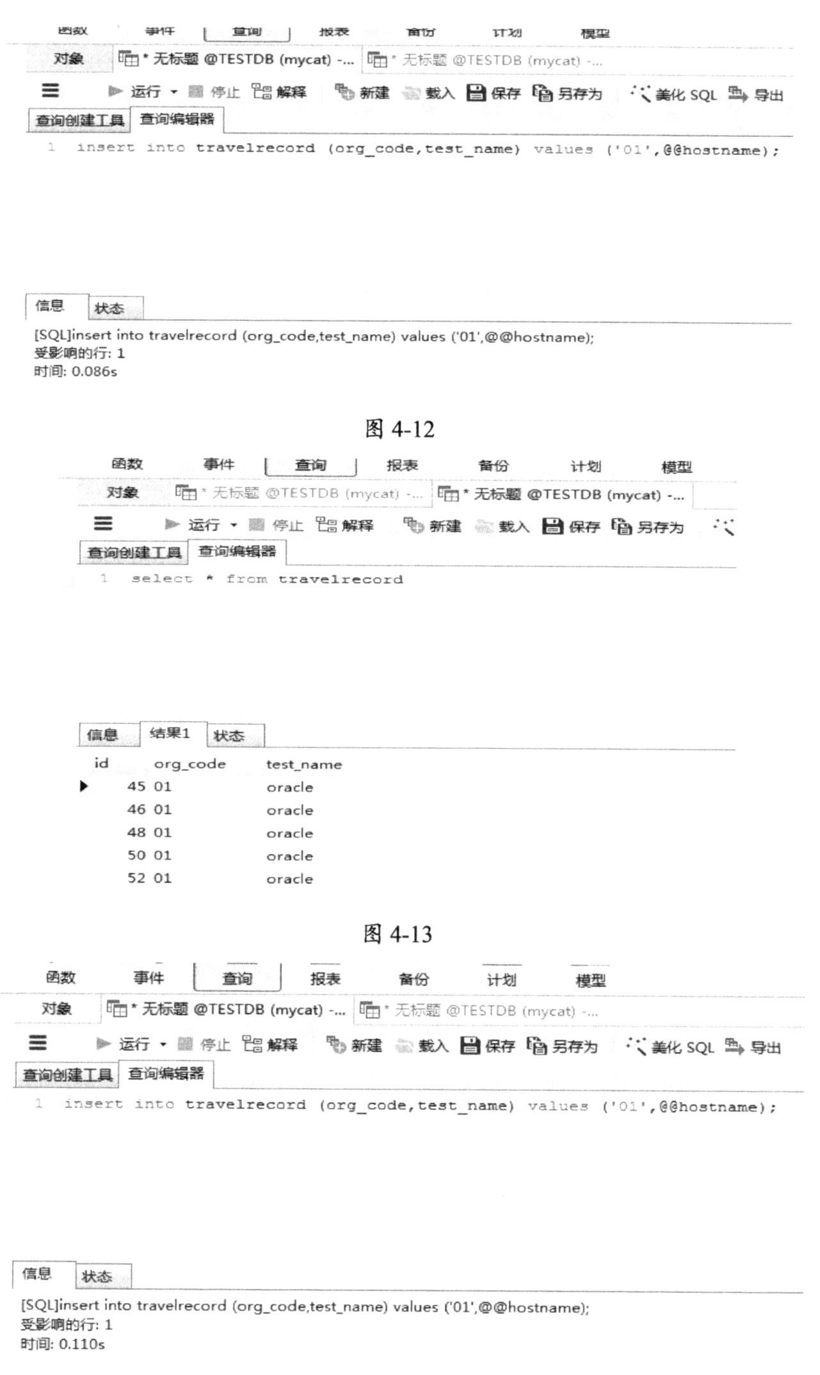

图 4-12

图 4-13

图 4-14

图 4-15

（4）启动 hostM1 节点的 MySQL 服务。

（5）执行如下语句：

```
select * from travelrecord
```

预期结果为：根据配置，选择 hostM1 节点作为读节点执行查询，结果如图 4-16 所示。

图 4-16

测试结论为符合预期。

下面测试服务器的 Mycat-MS-03 节点。经过前面的测试后，当前写节点为 hostM2，备用写节点为 hostM1。

测试预期为：当备用节点的同步延时超过配置值时，当前写节点宕机，将不发生切换，会导致所有 Mycat 的所有读写操作失败，再次启动当前的写节点服务后，继续使用原有的写节点作为写节点。

测试过程如下。

（1）将 hostM1 节点表加锁：

```
lock table travelrecord read;
```

（2）执行如下语句：

```
insert into travelrecord (org_code,test_name) values
('01',@@hostname);
```

预期结果为：hostM2 为当前写节点，根据 hostM2 自增长键步长配置，插入的记录 ID 值为偶数，结果如图 4-17 所示。

图 4-17

（3）等待 30 秒，确保同步延迟高于配置值，查看 hostM1slave status。

（4）执行如下语句：

```
select * from travelrecord
```

预期结果为：基于配置及同步延迟，选择 hostM2 节点查询，查询结果集 test_name 的值为 xbserver，结果如图 4-18 所示。

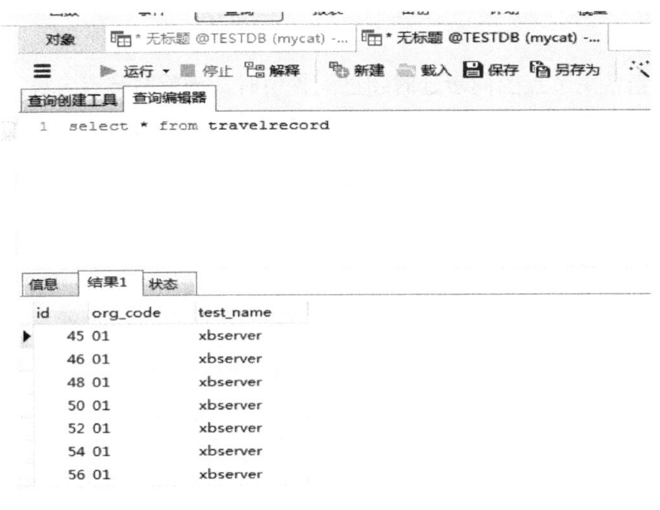

图 4-18

（5）停止 hostM2 节点的 MySQL 服务./mysql.server stop。

（6）执行如下语句：

```
insert into travelrecord (org_code,test_name) values
('01',@@hostname);
```

预期结果为：无法执行插入动作。结果如图 4-19 所示。

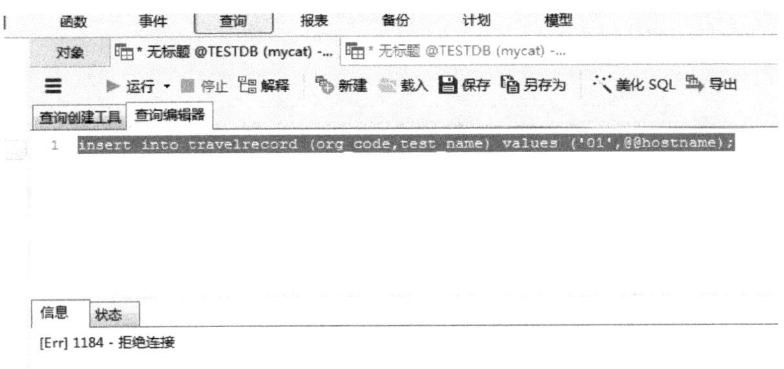

图 4-19

（7）执行如下语句：

```
select * from travelrecord
```

预期结果为：无法执行查询动作。结果如图 4-20 所示。

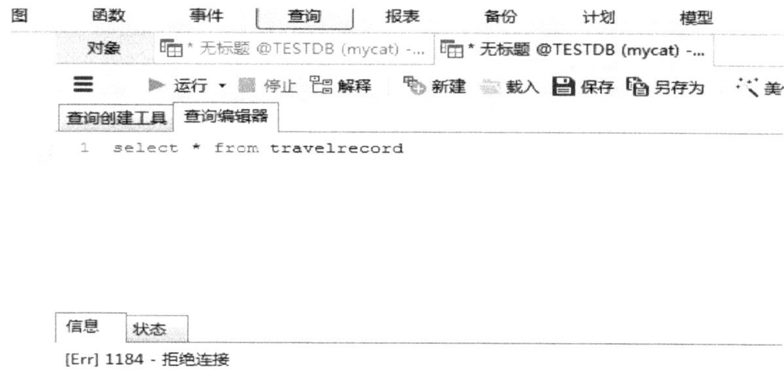

图 4-20

（8）启动 hostM2 节点的 MySQL 服务./mysql.server start。

（9）hostM1 节点表解锁：

```
unlock table;
```

（10）执行如下语句：

```
select * from travelrecord;
```

结果如图 4-21、图 4-22 所示。

信息　结果1　状态

id	org_code	test_name
45	01	oracle
46	01	oracle
48	01	oracle
50	01	oracle
52	01	oracle
54	01	oracle
56	01	oracle

图 4-21

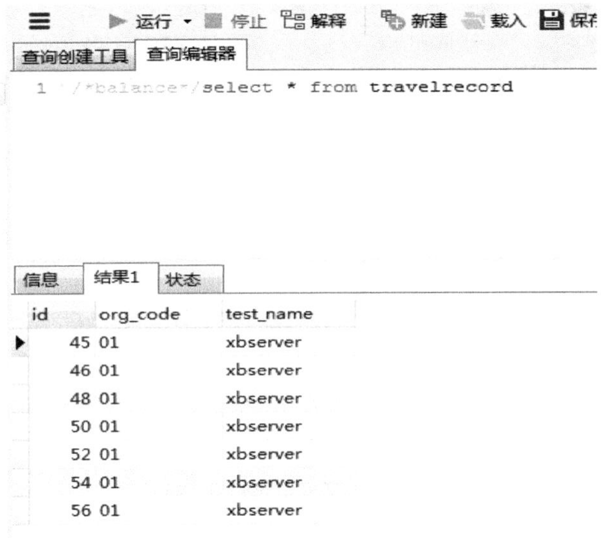

图 4-22

（11）执行如下语句：

```
insert into travelrecord (org_code,test_name) values
('01',@@hostname);
```

预期结果为：hostM2 为写节点，写入 ID 为偶数的记录。结果如图 4-23、图 4-24 所示。

图 4-23

信息　结果1　状态

id	org_code	test_name
45	01	oracle
46	01	oracle
48	01	oracle
50	01	oracle
52	01	oracle
54	01	oracle
56	01	oracle
58	01	oracle

图 4-24

测试结论为符合预期。

4.2.2　MySQL Galera 节点切换

从 1.4 版本开始，Mycat 新增了对 Galera Cluster 的心跳检测，这样就为 Galera Cluster 的节点切换提供了判断依据。下面通过实例来演示具体的节点切换过程。

方案一：schema.xml 配置

```
<schema name="pxc" checkSQLschema="false" sqlMaxLimit="100">
          <table name="travelrecord" primaryKey="id" dataNode="pxc" />
      </schema>
      <dataNode name="pxc" dataHost="pxc_Host" database="yf" />
      <dataHost name="pxc_Host" maxCon="1000" minCon="3" balance="1"
writeType="0" dbType="mysql" dbDriver="native" switchType="3">
            <heartbeat>show status like 'wsrep%'</heartbeat>
            <writeHost host="pxc_M1" url="10.35.12.26:3307" user="xjw_dba"
password="123">
          <readHost host="pxc_R1" url="10.35.12.27:3307" user="xjw_dba"
password="123" />
          <readHost host="pxc_R2" url="10.35.12.28:3307" user="xjw_dba"
password="123" />
            </writeHost>
            <writeHost host="pxc_M2" url="10.35.12.27:3307" user="xjw_dba"
password="123">
          <readHost host="pxc_R3" url="10.35.12.26:3307" user="xjw_dba"
```

```
password="123" />
            <readHost host="pxc_R4" url="10.35.12.28:3307" user="xjw_dba"
password="123" />
               </writeHost>
               <writeHost host="pxc_M3" url="10.35.12.28:3307" user="xjw_dba"
password="123">
            <readHost host="pxc_R5" url="10.35.12.26:3307" user="xjw_dba"
password="123" />
            <readHost host="pxc_R6" url="10.35.12.27:3307" user="xjw_dba"
password="123" />
            </writeHost>
         </dataHost>
```

对代码的说明如下。

- blance=1：所有 readHost 和 standby writeHost 都参与了读负载。
- writeType=0：所有写操作都被发送到配置的第 1 个 writeHost 上，第 1 个挂了则切换到第 2 个 writeHost。
- switchType=3：基于 MySQL Galary Cluster 的切换机制，心跳语句为 show status like 'wsrep%'。

配置为 writeHost 的节点为写节点，其他节点作为 writeHost 的 readHost 查看所有 host 的状态：

```
(xjw_dba@10.35.9.200) [(none)]> show @@heartbeat;
+--------+-------+-------------+------+---------+-------+--------+---------+
--------------+--------------------+-------+
| NAME   | TYPE  | HOST        | PORT | RS_CODE | RETRY | STATUS | TIMEOUT |
EXECUTE_TIME | LAST_ACTIVE_TIME    | STOP  |
+--------+-------+-------------+------+---------+-------+--------+---------+
--------------+--------------------+-------+
| pxc_M1 | mysql | 10.35.12.26 | 3307 |       1 |     0 | idle   |       0 | 43,43,43
| 2016-08-24 14:05:17 | false |
| pxc_M2 | mysql | 10.35.12.27 | 3307 |       1 |     0 | idle   |       0 | 68,68,68
| 2016-08-24 14:05:17 | false |
| pxc_M3 | mysql | 10.35.12.28 | 3307 |       1 |     0 | idle   |       0 | 71,71,71
| 2016-08-24 14:05:17 | false |
...
+--------+-------+-------------+------+---------+-------+--------+---------+
--------------+--------------------+-------+
9 rows in set (0.01 sec)
```

创建测试表并插入数据：

```
(xjw_dba@10.35.9.200) [(none)]> use pxc
Database changed
(xjw_dba@10.35.9.200) [pxc]> create table travelrecord (id int  not null
```

```
auto_increment, org_code varchar(10),host_name varchar(50),primary key (id));
    Query OK, 0 rows affected (0.21 sec)
    (xjw_dba@10.35.9.200) [pxc]> insert into travelrecord(org_code,host_name)
values('1',@@hostname);
    Query OK, 1 row affected (0.07 sec)
```

查看 dnindex.properties：

```
#update
#Wed Aug 24 10:35:35 CST 2016
pxc_Host=0
```

执行 create 语句的 Mycat 日志如下：

```
    08/24 14:06:57.345  DEBUG [$_NIOREACTOR-0-RW] (NonBlockingSession.java:113)
-ServerConnection [id=2, schema=pxc, host=10.35.9.200, user=xjw_dba,txIsolation=3,
autocommit=true, schema=pxc]create table travelrecord (id int  not null
auto_increment, org_code varchar(10),host_name varchar(50),primary key (id)),
route={
       1 -> pxc{create table travelrecord (id int  not null auto_increment, org_code
varchar(10),host_name varchar(50),primary key (id))}
    } rrs
    08/24 14:06:57.347  DEBUG [$_NIOREACTOR-0-RW] (MySQLConnection.java:459) -con
need syn ,total syn cmd 1 commands SET SESSION TRANSACTION ISOLATION LEVEL REPEATABLE
READ;schema change:false con:MySQLConnection [id=3, lastTime=1472018817346,
user=xjw_dba, schema=yf, old shema=yf, borrowed=true, fromslaveDB=false, threadId=47,
charset=utf8, txIsolation=0, autocommit=true, attachment=pxc{create table
travelrecord (id int  not null auto_increment, org_code varchar(10),host_name
varchar(50),primary key (id))}, respHandler=SingleNodeHandler [node=pxc{create
table travelrecord (id int  not null auto_increment, org_code varchar(10),host_name
varchar(50),primary key (id))}, packetId=0], host=10.35.12.26, port=3307,
statusSync=null, writeQueue=0, modifiedSQLExecuted=true]
    ...
```

执行 insert 语句的 Mycat 日志如下：

```
    08/24 14:07:11.495  DEBUG [$_NIOREACTOR-0-RW] (ServerQueryHandler.java:56)
-ServerConnection [id=2, schema=pxc, host=10.35.9.200, user=xjw_dba,txIsolation=3,
autocommit=true, schema=pxc]insert into travelrecord(org_code,host_name) values
('1',@@hostname)
    08/24 14:07:11.548  DEBUG [$_NIOREACTOR-0-RW] (NonBlockingSession.java:113)
-ServerConnection [id=2, schema=pxc, host=10.35.9.200, user=xjw_dba,txIsolation=3,
autocommit=true, schema=pxc]insert into travelrecord(org_code,host_name) values
('1',@@hostname), route={
       1 -> pxc{insert into travelrecord(org_code,host_name) values('1',@@hostname)}
    } rrs
    ...
```

下面执行 3 次 select 查询语句：

```
(xjw_dba@10.35.9.200) [pxc]> select * from travelrecord;
```

```
+----+----------+-----------+
| id | org_code | host_name |
+----+----------+-----------+
|  1 | 1        | mysqldb26 |
+----+----------+-----------+
1 row in set (0.06 sec)
```

3 个节点都负载了 select 操作：

```
08/24 14:20:16.391  DEBUG [$_NIOREACTOR-0-RW] (NonBlockingSession.java:113)
-ServerConnection [id=2, schema=pxc, host=10.35.9.200, user=xjw_dba,txIsolation=3,
autocommit=true, schema=pxc]select * from travelrecord, route={
   1 -> pxc{SELECT *
FROM travelrecord
LIMIT 100}
} rrs
08/24 14:20:16.391  DEBUG [$_NIOREACTOR-0-RW] (PhysicalDBPool.java:452) -select
read source pxc_R5 for dataHost:pxc_Host
08/24 14:20:17.629  DEBUG [$_NIOREACTOR-0-RW] (NonBlockingSession.java:113)
-ServerConnection [id=2, schema=pxc, host=10.35.9.200, user=xjw_dba,txIsolation=3,
autocommit=true, schema=pxc]select * from travelrecord, route={
   1 -> pxc{SELECT *
FROM travelrecord
LIMIT 100}
} rrs
08/24 14:20:17.629  DEBUG [$_NIOREACTOR-0-RW] (PhysicalDBPool.java:452) -select
read source pxc_R3 for dataHost:pxc_Host
08/24 14:20:19.005  DEBUG [$_NIOREACTOR-0-RW] (NonBlockingSession.java:113)
-ServerConnection [id=2, schema=pxc, host=10.35.9.200, user=xjw_dba,txIsolation=3,
autocommit=true, schema=pxc]select * from travelrecord, route={
   1 -> pxc{SELECT *
FROM travelrecord
LIMIT 100}
} rrs
08/24 14:20:19.005  DEBUG [$_NIOREACTOR-0-RW] (PhysicalDBPool.java:452) -select
read source pxc_M2 for dataHost:pxc_Host
```

关闭 10.35.12.26 上的 MySQL 进程，查看节点的状态：

```
(xjw_dba@10.35.9.200) [(none)]> show @@heartbeat;
+--------+-------+-------------+------+---------+-------+--------+---------+
--------------+---------------------+-------+
| NAME   | TYPE  | HOST        | PORT | RS_CODE | RETRY | STATUS | TIMEOUT |
EXECUTE_TIME | LAST_ACTIVE_TIME    | STOP  |
+--------+-------+-------------+------+---------+-------+--------+---------+
--------------+---------------------+-------+
| pxc_M1 | mysql | 10.35.12.26 | 3307 |      -1 |     0 | idle   |       0 | 7,15,18
| 2016-08-24 14:28:07 | false |
| pxc_M2 | mysql | 10.35.12.27 | 3307 |       1 |     0 | idle   |       0 | 12,15,17
```

```
| 2016-08-24 14:28:07 | false |
    ...
    | pxc_R6 | mysql | 10.35.12.27 | 3307 |      1 |     0 | idle   |      0 | 8,6,8
| 2016-08-24 14:28:07 | false |
    +--------+-------+------------+------+---------+-------+--------+---------+
--------------+--------------------+-------+
    9 rows in set (0.00 sec)
```

再次执行插入语句：

```
(xjw_dba@10.35.9.200) [(none)]> use pxc
Database changed
(xjw_dba@10.35.9.200) [pxc]> insert into travelrecord(org_code,host_name)
values('1',@@hostname);
Query OK, 1 row affected (0.01 sec)
```

查看 dnindex.properties：

```
#update
#Wed Aug 24 14:28:07 CST 2016
pxc_Host=1
```

通过 Mycat 日志可以看得出，已选择 10.35.12.27 作为写入的节点：

```
08/24 14:28:59.697  DEBUG [$_NIOREACTOR-0-RW] (NonBlockingSession.java:229)
-release connection MySQLConnection [id=13, lastTime=1472020139690, user=xjw_dba,
schema=yf, old shema=yf, borrowed=true, fromslaveDB=false, threadId=38, charset=utf8,
txIsolation=3, autocommit=true, attachment=pxc{insert into travelrecord(org_code,
host_name) values('1',@@hostname)}, respHandler=SingleNodeHandler [node=pxc{insert
into travelrecord(org_code,host_name) values('1',@@hostname)}, packetId=1], host=
10.35.12.27, port=3307, statusSync=null, writeQueue=0, modifiedSQLExecuted=true]
```

下面执行 3 次 select 查询语句：

```
(xjw_dba@10.35.9.200) [pxc]> select * from travelrecord;
+----+----------+-----------+
| id | org_code | host_name |
+----+----------+-----------+
|  1 | 1        | mysqldb26 |
|  3 | 1        | mysqldb27 |
+----+----------+-----------+
2 rows in set (0.00 sec)
```

可看出 alive 的节点都负载了 select 操作：

```
08/24 14:34:14.010  DEBUG [$_NIOREACTOR-0-RW] (NonBlockingSession.java:113)
-ServerConnection [id=4, schema=pxc, host=10.35.9.200, user=xjw_dba,txIsolation=3,
autocommit=true, schema=pxc]select * from travelrecord, route={
   1 -> pxc{SELECT *
FROM travelrecord
LIMIT 100}
} rrs
```

```
08/24 14:34:14.010  DEBUG [$_NIOREACTOR-0-RW] (PhysicalDBPool.java:452) -select read source pxc_R6 for dataHost:pxc_Host
......
08/24 14:34:14.786  DEBUG [$_NIOREACTOR-0-RW]  (NonBlockingSession.java:113) -ServerConnection [id=4, schema=pxc, host=10.35.9.200, user=xjw_dba,txIsolation=3, autocommit=true, schema=pxc]select * from travelrecord, route={
   1 -> pxc{SELECT *
FROM travelrecord
LIMIT 100}
} rrs
08/24 14:34:14.787  DEBUG [$_NIOREACTOR-0-RW] (PhysicalDBPool.java:452) -select read source pxc_M3 for dataHost:pxc_Host
......
08/24 14:34:15.626  DEBUG [$_NIOREACTOR-0-RW]  (NonBlockingSession.java:113) -ServerConnection [id=4, schema=pxc, host=10.35.9.200, user=xjw_dba,txIsolation=3, autocommit=true, schema=pxc]select * from travelrecord, route={
   1 -> pxc{SELECT *
FROM travelrecord
LIMIT 100}
} rrs
08/24 14:34:15.626  DEBUG [$_NIOREACTOR-0-RW] (PhysicalDBPool.java:452) -select read source pxc_R4 for dataHost:pxc_Host
```

关闭 10.35.12.27 上的 MySQL 进程，查看 dnindex.properties：

```
#update
#Wed Aug 24 15:03:47 CST 2016
pxc_Host=2
```

查看节点的状态：

```
(xjw_dba@10.35.9.200) [(none)]> show @@heartbeat;
+--------+-------+------------+------+---------+-------+--------+---------+--------------+---------------------+-------+
| NAME   | TYPE  | HOST        | PORT | RS_CODE | RETRY | STATUS | TIMEOUT | EXECUTE_TIME | LAST_ACTIVE_TIME    | STOP  |
+--------+-------+------------+------+---------+-------+--------+---------+--------------+---------------------+-------+
| pxc_M1   | mysql | 10.35.12.26 | 3307 |   -1  |    0  | idle |   0    | 1,1,1 | 2016-08-24 15:04:07 | false |
| pxc_M2   | mysql | 10.35.12.27 | 3307 |   -1  |    0  | idle |   0    | 4,15,17 | 2016-08-24 15:04:07 | false |
...
| pxc_R6   | mysql | 10.35.12.27 | 3307 |   -1  |    0  | idle |   0    | 1,7,8 | 2016-08-24 15:04:07 | false |
+--------+-------+------------+------+---------+-------+--------+---------+--------------+---------------------+-------+
9 rows in set (0.00 sec)
```

执行插入语句：

```
(xjw_dba@10.35.9.200) [(none)]> use pxc
Database changed
(xjw_dba@10.35.9.200) [pxc]> insert into travelrecord(org_code,host_name)
values('1',@@hostname);
Query OK, 1 row affected (0.00 sec)
```

将 alive 的 10.35.12.28 作为 writeHost：

```
08/24 15:04:43.845  DEBUG [$_NIOREACTOR-0-RW] (NonBlockingSession.java:229)
-release connection MySQLConnection [id=5, lastTime=1472022283840, user=xjw_dba,
schema=yf, old shema=yf, borrowed=true, fromslaveDB=false, threadId=37, charset=utf8,
txIsolation=3, autocommit=true, attachment=pxc{insert into travelrecord(org_code,
host_name) values('1',@@hostname)}, respHandler=SingleNodeHandler [node=pxc{insert
into travelrecord(org_code,host_name) values('1',@@hostname)}, packetId=1], host=
10.35.12.28, port=3307, statusSync=null, writeQueue=0, modifiedSQLExecuted=true]
```

重启关闭节点，此时节点加入成功：

```
(xjw_dba@10.35.9.200) [(none)]> show @@heartbeat;
+--------+-------+-------------+------+---------+-------+--------+---------+
--------------+--------------------+-------+
| NAME   | TYPE  | HOST        | PORT | RS_CODE | RETRY | STATUS | TIMEOUT |
EXECUTE_TIME | LAST_ACTIVE_TIME    | STOP  |
+--------+-------+-------------+------+---------+-------+--------+---------+
--------------+--------------------+-------+
| pxc_M1 | mysql | 10.35.12.26 | 3307 |       1 |     0 | idle   |       0 | 12,1,1
| 2016-08-24 15:14:37 | false |
| pxc_M2 | mysql | 10.35.12.27 | 3307 |       1 |     0 | idle   |       0 | 91,9,9
| 2016-08-24 15:14:37 | false |
...
| pxc_R6 | mysql | 10.35.12.27 | 3307 |       1 |     0 | idle   |       0 | 97,9,9
| 2016-08-24 15:14:37 | false |
+--------+-------+-------------+------+---------+-------+--------+---------+
--------------+--------------------+-------+
9 rows in set (0.01 sec)
```

执行插入语句：

```
(xjw_dba@10.35.9.200) [(none)]> use pxc
Database changed
(xjw_dba@10.35.9.200) [pxc]> insert into travelrecord(org_code,host_name)
values('1',@@hostname);
Query OK, 1 row affected (0.00 sec)
```

查看 dnindex.properties：

```
#update
#Wed Aug 24 15:03:47 CST 2016
pxc_Host=2
```

没有变化，仍然将最后一次的 10.35.12.28 作为 writeHost：

```
08/24 15:15:13.970 DEBUG [$_NIOREACTOR-0-RW] (NonBlockingSession.java:229)
-release connection MySQLConnection [id=5, lastTime=1472022913956, user=xjw_dba,
schema=yf, old shema=yf, borrowed=true, fromslaveDB=false, threadId=37, charset=utf8,
txIsolation=3, autocommit=true, attachment=pxc{insert into travelrecord(org_code,
host_name) values('1',@@hostname)}, respHandler=SingleNodeHandler [node=pxc{insert
into travelrecord(org_code,host_name) values('1',@@hostname)}, packetId=1], host=
10.35.12.28, port=3307, statusSync=null, writeQueue=0, modifiedSQLExecuted=true]
```

总结如下。

- Mycat 选取一个 writeHost 作为写节点，所有节点都作为 readHost。
- 如果 writeHost 失效，则选择另一个 writeHost 作为写节点，其余节点作为 readHost。
- 除非全部 writeHost 失效，否则业务不会中断，安全性较高。

但是在这种配置下，在同一个节点心跳会发生三次。在笔者的生产环境中，Mycat 曾多次发生莫名其妙的 switch 现象，所以在 Galera 环境下不推荐上述配置。用 Mycat 搭配 Galera，schema 配置如下：

```
<?xml version="1.0"?>
<!DOCTYPE mycat:schema SYSTEM "schema.dtd">
<mycat:schema xmlns:mycat="http://org.opencloudb/">
        <schema name="pxc" checkSQLschema="false" sqlMaxLimit="100">
            <table name="travelrecord" primaryKey="id" dataNode="pxc" />
        </schema>
        <dataNode name="pxc" dataHost="pxc_Host" database="yf" />
        <dataHost name="pxc_Host" maxCon="1000" minCon="3" balance="1"
writeType="0" dbType="mysql" dbDriver="native" switchType="3">
                <heartbeat>show status like 'wsrep%'</heartbeat>
                <writeHost host="pxc_M1" url="10.35.12.26:3307" user="xjw_dba"
password="123">
                </writeHost>
                <writeHost host="pxc_M2" url="10.35.12.27:3307" user="xjw_dba"
password="123">
                </writeHost>
                <writeHost host="pxc_M3" url="10.35.12.28:3307" user="xjw_dba"
password="123">
                    </writeHost>
        </dataHost>
</mycat:schema>
```

对代码的说明如下。

- balance=1：所有 readHost 和 standby writeHost 都参与读负载，既然没有 readhost，那么只有其他 writehost 节点做读节点了。

- writeType=0：所有写操作都发送到配置的第 1 个 writeHost 上，第 1 个挂了就切换到第 2 个 writeHost。
- switchType=3：基于 MySQL Galary Cluster 的切换机制，心跳语句为 show status like 'wsrep%'。

查看节点的状态：

```
(xjw_dba@10.35.9.200) [(none)]> show @@heartbeat;
+--------+-------+-------------+------+---------+-------+--------+---------+
--------------+--------------------+-------+
| NAME   | TYPE  | HOST        | PORT | RS_CODE | RETRY | STATUS | TIMEOUT |
EXECUTE_TIME | LAST_ACTIVE_TIME    | STOP  |
+--------+-------+-------------+------+---------+-------+--------+---------+
--------------+--------------------+-------+
| pxc_M1 | mysql | 10.35.12.26 | 3307 |       1 |     0 | idle   |       0 | 44,44,44
| 2016-08-24 15:34:32 | false |
| pxc_M2 | mysql | 10.35.12.27 | 3307 |       1 |     0 | idle   |       0 | 40,40,40
| 2016-08-24 15:34:32 | false |
| pxc_M3 | mysql | 10.35.12.28 | 3307 |       1 |     0 | idle   |       0 | 16,16,16
| 2016-08-24 15:34:32 | false |
+--------+-------+-------------+------+---------+-------+--------+---------+
--------------+--------------------+-------+
3 rows in set (0.01 sec)
```

执行建表语句及插入语句：

```
(xjw_dba@10.35.9.200) [(none)]> use pxc
Database changed
(xjw_dba@10.35.9.200) [pxc]> drop table travelrecord;
Query OK, 0 rows affected (0.02 sec)
(xjw_dba@10.35.9.200) [pxc]> create table travelrecord (id int not null
auto_increment, org_code varchar(10),host_name varchar(50),primary key (id));
Query OK, 0 rows affected (0.01 sec)
(xjw_dba@10.35.9.200) [pxc]> insert into travelrecord(org_code,host_name)
values('1',@@hostname);
Query OK, 1 row affected (0.04 sec)
```

查看 dnindex.properties：

```
#update
#Wed Aug 24 15:39:22 CST 2016
pxc_Host=0
```

将 10.35.12.26 作为 writeHost，执行 create 语句的 Mycat 日志如下：

```
08/24 15:39:58.088 DEBUG [$_NIOREACTOR-0-RW] (NonBlockingSession.java:229)
-release connection MySQLConnection [id=5, lastTime=1472024398065, user=xjw_dba,
schema=yf, old shema=yf, borrowed=true, fromslaveDB=false, threadId=39, charset=utf8,
txIsolation=3, autocommit=true, attachment=pxc{create table travelrecord (id int
```

```
not null auto_increment, org_code varchar(10),host_name varchar(50),primary key
(id))}, respHandler=SingleNodeHandler [node=pxc{create table travelrecord (id int
not null auto_increment, org_code varchar(10),host_name varchar(50),primary key
(id))}, packetId=1], host=10.35.12.26, port=3307, statusSync=null, writeQueue=0,
modifiedSQLExecuted=true]
```

执行 insert 语句的 Mycat 日志如下：

```
08/24 15:40:04.704 DEBUG [$_NIOREACTOR-0-RW] (NonBlockingSession.java:229)
-release connection MySQLConnection [id=5, lastTime=1472024404698, user=xjw_dba,
schema=yf, old shema=yf, borrowed=true, fromslaveDB=false, threadId=39, charset=utf8,
txIsolation=3, autocommit=true, attachment=pxc{insert into travelrecord(org_code,
host_name) values('1',@@hostname)}, respHandler=SingleNodeHandler [node=pxc{insert
into travelrecord(org_code,host_name) values('1',@@hostname)}, packetId=1], host=
10.35.12.26, port=3307, statusSync=null, writeQueue=0, modifiedSQLExecuted=true]
```

下面执行 3 次 select 查询语句：

```
(xjw_dba@10.35.9.200) [pxc]> select * from travelrecord;
+----+----------+-----------+
| id | org_code | host_name |
+----+----------+-----------+
|  1 | 1        | mysqldb26 |
+----+----------+-----------+
1 row in set (0.03 sec)
```

分析 Mycat 日志可以看出，所有备 writeHost 节点都分担了读操作：

```
08/24 15:43:57.344 DEBUG [$_NIOREACTOR-0-RW] (NonBlockingSession.java:113)
-ServerConnection [id=1, schema=pxc, host=10.35.9.200, user=xjw_dba,txIsolation=3,
autocommit=true, schema=pxc]select * from travelrecord, route={
   1 -> pxc{SELECT *
FROM travelrecord
LIMIT 100}
} rrs
08/24 15:43:57.344 DEBUG [$_NIOREACTOR-0-RW] (PhysicalDBPool.java:452) -select
read source pxc_M3 for dataHost:pxc_Host
08/24 15:43:57.970 DEBUG [$_NIOREACTOR-0-RW] (NonBlockingSession.java:113)
-ServerConnection [id=1, schema=pxc, host=10.35.9.200, user=xjw_dba,txIsolation=3,
autocommit=true, schema=pxc]select * from travelrecord, route={
   1 -> pxc{SELECT *
FROM travelrecord
LIMIT 100}
} rrs
08/24 15:43:57.970 DEBUG [$_NIOREACTOR-0-RW] (PhysicalDBPool.java:452) -select
read source pxc_M2 for dataHost:pxc_Host

08/24 15:43:58.650 DEBUG [$_NIOREACTOR-0-RW] (NonBlockingSession.java:113)
-ServerConnection [id=1, schema=pxc, host=10.35.9.200, user=xjw_dba,txIsolation=3,
autocommit=true, schema=pxc]select * from travelrecord, route={
```

```
    1 -> pxc{SELECT *
FROM travelrecord
LIMIT 100}
} rrs
08/24 15:43:58.650  DEBUG [$_NIOREACTOR-0-RW] (PhysicalDBPool.java:452) -select
read source pxc_M3 for dataHost:pxc_Host
```

关闭 10.35.12.26 上的 mysql 进程，查看节点状态：

```
(xjw_dba@10.35.9.200) [(none)]> show @@heartbeat;
+--------+-------+-------------+------+---------+-------+--------+---------+
--------------+---------------------+-------+
| NAME   | TYPE  | HOST        | PORT | RS_CODE | RETRY | STATUS | TIMEOUT |
EXECUTE_TIME | LAST_ACTIVE_TIME    | STOP  |
+--------+-------+-------------+------+---------+-------+--------+---------+
--------------+---------------------+-------+
| pxc_M1 | mysql | 10.35.12.26 | 3307 |      -1 |     0 | idle   |       0 | 6,9,10
| 2016-08-25 10:38:48 | false |
| pxc_M2 | mysql | 10.35.12.27 | 3307 |       1 |     0 | idle   |       0 | 9,7,7
| 2016-08-25 10:38:48 | false |
| pxc_M3 | mysql | 10.35.12.28 | 3307 |       1 |     0 | idle   |       0 | 9,6,7
| 2016-08-25 10:38:48 | false |
+--------+-------+-------------+------+---------+-------+--------+---------+
--------------+---------------------+-------+
3 rows in set (0.00 sec)
```

执行插入语句：

```
(xjw_dba@10.35.9.200) [(none)]> use pxc
Database changed
(xjw_dba@10.35.9.200) [pxc]> insert into travelrecord(org_code,host_name)
values('1',@@hostname);
Query OK, 1 row affected (0.01 sec)
```

查看 dnindex.properties：

```
#update
#Thu Aug 25 10:38:48 CST 2016
pxc_Host=1
```

分析 Mycat 日志可以看出，已选择 10.35.12.27 作为写节点：

```
08/25 10:39:17.200  DEBUG [$_NIOREACTOR-0-RW] (NonBlockingSession.java:229)
-release connection MySQLConnection [id=4, lastTime=1472092757185, user=xjw_dba,
schema=yf, old shema=yf, borrowed=true, fromslaveDB=false, threadId=38, charset=utf8,
txIsolation=3, autocommit=true, attachment=pxc{insert into travelrecord(org_code,
host_name) values('1',@@hostname)}, respHandler=SingleNodeHandler [node=pxc{insert
into travelrecord(org_code,host_name) values('1',@@hostname)}, packetId=1], host=
10.35.12.27, port=3307, statusSync=null, writeQueue=0, modifiedSQLExecuted=true]
```

下面执行 3 次 select 查询语句：

```
(xjw_dba@10.35.9.200) [pxc]> select * from travelrecord;
+----+----------+-----------+
| id | org_code | host_name |
+----+----------+-----------+
|  1 | 1        | mysqldb26 |
|  3 | 1        | mysqldb27 |
+----+----------+-----------+
2 rows in set (0.00 sec)
```

可看出 10.35.12.27 负载了读操作。这种情况有点类似于传统的主从读写分离模式。接下来继续关闭 10.35.12.27 上的 mysql 进程，查看节点的状态：

```
(xjw_dba@10.35.9.200) [(none)]> show @@heartbeat;
+--------+-------+-------------+------+---------+-------+--------+---------+
--------------+---------------------+-------+
| NAME   | TYPE  | HOST        | PORT | RS_CODE | RETRY | STATUS | TIMEOUT |
EXECUTE_TIME | LAST_ACTIVE_TIME    | STOP  |
+--------+-------+-------------+------+---------+-------+--------+---------+
--------------+---------------------+-------+
| pxc_M1 | mysql | 10.35.12.26 | 3307 |      -1 |     0 | idle   |       0 | 0,2,3
| 2016-08-25 10:44:58 | false |
| pxc_M2 | mysql | 10.35.12.27 | 3307 |      -1 |     0 | idle   |       0 | 5,8,8
| 2016-08-25 10:44:58 | false |
| pxc_M3 | mysql | 10.35.12.28 | 3307 |       1 |     0 | idle   |       0 | 7,7,7
| 2016-08-25 10:44:58 | false |
+--------+-------+-------------+------+---------+-------+--------+---------+
--------------+---------------------+-------+
3 rows in set (0.00 sec)
```

查看 dnindex.properties：

```
#update
#Thu Aug 25 10:44:58 CST 2016
pxc_Host=2
```

执行插入语句：

```
(xjw_dba@10.35.9.200) [(none)]> use pxc
Database changed
(xjw_dba@10.35.9.200) [pxc]> insert into travelrecord(org_code,host_name)
values('4',@@hostname);
Query OK, 1 row affected (0.00 sec)
```

分析 Mycat 日志可以看出，已选择 10.35.12.28 作为写节点：

```
08/25 10:45:57.293 DEBUG [$_NIOREACTOR-0-RW] (NonBlockingSession.java:229)
-release connection MySQLConnection [id=6, lastTime=1472093157281, user=xjw_dba,
schema=yf, old shema=yf, borrowed=true, fromslaveDB=false, threadId=34, charset=utf8,
txIsolation=3, autocommit=true, attachment=pxc{insert into travelrecord(org_code,
host_name) values('4',@@hostname)}, respHandler=SingleNodeHandler [node=pxc{insert
```

```
into travelrecord(org_code,host_name) values('4',@@hostname)}, packetId=1], host=
10.35.12.28, port=3307, statusSync=null, writeQueue=0, modifiedSQLExecuted=true]
```

下面执行 3 次 select 查询语句：

```
(xjw_dba@10.35.9.200) [pxc]> select * from travelrecord;
+----+----------+-----------+
| id | org_code | host_name |
+----+----------+-----------+
|  1 | 1        | mysqldb26 |
|  3 | 1        | mysqldb27 |
|  4 | 4        | mysqldb28 |
+----+----------+-----------+
3 rows in set (0.01 sec)
```

将 10.35.12.28 同时作为读节点。

启动关闭的 mysql 进程，可发现节点加入成功：

```
(xjw_dba@10.35.9.200) [(none)]> show @@heartbeat;
+--------+-------+-------------+------+---------+-------+--------+---------+
--------------+---------------------+-------+
| NAME   | TYPE  | HOST        | PORT | RS_CODE | RETRY | STATUS | TIMEOUT |
EXECUTE_TIME | LAST_ACTIVE_TIME    | STOP  |
+--------+-------+-------------+------+---------+-------+--------+---------+
--------------+---------------------+-------+
| pxc_M1 | mysql | 10.35.12.26 | 3307 |    1 |     0 | idle   |    0 | 11,11,4
| 2016-08-25 11:18:58 | false |
| pxc_M2 | mysql | 10.35.12.27 | 3307 |    1 |     0 | idle   |    0 | 10,8,2
| 2016-08-25 11:18:58 | false |
| pxc_M3 | mysql | 10.35.12.28 | 3307 |    1 |     0 | idle   |    0 | 9,6,9
| 2016-08-25 11:18:58 | false |
+--------+-------+-------------+------+---------+-------+--------+---------+
--------------+---------------------+-------+
3 rows in set (0.00 sec)
```

查看 dnindex.properties，pxc_host 写库仍然是 2（pxc_M3）：

```
#update
#Thu Aug 25 10:44:58 CST 2016
pxc_Host=2
```

执行插入语句：

```
(xjw_dba@10.35.9.200) [(none)]> use pxc
Database changed
(xjw_dba@10.35.9.200) [pxc]> insert into travelrecord(org_code,host_name)
values('6',@@hostname);
Query OK, 1 row affected (0.01 sec)
```

分析 Mycat 日志可以看出，已选择 10.35.12.28 作为写节点：

```
08/25 11:20:46.597  DEBUG [$_NIOREACTOR-0-RW] (NonBlockingSession.java:229)
-release connection MySQLConnection [id=6, lastTime=1472095246588, user=xjw_dba,
schema=yf, old shema=yf, borrowed=true, fromslaveDB=false, threadId=34, charset=utf8,
txIsolation=3, autocommit=true, attachment=pxc{insert into travelrecord(org_code,
host_name) values('6',@@hostname)}, respHandler=SingleNodeHandler [node=pxc{insert
into travelrecord(org_code,host_name) values('6',@@hostname)}, packetId=1], host=
10.35.12.28, port=3307, statusSync=null, writeQueue=0, modifiedSQLExecuted=true]
```

总结如下：Percona Cluster 通过 Mycat 读写分离机制实现了读写分离的效果；当 writeHost 失效时，其他备 writeHost 会提升为 writeHost，只有当所有节点失效时，服务才会中断，可靠性高。当所有节点均正常的时候，writeHost 负责写操作，备 writeHost 负责读操作。如图 4-25 所示。

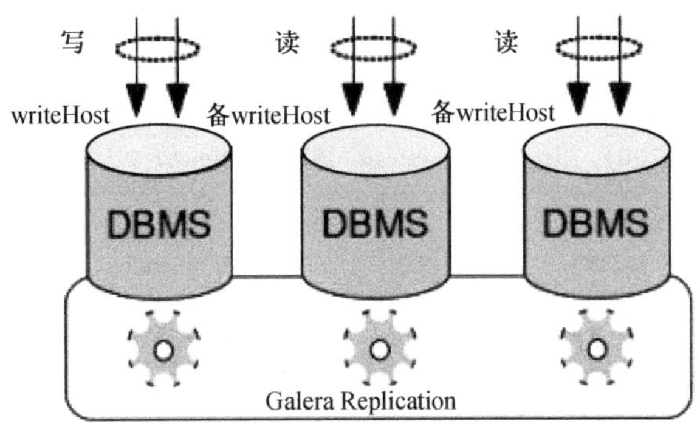

图 4-25

当第 1 个 writeHost 失效时，其中一个备 writeHost 负责写操作，其他备 writeHost 负责读操作。如图 4-26 所示。

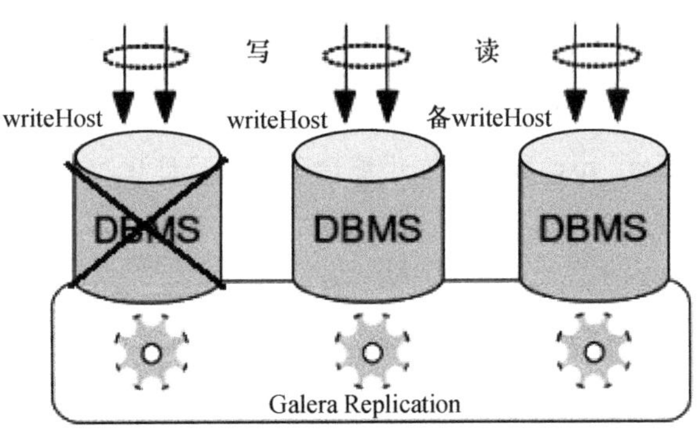

图 4-26

当只有一个 writeHost 时，writeHost 同时负担读和写操作。如图 4-27 所示。

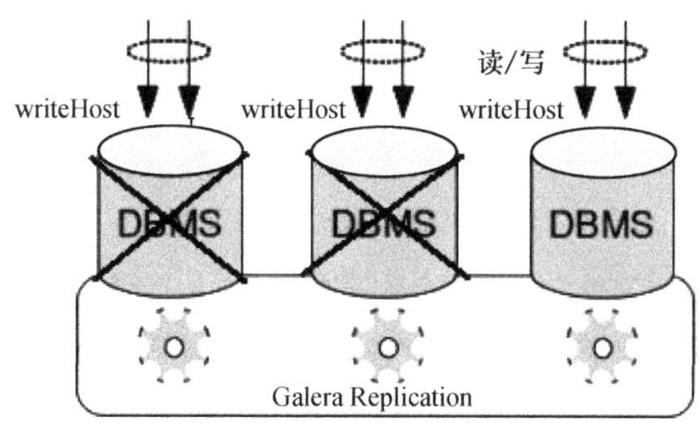

图 4-27

当所有 writeHost 都失效时，无法对外提供服务。

4.3 Mycat+Percona+HAProxy+Keepalived

本节讲解 Percona 集群技术，主要是为了处理解决 Mycat 路由后，线上压力集中在后端某一台实例上而导致 CPU 负荷过高、复制延迟、数据不一致等问题。

4.3.1 Mycat

1. 配置环境

```
10.1.166.99: Keepalived 连接 HAProxy 的 vip
10.1.166.22 haproxy1  percona-server
10.1.166.23 haproxy2  percona-server
10.1.166.24 percona-server
10.1.166.25 percona-server
10.1.166.26 percona-server
10.1.166.27  mysql-server
10.1.166.28  mycat mysql-server
10.1.166.29  mysql-server
10.1.166.30  mysql-server
```

2. 配置 hosts 文件

```
[root@node22 src]#cat /etc/hosts
127.0.0.1   localhost localhost.localdomain localhost4 localhost4.localdomain4
::1         localhost localhost.localdomain localhost6 localhost6.localdomain6
10.1.166.99       keepalivehost
```

```
10.1.166.22 dn1    node22  haproxy1
10.1.166.23        node23  haproxy2
10.1.166.27 dn2
10.1.166.28 dn3
10.1.166.29 dn4
10.1.166.30 dn5
```

将已配置好的/etc/hosts 文件上传到各个数据库节点，执行脚本如下：

```
for i in 22 23 27 28 29 30;do scp /etc/hosts 10.1.166.$i:/etc;done
```

安装 Mycat 软件：

```
[root@node28 src]#tar -xf
Mycat-server-1.4-RELEASE-20150901112004-linux.tar.gz
[root@node28 src]#mv mycat/ /usr/local/
[root@node28 src]#useradd mycat
[root@node28 local]#chown mycat:root mycat/  -R
```

下载和安装 Java：

```
[root@node28 src]#rpm -ivh jre-8u51-linux-x64.rpm
[root@node28 src]#java -version
```

此处注意 Java 版本不得低于 1.6，至此 Mycat 安装就算完成了，为你的 Mycat 用户设置密码。

```
[root@node24 src]#passwd mycat
Changing password for user mycat.
New password:
BAD PASSWORD: it is WAY too short
BAD PASSWORD: is too simple
Retype new password:
passwd: all authentication tokens updated successfully.
```

3. 原始 Mycat 测试配置

```
[root@node28 conf]#>../logs/wrapper.log
```

为了在后面更清楚地观察启动后的 Mycat 是否存在错误，清空日志：

```
[root@node28 conf]#pwd
/usr/local/mycat/conf
[root@node28 conf]#../bin/mycat start
Starting Mycat-server...
```

4. 报错处理方案

```
[root@node28 conf]#cat ../logs/wrapper.log
...
ERROR  | wrapper  | 2015/09/18 01:27:36 | JVM exited while loading the application.
INFO   | jvm 1    | 2015/09/18 01:27:36 | Error: Exception thrown by the agent :
```

```
java.net.MalformedURLException: Local host name unknown:
java.net.UnknownHostException: node28: node28: unknown error
```

可以看到，域名的报错可能是因为刚刚修改 hosts 文件时没有写本机域名：

```
[root@node28 conf]#cat /etc/hosts
127.0.0.1   localhost localhost1 localhost.localdomain localhost4
localhost4.localdomain4
::1         localhost localhost.localdomain localhost6 localhost6.localdomain6
10.1.166.99 dn1     keepalivehost
10.1.166.22  node22  haproxy1
10.1.166.23  node23  haproxy2
10.1.166.27 dn2
10.1.166.28 dn3
10.1.166.29 dn4
10.1.166.30 dn5

[root@node28 conf]#hostname
node28
[root@node28 conf]#cat /etc/sysconfig/network
NETWORKING=yes
HOSTNAME=node28
```

所以，需要将 hosts 文件中 IP 地址为 10.1.166.28 的行定义一个域名，即 node28。配置如下：

```
[root@node28 conf]#cat /etc/hosts
127.0.0.1  node24 localhost localhost1 localhost.localdomain localhost4
localhost4.localdomain4
::1         localhost localhost.localdomain localhost6 localhost6.localdomain6
10.1.166.99 dn1     keepalivehost
10.1.166.22  node22  haproxy1
10.1.166.23  node23  haproxy2
10.1.166.27 dn2    node27
10.1.166.28 dn3    node28
10.1.166.29 dn4    node29
10.1.166.30 dn5    node30

[root@node24 conf]#../bin/mycat restart
Stopping Mycat-server...
Starting Mycat-server...
[root@node28 conf]#cat ../logs/wrapper.log
...
INFO   | jvm 1    | 2015/09/18 02:17:32 | MyCAT Server startup successfully. see
logs in logs/mycat.log
```

5. 基于案例配置

如下所示为创建一张日志表：

```
CREATE TABLE `zlog` (
  `id` bigint(15) NOT NULL COMMENT ,
  `xiha` varchar(30) NOT NULL COMMENT ,
  `qiqi` varchar(300) DEFAULT NULL,
  `note1` varchar(2) DEFAULT NULL,
  `note2` varchar(2) DEFAULT NULL,
  `stat` tinyint(1) NOT NULL DEFAULT '1' COMMENT ',
  `isno` tinyint(1) NOT NULL DEFAULT '0'
) ENGINE=InnoDB DEFAULT CHARSET=utf8 COMMENT='日志';
```

假设这是一张生产表，那么现在的数据已经有 3 亿条了，且为线上实时访问数据。根据日期分片，最后一个实例为本月数据，客户端访问也集中在了本月的后端实例中，因此仅仅采用 Mycat 仍然无法解决后端中本月所分配的实例负载，由此引进 Percona 集群项目为后端实例，一组 Percona 集群为 Mycat 路由的一个分片地址。这里手绘一下处理思路（图 4-28 中仅用一套 Percona 集群编辑解释，实际上至少需要两套 Percona 集群或者一套 Percona 外加一至多个单独的 MySQL-SERVER 或 MySQL 分支实例或 MySQL 主从等来实现分片）。

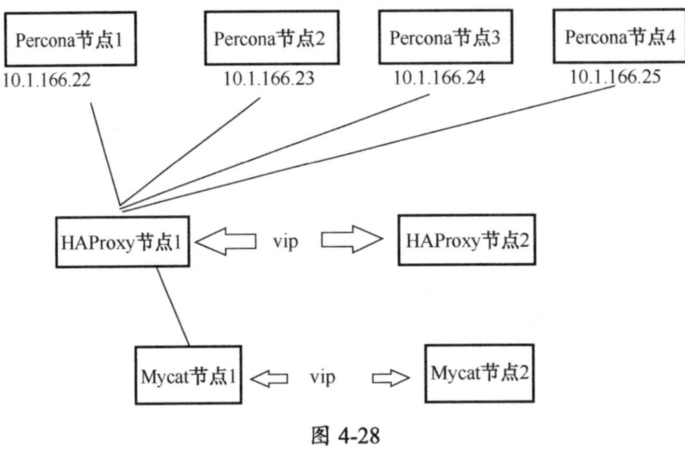

图 4-28

在图 4-28 中，vip 所使用的软件为 Keepalived 实现了高可用目的，通过配置 HAProxy 软件实现了负载均衡，Mycat 访问 HAProxy 则等同于轮询访问后端的 Percona 集群节点，可以解决发往单台实例的 CPU 负载、主从复制导致的延迟等问题。后面将会对 Percona 做详细的介绍（此处以 HAProxy 配置的方式达到负载均衡的效果，也可以通过 Mycat 对 Percona 读写分离的配置方式进行处理）。

6. 配置 schema.xml 文件

```
[root@node24 conf]#cat schema.xml
<?xml version="1.0"?>
<!DOCTYPE mycat:schema SYSTEM "schema.dtd">
```

```
    <mycat:schema xmlns:mycat="http://org.opencloudb/">
            <schema name="zyz" checkSQLschema="false" sqlMaxLimit="100">
                    <table name="employee" primaryKey="ID" autoIncrement="true"
dataNode="dn1,dn2"
                            rule="sharding-by-intfile" />
            <table name="zlog" primaryKey="id" autoIncrement="true"
dataNode="dn1,dn2,dn3,dn4,dn5"
            rule="mod-long" />
            <table name="tt2" primaryKey="id" autoIncrement="true"
dataNode="dn1,dn2,dn3,dn4,dn5" rule="mod-long" />
            <table name="MYCAT_SEQUENCE" primaryKey="name"   dataNode="dn1" />
            </schema>
            <dataNode name="dn1" dataHost="10.1.166.99" database="db1" /><!--此处为
给 HAProxy 做高可用的 Keepalived 所产生的 vip-->
            <dataNode name="dn2" dataHost="10.1.166.27" database="db2" />
            <dataNode name="dn3" dataHost="10.1.166.28" database="db3" />
            <dataNode name="dn4" dataHost="10.1.166.29" database="db4" />
            <dataNode name="dn5" dataHost="10.1.166.30" database="db5" />

            <dataHost name="10.1.166.99" maxCon="1000" minCon="10" balance="0"
writeType="0" dbType="MySQL" dbDriver="native" switchType="1"
slaveThreshold="100">
                    <heartbeat>select user()</heartbeat>
                    <!-- can have multi write hosts -->
                    <writeHost host="hostM1" url="10.1.166.99:3399" user="root"
password="123456">
                    <!-- can have multi read hosts -->
                    </writeHost>
                    <writeHost host="hostS1" url="10.1.166.99:3399" user="root"
password="123456" />
                    <!-- <writeHost host="hostM2" url="localhost:3399" user="root"
password="123456"/>  此处配置的端口为 HAProxy 配置文件中配置的访问后端数据库实例的绑定端口
-->
            </dataHost>
            <dataHost name="10.1.166.27" maxCon="1000" minCon="10" balance="0"
                    writeType="0" dbType="MySQL" dbDriver="native" switchType="1"
slaveThreshold="100">
                    <heartbeat>select user()</heartbeat>
                    <!-- can have multi write hosts -->
                    <writeHost host="hostM1" url="10.1.166.27:3306" user="root"
password="123456">
                <!-- can have multi read hosts -->              </writeHost>
                    <writeHost host="hostS1" url="10.1.166.27:3306" user="root"
password="123456" />
                    <!-- <writeHost host="hostM2" url="localhost:3306" user="root"
password="123456"/> -->
            </dataHost>
```

```
        ....余下配置省略
    </mycat:schema>
```

注意：10.1.166.99 为图 4-28 中 Keepalived 所产生的 vip 地址，连接的是 HAProxy 服务，HAProxy 则配置连接 Percona 各节点的地址。因此，客户端实际上是通过访问 10.1.166.99 这个 vip 去连接后端的多个 Percona 实例。至于 Percona 集群读写分离的具体配置可参考 4.1.2 和 4.2.2 节。本章的部分配置可以通过 HAProxy 实现。

对其中的一部分参数解释如下。

（1）balance

参数 balance 决定了哪些 MySQL 服务器参与到读 SQL 的负载均衡中，0 为不开启读写分离，1 为全部 readHost 与 standby writeHost 参与 select 语句的负载均衡，比如我们配置了 1 主 3 从（级联复制，第 2 个从节点为其他从节点的主节点）的 MySQL 主从环境，并把第 1 个从节点 MySQL 配置为 dataHost 中的第 2 个 writeHost，以便主节点宕机后，Mycat 自动切换到这个 writeHost 上来执行写操作，此时 balance="1"意味着第 1 个 writeHost 不参与读 SQL 的负载均衡，其他 3 个 writeHost 都参与；balance="2"意味着所有读操作都随机地在 writeHost、readHost 上分发；balance="3"意味着所有读请求都随机地分发至 writeHost 对应的 readHost 上执行，writerHost 不负担读压力。

（2）writeType

writeType="0"意味着所有写操作都发送到配置的第 1 个 writeHost 上，第 1 个 writeHost 服务宕机了便切到正常运行的第 2 个 writeHost 上；writeType="1"对于 galera for MySQL 类集群这种多主多节点都能写入的集群有效，Mycat 会随机选择一个 writeHost 并写入数据，对于非 galera for MySQL 集群，最好不要配置 writeType="1"，可能会导致数据库不一致的严重问题。

Mycat 1.3 支持自动方式及编程指定的两种读写分离方式。

自动方式为一个查询 SQL 语句是自动提交的模式；connection.setAutocommit(true)或者 set autocommit=1 的编程指定方式为一个查询 SQL 语句以/*balance*/注解来确定其是走读节点还是走写节点。在 Mycat 1.3 中，若事务内的查询语句增加了此注解，则强制其走读节点，在 Mycat 1.4 中可以在非事务内的查询语句前增加此注解，强制其走写节点，这个增强是为了避免在主从不同步的情况下要求查询刚写入的数据。

另外，Mycat 从 1.4 版本开始支持 MySQL 主从复制状态绑定的读写分离机制，让读更加安全可靠，配置如下。

将 Mycat 心跳检查语句配置为 show slave status，在 dataHost 上定义两个新属性：switchType="2"与 slaveThreshold="100"，此时意味着开启 MySQL 主从复制状态绑定的读写分离与切换机制。Mycat 心跳机制通过检测 show slave status 中的"Seconds_Behind_master" "slave_IO_Running""slave_SQL_Running"三个字段来确定当前主从同步的状态及 Seconds_Behind_master 主从复制时延，当 Seconds_Behind_master>slaveThreshold 时，读写分离筛选器会过滤掉此 slave 机器，防止读到很久之前的旧数据，而当主节点宕机后，切换逻辑会检查 slave 上的 Seconds_Behind_master 是否为 0，为 0 则表示主从同步，可以安全切换，否则不会切换。

（3）switchType

switchType 指的是切换模式，目前的取值也有 4 种。

- switchType='-1'：表示不自动切换。
- switchType='1'：默认值，表示自动切换。
- switchType='2'：基于 MySQL 主从同步的状态决定是否切换，心跳语句为 show slave status。
- switchType='3'：基于 Cluster 的切换机制（适合集群，Mycat 从 1.4.1 版本开始支持），心跳语句为 show status like 'wsrep%'。

7. MySQL 授权及配置文件

在 MySQL 配置文件中加入如下两句：

```
log_bin_trust_function_creators=1
lower_case_table_names = 1
```

在 schema.xml 配置文件中定义 MySQL-SERVER，在所在的服务器上授权如下：

```
GRANT ALL PRIVILEGES ON *.* TO 'root'@'10.1.166.%' IDENTIFIED BY '123456' WITH GRANT OPTION ;
GRANT ALL PRIVILEGES ON *.* TO 'root'@'127.0.0.1' IDENTIFIED BY '123456' WITH GRANT OPTION;
GRANT ALL PRIVILEGES ON *.* TO 'root'@'localhost' IDENTIFIED BY '123456' WITH GRANT OPTION ;
flush privileges;
```

这里的密码与上面配置文件的密码相对应，如果授权错误，则 Mycat 将无法连接后端的数据库服务。

8. 配置 server.xml 文件

```
[root@node28 conf]#vi server.xml #需要改的地方如下
 </system>
        <user name="adminz">
```

```
                <property name="password">keymypass</property>
                <property name="schemas">zyz</property>
        </user>
<user name="user">
<property name="password">user</property>
<property name="schemas">zyz</property>
<property name="readOnly">true</property>

[root@node28 conf]#tail -f  server.xml
        </system>
        <user name="adminz">
                <property name="password">keymypass</property>
                <property name="schemas">zyz</property>
        </user>
        <user name="user">
                <property name="password">user</property>
                <property name="schemas">zyz</property>
                <property name="readOnly">true</property>
        </user>
</mycat:server>
```

9. 配置 rule.xml

```
[root@node28 conf]#vi rule.xml
<property name="count">5</property><!--  这一行改成实际分片的节点数量，必须指定要分片的数据库节点的数量，否则没法分片 -->
```

10. 启动测试

```
[root@node24 conf]#rm -rf dnindex.properties  #首先要删除 dnindex.properties 文件，因为它会记录上次测试保存的节点信息,然后启动 Mycat
[root@node28 conf]#../bin/mycat start
[root@node28 conf]#cat dnindex.properties
#update
#Fri Sep 18 03:58:06 CST 2015
10.1.166.30=0
10.1.166.29=0
10.1.166.28=0
10.1.166.27=0
10.1.166.99=0
```

生成这个新文件

10. 创建表

```
MySQL -uadminz -pkeymypass -h127.0.0.1 -P8066 -Dzyz
MySQL> show databases;
```

```
+----------+
| DATABASE |
+----------+
| zyz     |
+----------+
1 row in set (0.00 sec)
explain create table employee (id int not null primary key,name
varchar(100),sharding_id int not null);
create table employee (id int not null primary key,name varchar(100),sharding_id
int not null);
insert into employee(id,name,sharding_id) values(1,'leader us',10000);
insert into employee(id,name,sharding_id) values(2, 'me',10010);
insert into employee(id,name,sharding_id) values(3, 'mycat',10000);
insert into employee(id,name,sharding_id) values(4, 'mydog',10010);
```

如果 db1 和 db2 都有数据，则说明以上配置没问题：

```
CREATE TABLE `zlog` (
  `id` bigint(15) NOT NULL AUTO_INCREMENT ,
  `xiha` varchar(30) NOT NULL ,
  `nobai` decimal(2,0) NOT NULL ,
  `shopvalue` decimal(11,1) NOT NULL ,
  `shoptype` tinyint(2) NOT NULL ,
  `hatype` tinyint(2) NOT NULL ,
  `targetid` decimal(11,0) NOT NULL ,
  `consumetime` datetime NOT NULL DEFAULT '1753-01-01 12:00:00' ,
  `opid` bigint(20) NOT NULL,
  `qiqi` varchar(300) DEFAULT NULL,
  `note1` varchar(2) DEFAULT NULL,
  `note2` varchar(2) DEFAULT NULL,
  `stat` tinyint(1) NOT NULL DEFAULT '1' ,
  `isno` tinyint(1) NOT NULL DEFAULT '0' ,
  PRIMARY KEY (`id`)
) ENGINE=InnoDB AUTO_INCREMENT=1564262 DEFAULT CHARSET=utf8 COMMENT='消费日志';
MySQL> INSERT INTO `zlog` (`id`, `xiha`, `nobai`, `shopvalue`, `shoptype`,
`hatype`, `targetid`, `consumetime`, `opid`, `qiqi`, `note1`, `note2`, `stat`, `isno`)
VALUES ('1945', '1182714042231465669201901', '0', '-0.5', '0', '0', '10214131',
'2014-07-03 09:43:13', '347', NULL, NULL, NULL, '1', '0');
Query OK, 1 row affected (0.01 sec)
INSERT INTO `zlog` (`xiha`, `nobai`, `shopvalue`, `shoptype`, `hatype`,
`targetid`, `consumetime`, `opid`, `qiqi`, `note1`, `note2`, `stat`, `isno`) VALUES
('1182714042231465669201901', '0', '-0.5', '0', '0', '10214131', '2014-07-03
09:43:13', '347', NULL, NULL, NULL, '1', '0');
ERROR 1003 (HY000): mycat sequnce err.java.lang.NumberFormatException: null
MySQL>
```

若执行 insert 语句时报错，则说明自增主键不能自增。

11. 自增主键

（1）配置 server.xml，开启数据库层面设计的自增主键方式：

```
<property name="sequnceHandlerType">1</property>
```

（2）配置 sequence_db_conf.properties：

```
[root@node24 conf]#vi sequence_db_conf.properties
#sequence stored in dataNode
GLOBAL=dn1
COMPANY=dn1
CUSTOMER=dn1
ORDERS=dn1
tt2=dn1
TT2=dn1
zlog=dn1
ZLOG=dn1
~
```

（3）重启 Mycat 并实现自增主键功能，通过本地方式连接每一套数据库实例（为了更安全），执行以下 SQL 语句。

```
[root@node24 conf]#mysql -p123456
MySQL>
DROP TABLE IF EXISTS MYCAT_SEQUENCE;
DROP FUNCTION IF EXISTS `mycat_seq_nextval`;
DROP FUNCTION IF EXISTS `mycat_seq_setval`;
DROP FUNCTION IF EXISTS `mycat_seq_setval`;
CREATE TABLE MYCAT_SEQUENCE (  name VARCHAR(50) NOT NULL,  current_value INT NOT NULL,  increment INT NOT NULL DEFAULT 100, PRIMARY KEY (name) ) ENGINE=InnoDB;
-- ----------------------------
-- Function structure for `mycat_seq_currval`
-- ----------------------------
DROP FUNCTION IF EXISTS `mycat_seq_currval`;
DELIMITER ;;
CREATE FUNCTION `mycat_seq_currval`(seq_name VARCHAR(50)) RETURNS varchar(64) CHARSET latin1
    DETERMINISTIC
BEGIN
        DECLARE retval VARCHAR(64);
        SET retval="-999999999,null";
        SELECT concat(CAST(current_value AS CHAR),",",CAST(increment AS CHAR) ) INTO retval   FROM MYCAT_SEQUENCE  WHERE name = seq_name;
        RETURN retval ;
END
;;
DELIMITER ;
-- ----------------------------
```

```
-- Function structure for `mycat_seq_nextval`
-- ----------------------------
DROP FUNCTION IF EXISTS `mycat_seq_nextval`;
DELIMITER ;;
CREATE FUNCTION `mycat_seq_nextval`(seq_name VARCHAR(50)) RETURNS varchar(64)
CHARSET latin1
    DETERMINISTIC
BEGIN
        UPDATE MYCAT_SEQUENCE
                SET current_value = current_value + increment  WHERE name =
seq_name;
        RETURN mycat_seq_currval(seq_name);
END
;;
DELIMITER ;
-- ----------------------------
-- Function structure for `mycat_seq_setval`
-- ----------------------------
DROP FUNCTION IF EXISTS `mycat_seq_setval`;
DELIMITER ;;
CREATE  FUNCTION `mycat_seq_setval`(seq_name VARCHAR(50), value INTEGER)
RETURNS varchar(64) CHARSET latin1
    DETERMINISTIC
BEGIN
        UPDATE MYCAT_SEQUENCE
                 SET current_value = value
                 WHERE name = seq_name;
        RETURN mycat_seq_currval(seq_name);
END
;;
DELIMITER ;
```

登录 Mycat 后查出来是如下结果（若每次可以按照 MYCAT_SEQUENCE 表中定义的方式增长，则表示正常）：

```
[root@node24 conf]#MySQL -uadminz -pkeymypass -h127.0.0.1 -P8066 -Dzyz
MySQL> SELECT MYCAT_SEQ_NEXTVAL('GLOBAL');
+-----------------------------+
| MYCAT_SEQ_NEXTVAL('GLOBAL') |
+-----------------------------+
| 201,100                   |
+-----------------------------+
1 row in set (0.00 sec)
MySQL> SELECT MYCAT_SEQ_NEXTVAL('GLOBAL');
+-----------------------------+
| MYCAT_SEQ_NEXTVAL('GLOBAL') |
+-----------------------------+
```

```
| 301,100                    |
+-----------------------------+
1 row in set (0.00 sec)
```

（4）通过测试表 TT2 进行测试：

```
MySQL> insert into mycat_sequence values('TT2',0,1);  /*这一条的目的是定义为 TT2
表从 0 开始自增长，每次增长数为 1，如果没有这个就会报错，如下所示：*/
MySQL> delete from  mycat_sequence;
MySQL> insert into tt2(name) values('df') ;
ERROR 1003 (HY000): mycat sequnce err.java.lang.RuntimeException:
sequnce not found in db table
MySQL> create table tt2(id int auto_increment primary key,name
varchar(10));
Query OK, 0 rows affected (0.02 sec)
MySQL> insert into mycat_sequence values('TT2',0,1);
/*如果执行此步后仍报错话，则建议重启 Mycat（配置好自增长后需要重启 Mycat 以生效或者重新加载
配置文件以生效），不报错就不需要重启了*/
MySQL> insert into tt2(name) values('123');
Query OK, 1 row affected (0.11 sec)
MySQL> insert into tt2(name) values('123');
Query OK, 1 row affected (0.11 sec)
MySQL> select * from tt2 order by id asc;
+----+------+
| id | name |
+----+------+
|  1 | 123  |
|  2 | 123  |
...
| 13 | 123  |
+----+------+
13 rows in set (0.03 sec)
```

4.3.2 Percona 集群

1. Percona 集群介绍

本节中 Percona Cluster 的搭建过程与 4.3.1 节无关。项目地址为 http://www.percona.com/doc/percona-xtradb-cluster/intro.htmlPercona。Percona Cluster 是 MySQL 高可用性和可扩展性的解决方案。Percona XtraDB Cluster 的特性如下。

- 同步复制，事务要么在所有节点提交，要么不提交。
- 多主复制，可以在任意节点进行写操作。
- 并行复制。

- 节点自动配置。
- 最大化保证数据的一致性。

Percona XtraDB Cluster 完全兼容 MySQL 和 Percona Server，表现如下。

- 数据的兼容性。
- 应用程序的兼容性：无须更改应用程序。
- 集群是由节点组成的，推荐配置至少有 3 个节点，但也可以运行在两个节点上。
- 每个节点都是普通的 MySQL、Percona 服务器，可以将现有的数据库服务器组成集群，反之，也可以将集群拆分成单独的服务器。
- 每个节点都包含完整的数据副本。

优点如下。

- 当执行一个查询时，在本地节点上执行。所有数据都在本地，所以无须远程访问。
- 无须集中管理。可以在任何时间点失去任何节点，但是集群将照常工作。
- 良好的读负载扩展，任意节点都可以查询。

缺点如下。

- 加入新节点的开销大，需要复制完整的数据。
- 不能有效地解决写缩放的问题，所有的写操作都将发生在所有节点上。
- 有多少个节点就有多少重复的数据。

Percona XtraDB Cluster 与 MySQL Replication 两种复制模式基于 CAP 理论的主要区别在于：Percona XtraDB Cluster 支持一致性和可用性；MySQL Replication 支持可用性和分区容忍性。

分布式系统的 CAP 理论如下。

- C：一致性，所有节点的数据一致。
- A：可用性，一个或多个节点的失效不影响请求服务。
- P：分区容忍性，节点间的连接失效时仍然可以处理请求。

任何一个分布式系统都需要满足其中两个理论。

因此 MySQL Replication 并不保证数据的一致性，而 Percona XtraDB Cluster 保证数据的一致性。

Percona Cluster 包含写复制集补丁，使用 Galera 2.x library 多主同步复制插件。Galera 2.x 的新特性如下。

- IST（Incremental State Transfer）：增量状态传输，对于 WAN 特别有用。
- RSU（Rolling Schema Update）：旋转更新架构，不会阻止对表的操作。

2. 初始配置

为了使用 XtraDB 集群，需要在 my.cnf 文件中配置以下选项：

```
wsrep_provider -- a path to Galera library.
wsrep_cluster_address -- cluster connection URL.
binlog_format=ROW
default_storage_engine=InnoDB
innodb_autoinc_lock_mode=2
innodb_locks_unsafe_for_binlog=1
wsrep_sst_method=xtrabackup
```

3. 局限性

（1）目前的复制仅仅支持 InnoDB 存储引擎。任何写入其他引擎的表，包括 MySQL 库下的 MYISAM 引擎表将不会被复制，但是 DDL 语句会被复制，因此采用 GRANT 方式创建用户将会被复制，但采用 insert into MySQL.user...方式不会被复制。

（2）delete 操作不支持没有主键的表。没有主键的表在不同节点的顺序不同，如果执行 SELECT...LIMIT...，则将出现不同的结果集。

（3）在多主环境下不支持 LOCK/UNLOCK TABLES 及锁函数 GET_LOCK()、RELEASE_LOCK()等。

（4）查询日志不能保存在表中。如果开启查询日志，则只能保存到文件中。

（5）所允许的最大的事务大小由 wsrep_max_ws_rows 和 wsrep_max_ws_size 定义。任何大型操作都将被拒绝，例如大型的 LOAD DATA 操作。

（6）由于集群是乐观的并发控制，事务 commit 可能在该阶段中止，所以如果有两个事务向集群中的不同节点的同一行写入并提交，则失败的节点将中止。对于集群级别的中止，集群返回死锁错误如 Error: 1213 SQLSTATE: 40001 (ER_LOCK_DEADLOCK)。

（7）不支持 XA 事务，在提交上可能回滚。

（8）整个集群的写入吞吐量由最弱的节点限制，如果有一个节点变得缓慢，那么整个集群将变得缓慢。为了满足稳定及高性能的要求，所有节点应使用统一的硬件。

（9）集群节点建议有最少 3 个或者通过 Percona Arbitrator 技术预防脑裂问题。

（10）如果 DDL 语句有问题，则将破坏集群。

4. id 自增

集群内部按照 id 自增长机制写入数据，比如集群中有三台可能是 3、6、9 这样的递增。

5. 数据备份

刚搭建集群时，每个节点的日志尽量保留得久一些，最好是对每个节点都做日志备份。数据备份可以防止某个节点的错误配置导致数据不一致，比如存在 MyISAM 表。

6. 名词解释

（1）ST（Increment Snapshot Transfer）：用于在节点间传送数据，发生在初始化节点或节点发生故障时，但是能够从 galera.dat 中获得增量同步点的情况，仅仅做必要的增量同步是最理想的数据恢复方法。

（2）SST（State Snapshot Transfer）：用于在节点之间传送数据，发生在初始化节点或者节点发生故障且需要全部重置数据时，相当于复制整份数据到新的节点，这个过程的影响非常大，会导致 Donor 节点无法访问。

（3）DDL--TOI Schema Upgrade：指任何修改数据库结构的 DDL 语句，这种语句不具有事务性。

（4）TOI（Total Order Isolation）：为默认的工作模式，在这种模式下，DDL 语句的表现将和在一个单机数据库上一样，会锁定整个集群库，并且这个语句会被同时发送到所有的集群节点去执行。

（5）DDL RSU wsrep：通过设置 wsrep_OSU_method 参数，DDL 语句将会在当前节点执行，并且在执行的过程中不锁定其他节点，当这个节点的 DDL 操作完成时，将会应用操作过程中延迟的 replication，然后对每个节点进行人为的 DDL 操作。在这个滚动、升级的过程中，集群中会有部分服务器有新的表结构，部分有旧的表结构。

（6）Percona Arbitrator：通常单纯双节点的 Percona 集群无法保证环境的可靠性，因此引入了仲裁技术 Arbitrator 来防止双节点的集群环境中由于通信故障而导致节点无法正常同步的情况发生。

（7）Donor：当发生 SST 时，可以手动指定或自动选择集群中的数据源节点为 Donor 节点，但是新的数据变化情况无法被应用，会被缓存在当前节点的 cache 文件中。当 SST 过程结束后，Donor 节点将变为 JOINED 状态，并应用这些缓存的内容，从而返回 SYNCED 状态。

7. 检查现有环境是否可以搭建 Percona Cluster

```
SELECT DISTINCT
       CONCAT(t.table_schema,'.',t.table_name) as tablename,
       t.engine,
       IF(ISNULL(c.constraint_name),'no','yes') AS primarykey,
       IF(s.index_type = 'FULLTEXT','FULLTEXT','') as fulltextindex,
       IF(s.index_type = 'SPATIAL','SPATIAL','') as spatialindex
  FROM information_schema.tables AS t
  LEFT JOIN information_schema.key_column_usage AS c
    ON (t.table_schema = c.constraint_schema AND t.table_name = c.table_name
       AND c.constraint_name = 'PRIMARY')
  LEFT JOIN information_schema.statistics AS s
    ON (t.table_schema = s.table_schema AND t.table_name = s.table_name
       AND s.index_type IN ('FULLTEXT','SPATIAL'))
  WHERE t.table_schema NOT IN
('information_schema','performance_schema','mysql')
    AND t.table_type = 'BASE TABLE'
    AND (t.engine <> 'InnoDB' OR c.constraint_name IS NULL OR s.index_type IN
('FULLTEXT','SPATIAL'))
  ORDER BY t.table_schema,t.table_name;
```

可以通过上面的代码检测不符合条件的表：表引擎是否有全文索引、空间索引是否存在、是否存在主键，引擎是否为 InnoDB。若上述查询没有结果，则表示可以将现有环境迁移至 Percona Cluster。

8. 安装配置 Percona

服务器的环境如下：

```
[root@zyz_dba_test01 ~]# cat /proc/cpuinfo | grep name | cut -f2 -d: | uniq
-c
2 Intel(R) Xeon(R) CPU E5-2680 0 @ 2.7GHZ
[root@zyz_dba_test01 ~]# cat /proc/cpuinfo | grep physical | uniq -c
1 physical id : 0
1 address sizes : 40 bits physical, 48 bits virtal
1 physical id : 0
1 address sizes : 40 bits physical, 48 bits virtal
[root@zyz_dba_test01 ~]# head -4 /proc/meminfo
MemTotal:       1922464 kB
MemFree:          79904 kB
Buffers:         159296 kB
Cached:         1366644 kB
```

系统环境为 Red Hat Enterprise Linux Server release 6.4（Santiago），服务器的地址如下：

```
10.21.3.106 node1
```

```
10.21.3.107 node2
10.21.3.108 node3
10.21.3.109 haproxy
```

这些将配置到每台测试服务器的/etc/hosts 文件中。

安装操作如下。

node1、node2、node3 执行以下操作：

```
ln -sf /usr/lib64/libssl.so.10 /usr/lib64/libssl.so.6
ln -sf /usr/lib64/libcrypto.so.10 /usr/lib64/libcrypto.so.6
useradd mysql
mkdir /opt/data
chown mysql:mysql /opt/data
chown 755 /opt
chown mysql:mysql /opt
yum -y install nmap
yum -y install perl-DBD-MySQL perl-Time-HiRes nc
yum -y install boost-program-options scoat
iptables -F
setenforce 0
 tar -xf /opt/Percona-XtraDB-Cluster-5.6.24-rel72.2-25.11..Linux.x86_64.tar.gz
mv /opt/Percona-XtraDB-Cluster-5.6.24-rel72.2-25.11..Linux.x86_64
/opt/percona
rpm -ivh Percona-XtraDB-Cluster-galera-3-3.11-1.rhel6.x86_64.rpm
[root@zyz_dba_test03 opt]# cat /etc/my.cnf
# The follo#wing options will be passed to all MySQL clients
[client]
#password       = your_password
port            = 3306
socket          = /opt/mysql.sock
[mysqld]
port            = 3306
user=mysql
datadir=/opt/data
basedir=/opt/percona
socket          = /opt/mysql.sock
skip-external-locking
key_buffer_size = 16K
max_allowed_packet = 1M
table_open_cache = 4
sort_buffer_size = 64K
read_buffer_size = 256K
innodb_buffer_pool_size=1G
read_rnd_buffer_size = 256K
net_buffer_length = 2K
thread_stack = 240K
```

```
max_connections=10000
log-bin=mysql-bin
binlog_format = ROW
server-id       = 3  #各节点的配置不同
innodb_autoinc_lock_mode = 2
wsrep_provider = /usr/lib64/galera3/libgalera_smm.so
#wsrep_cluster_name = "my_mariadb_cluster"
wsrep_cluster_address="gcomm://10.21.3.106,10.21.3.107"#此处非主节点填写另外两台
服务器 ip 地址,主节点 10.21.3.106 上写"gcomm://"
wsrep_cluster_name='example_cluster' #各节点此处的配置均应相同
wsrep_node_name = "cluster_node3" #各节点的节点命名，各节点此处的配置不同
wsrep_node_address = 10.21.3.108:4406 #此处填写本机的 IP 地址:4406
wsrep_sst_auth=tt:123
#wsrep_node_name='node3'
wsrep_sst_method=rsync
[mysqldump]
quick
max_allowed_packet = 16M
[mysql]
no-auto-rehash
[myisamchk]
key_buffer_size = 8M
sort_buffer_size = 8M
[mysqlhotcopy]
interactive-timeout
#其他节点配置文件与上面的 my.cnf 配置文件的内容基本相同，不同的点和必须相同的点已经在注释中
说明
```

按照从 node1 到 node3 的顺序启动。

node1：

```
/opt/percona/bin/mysqld --defaults-file=/opt/my.cnf
--wsrep-new-cluster & 或者
/opt/percona/bin/mysqld_safe --defaults-file=/opt/my.cnf
--wsrep-new-cluster &
```

node2、node3：

```
cp /opt/percona/support-files/mysqld  /etc/init.d/percona
vi /etc/init.d/percona
```

找到 basedir 和 datadir 关键字所在的行，修改如下：

```
basedir=/opt/percona
datadir=/opt/data
```

如果无法修改，则可以通过如下方式启动：

```
/opt/percona/bin/mysqld --defaults-file=/opt/my.cnf &
```

或者

```
/opt/percona/bin/mysqld_safe --defaults-file=/opt/my.cnf &
```

4.3.3　HAProxy

HAProxy 是一个开源的、高性能的基于 TCP（第四层）和 HTTP（第 7 层）应用的负载均衡软件。使用 HAProxy 可以快速、可靠地实现基于 TCP 和 HTTP 应用的负载均衡解决方案。作为一个专业的负载均衡软件，它有如下优点。

（1）可靠性和稳定性非常好，可以与硬件级的 F5 负载均衡设备相媲美。

（2）最高可以同时维护 40000～50000 个并发连接，单位时间内处理的最大请求数达 20000 个，最大数据处理能力可达 10Gbps。

（3）支持多达 8 种负载均衡算法，同时支持会话保持。

（4）支持虚拟主机功能，使得实现 Web 负载均衡更加灵活。

（5）从 HAProxy 1.3 版本开始支持连接拒绝、全透明代理等功能，这些功能是其他负载均衡器所不具备的。

（6）有功能强大的监控页面，通过此页面可以实时了解系统的运行状况。

（7）拥有功能强大的 ACL 支持。

HAProxy 是借助操作系统的技术特性来实现性能最大化的，因此要想发挥 HAProxy 的最大性能，需要对操作系统的性能进行优化。

HAProxy 适用于并发量特别大且需要持久连接或有四层、七层处理机制的 Web 系统，例如门户网站或电子商务网站等，也适用于 MySQL 数据库（读操作）的负载均衡。HAProxy 的官方文档地址为 http://cbonte.github.io/haproxy-dconv/。

安装配置如下：

```
yum -y install gcc
[root@zyz_dba_test04 ~]# tar -xf haproxy-1.4.25 -C /usr/local/
[root@zyz_dba_test04 ~]# cd /usr/local/haproxy/
[root@zyz_dba_test04 haproxy]# useradd haproxy
[root@zyz_dba_test04 haproxy]# make TARGET=linux26
PREFIX=/usr/local/haproxy
[root@zyz_dba_test04 haproxy]# make install PREFIX=/usr/local/haproxy
[root@zyz_dba_test04 haproxy]# mkdir /usr/local/haproxy/logs/ -p &&
mkdir /usr/local/haproxy/var/run/ -p && mkdir
/usr/local/haproxy/var/chroot/ -p && mkdir
/usr/local/haproxy/conf/ -p
```

```
[root@zyz_dba_test04 haproxy]#  cat  > /usr/local/haproxy/haproxy.cfg
<< OO
global
        log           127.0.0.1        local0
        log           127.0.0.1        local1 notice
        maxconn        94096
        user           haproxy      #所属运行的用户
        group          haproxy      #所属运行的组
        nbproc         1
        pidfile        /usr/local/haproxy/var/run/haproxy1.pid
defaults
        log           global
        option         tcplog
        option         dontlognull
        retries         3
        option          redispatch
        maxconn         94096
        timeout         connect  50000ms
        timeout         client   50000ms
        timeout         server   50000ms
listen  mariadb-galera
        bind 10.21.3.109:3399  #客户端监听端口
        mode tcp
        balance  leastconn  #最少连接的负载均衡算法
        server   db1  10.21.3.106:3306 check
        server   db2  10.21.3.107:3306 check
        server   db3  10.21.3.108:3306 check
OO
```

加入页面监控：

```
global
log 127.0.0.1 local0 info #[err warning info debug]
maxconn 40960
user admin
group admin
daemon
nbproc 1
chroot  /usr/local/haproxy/var/chroot
pidfile /usr/local/haproxy/var/run/haproxy.pid
defaults
mode http
option httplog
retries 3
option redispatch
option abortonclose
maxconn 496
contimeout 5000
```

```
clitimeout 30000
srvtimeout 30000
timeout check 2000
listen admin_stats
bind 0.0.0.0:1080
mode http
log 127.0.0.1 local0 err #[err warning info debug]
stats refresh 3s
stats uri /admin?stats
stats realm Gemini\ Haproxy
stats auth admin:admin
stats auth admin1:admin1
listen  mariadb-galera
       bind 10.21.3.109:3399  #客户端监听端口
       mode tcp
#       stats   uri admin?stats
       balance  leastconn  #最少连接的负载均衡算法
       server   db1  10.21.3.106:3306 check
       server   db2  10.21.3.107:3306 check
       server   db3  10.21.3.108:3306 check
```

如上配置的 defaults 选项可能根据 HAProxy 版本的不同对参数有所调整，如果升级导致参数无效，则可将 defaults 选项改写如下：

```
defaults
       log           global
       option         tcplog
       option         dontlognull
       retries         3
       option          redispatch
       maxconn         94096
       timeout         connect  50000ms
       timeout         client   50000ms
       timeout         server   50000ms
       timeout check 2000
```

启动测试：

```
[root@zyz_dba_test04~]# /usr/local/haproxy/sbin/haproxy -f
/usr/local/haproxy/haproxy.cfg &
```

集群节点开放权限给 HAProxy 服务所在的机器：

```
MariaDB [(none)]> grant all privileges on *.* to zuo@'%' identified by '123';
Query OK, 0 rows affected (0.00 sec)
[root@zyz_dba_test04 haproxy]# MySQL -h 10.21.3.109  -uzuo -p123  --port 3399
```

将 HAProxy 写入服务：

```
cat > /etc/init.d/haproxyd <<OO
```

```
#!/bin/bash
#chkconfig: 35 35 -
. /etc/init.d/functions
BASE="/usr/local/haproxy"
PROG=$BASE/sbin/haproxy
PIDFILE=$BASE/var/run/haproxy.pid
CONFFILE=$BASE/conf/haproxy.conf
case "$1" in
start)
        #$PROG  -f $CONFFILE >/dev/null 2>&1
        $PROG  -f $CONFFILE
       [ $?  -eq 0 ] && {
        action  "haproxy start is  OK..."  /bin/true
 } || action  "haproxy start is error..." /bin/false
        ;;
status)
        if [ !  -f $PIDFILE ]; then
                echo  "pid not found"
                exit  1
        fi
        for  pid in $(cat $PIDFILE); do
                kill  -0 $pid
                RETVAL="$?"
                if  [ ! "$RETVAL" = "0" ]; then
                        echo  "process $pid died"
                        exit  1
                fi
        done
        echo  "process is running"
        ;;
restart)
        kill  $(cat $PIDFILE)
        [ $? -eq 0 ] && {
        action  "haproxy stop is  OK..."  /bin/true
 } || action  "haproxy stop is error..." /bin/false
        #$PROG  -f $CONFFILE -sf $(cat $PIDFILE) >/dev/null 2>&1
        $PROG  -f $CONFFILE
        [ $?  -eq 0 ] && {
        action  "haproxy start is  OK..."  /bin/true
 } || action  "haproxy start is error..." /bin/false
        ;;
stop)
        kill  $(cat $PIDFILE)
 [ $? -eq 0 ] && {
        action  "haproxy stop is  OK..."  /bin/true
 } || action  "haproxy stop is error..." /bin/false
        ;;
```

```
*)
        echo  "USAGE: $0 start|restart|status|stop"
        exit 1
        ;;
esac
OO
```

执行 chkconfig -add haproxy 命令，如果报 service haproxyd does not support chkconfig 错误，则解决办法为在/etc/init.d/haproxyd 中的#!/bin/bash 后添加如下内容：

```
# chkconfig: 2345 10 90
# description:haproxy
```

其中 2345 是默认的启动级别，有 0～6 共 7 个级别：等级 0 表示关机；等级 1 表示单用户模式；等级 2 表示无网络连接的多用户命令行模式；等级 3 表示有网络连接的多用户命令行模式；等级 4 表示不可用；等级 5 表示带图形界面的多用户模式；等级 6 表示重新启动；10 是启动优先级；90 是停机优先级，优先级范围是 0～100，数字越大，优先级越低。

为了验证客户端连接 HAProxy 后，查询语句是否成功分发至后端数据库服务，这里将集群中一个 MySQL 节点的配置内容进行了注释。重启该节点的 MySQL 服务：

```
[root@zyz_dba_test02 ~]# vi /etc/my.cnf
#wsrep_provider = /usr/local/MySQL/lib/libgalera_smm.so
#wsrep_cluster_name = "my_percona_cluster"
#wsrep_cluster_address="gcomm://10.21.3.106,10.21.3.108"
#wsrep_cluster_name='example_cluster'
#wsrep_node_name = "cluster_node2"
#wsrep_node_address = 10.21.3.107:4406
#wsrep_sst_auth=tt:123
#wsrep_node_name='node2'
#wsrep_sst_method=rsync
```

注释掉 10.21.3.107 中与集群相关的参数，使 10.21.3.107 脱离集群：

```
[root@zyz_dba_test02 ~]# service mysqld restart
Shutting down MySQL..
percna> drop database zyz;
percona> show databases;
+--------------------+
| Database          |
+--------------------+
| sbtest           |
| test             |
+--------------------+
6 rows in set (0.01 sec)
```

此时 10.21.3.108 和 10.21.3.106 服务器上的数据库节点还保留在集群中：

```
percona> show databases;
```

```
+---------------------+
| Database           |
+---------------------+
| test               |
| zyz                |
+---------------------+
7 rows in set (0.00 sec)
```

登录 10.21.3.109 即 HAProxy 所在的服务器，重复执行上面的命令三次，通过对比结果可以得知 HAProxy 配置已经生效。

```
[root@zyz_dba_test04 haproxy]# mysql -h 10.21.3.109  -uzuo -p123  --port 3399 -e
"show databases;"
+---------------------+
| Database           |
+---------------------+
| test               |
| zyz                |
+---------------------+
```

1. 日志支持功能的添加

```
# vim /etc/rsyslog.conf
```

在文件最后增加：

```
local0.* /usr/local/haproxy/logs/haproxy.log
# 重启日志服务
/etc/init.d/rsyslog restart
```

2. 压力测试

以下实验中使用的配置为单核 CPU、2GB 内存的 VMware 虚拟机进行压力测试。如果集群中的任意节点同时安装了 MySQL-server 和 Percona Cluster，则其压测过程中 tps 存在降低到 0 的情况且 tps 值的波动幅度较大。同样环境下的纯 Percona Cluster 则不会出现这种情况。如图 4-29 所示。

```
response time:
        min:                              226.48ms
        avg:                             1265.01ms
        max:                             9628.06ms
        approx.  95 percentile:              3422.12ms

Threads fairness:
    events (avg/stddev):           47.7656/7.15
    execution time (avg/stddev):   60.4241/0.28
RECORD LOCKS space id 26 page no 7 n bits 144 index `GEN_CLUST_INDEX` of table
`sbtest`.`sbtest9` trx id 166278 lock_mode X locks rec but not gap
```

```
[root@zyz_dba_test03 opt]# sysbench --test=/usr/share/doc/sysbench/tests/db/oltp.lua --oltp_tables_count=9 --oltp-table-size=100000 --rand-init=on --num-threads=64
oltp-read-only=off --report-interval=1 --rand-type=gaussian --max-time=60 --max-requests=0 --mysql-host=10.21.3.109 --mysql-port=3399 --mysql-user=zuo --mysql-passwo
=123 run
sysbench 0.5:  multi-threaded system evaluation benchmark

Running the test with following options:
Number of threads: 64
Report intermediate results every 1 second(s)
Initializing random number generator from timer.

Random number generator seed is 0 and will be ignored


Threads started!

[   1s] threads: 64, tps: 11.00, reads/s: 1189.85, writes/s: 173.98, response time: 883.69ms (95%)
[   2s] threads: 64, tps: 26.00, reads/s: 404.00, writes/s: 135.00, response time: 1954.04ms (95%)
[   3s] threads: 64, tps: 10.00, reads/s: 486.99, writes/s: 107.00, response time: 2913.09ms (95%)
[   4s] threads: 64, tps: 2.00, reads/s: 158.98, writes/s: 40.00, response time: 3744.78ms (95%)
[   5s] threads: 64, tps: 15.00, reads/s: 476.03, writes/s: 154.01, response time: 4243.91ms (95%)
[   6s] threads: 64, tps: 26.00, reads/s: 657.97, writes/s: 156.99, response time: 3848.18ms (95%)
[   7s] threads: 64, tps: 38.00, reads/s: 899.11, writes/s: 244.03, response time: 6140.28ms (95%)
[   8s] threads: 64, tps: 58.00, reads/s: 1409.00, writes/s: 321.00, response time: 7472.55ms (95%)
[   9s] threads: 64, tps: 76.00, reads/s: 1755.93, writes/s: 378.98, response time: 8382.80ms (95%)
[  10s] threads: 64, tps: 78.00, reads/s: 1874.09, writes/s: 424.02, response time: 2084.55ms (95%)
[  11s] threads: 64, tps: 52.00, reads/s: 1311.00, writes/s: 362.00, response time: 1955.21ms (95%)
[  12s] threads: 64, tps: 75.00, reads/s: 1739.94, writes/s: 362.99, response time: 1920.41ms (95%)
[  13s] threads: 64, tps: 76.01, reads/s: 1828.16, writes/s: 394.03, response time: 2092.68ms (95%)
[  14s] threads: 64, tps: 63.00, reads/s: 1577.01, writes/s: 392.00, response time: 1742.38ms (95%)
[  15s] threads: 64, tps: 51.00, reads/s: 1491.90, writes/s: 318.98, response time: 2516.43ms (95%)
[  16s] threads: 64, tps: 59.00, reads/s: 1540.98, writes/s: 346.00, response time: 2328.00ms (95%)
[  17s] threads: 64, tps: 71.00, reads/s: 1617.04, writes/s: 376.01, response time: 2350.41ms (95%)
[  18s] threads: 64, tps: 74.00, reads/s: 1752.93, writes/s: 402.98, response time: 2607.68ms (95%)
[  19s] threads: 64, tps: 74.00, reads/s: 1837.07, writes/s: 412.02, response time: 2092.06ms (95%)
[  20s] threads: 64, tps: 66.00, reads/s: 1657.96, writes/s: 369.99, response time: 1826.76ms (95%)
[  21s] threads: 64, tps: 47.00, reads/s: 1205.09, writes/s: 303.02, response time: 2536.85ms (95%)
[  22s] threads: 64, tps: 12.00, reads/s: 674.00, writes/s: 144.00, response time: 2325.91ms (95%)
[  23s] threads: 64, tps: 72.00, reads/s: 1527.93, writes/s: 334.98, response time: 3664.93ms (95%)
```

图 4-29

压测时，连接数据库服务执行 show engine innodb status，会看到存在大量的 lock_mode X 锁信息，tps 波动的幅度较大。

全局使用 Percona Cluster 成员进行压力测试，CPU 负载到达 1.5 时的 tps 情况如下：

```
    [root@zyz_dba_test03 ~]# sysbench
--test=/usr/share/doc/sysbench/tests/db/oltp.lua --oltp_tables_count=9
--oltp-table-size=100000 --rand-init=on -num -threads=12  --oltp-read-only=off
--report-interval=1 --rand-type=gaussian --max-time=3000 --max-requests=0
--mysql-host=10.21.3.109 --MySQL-port=3399 --mysql-user=zuo --mysql-password=123
run
    sysbench 0.5:  multi-threaded system evaluation benchmark
    Running the test with following options:
    Number of threads: 12
    Report intermediate results every 1 second(s)
    Initializing random number generator from timer.
    Random number generator seed is 0 and will be ignored
    Threads started!

    [   1s] threads: 12, tps: 77.96, reads/s: 1175.45, writes/s: 315.88, response time:
499.33ms (95%)
    [   2s] threads: 12, tps: 128.00, reads/s: 1843.01, writes/s: 522.00, response
time: 123.21ms (95%)
    [   3s] threads: 12, tps: 114.00, reads/s: 1584.98, writes/s: 455.99, response
time: 235.69ms (95%)
    ...
```

全局使用 Percona Cluster 成员进行更多线程的压力测试，CPU 负载到 18 时的 tps 情况如下：

```
    [root@zyz_dba_test03 ~]# sysbench
--test=/usr/share/doc/sysbench/tests/db/oltp.lua --oltp_tables_count=9
--oltp-table-size=10 --rand-init=on --num-threads=64  --oltp-read-only=off
```

```
--report-interval=1 --rand-type=gaussian --max-time=3000 --max-requests=0
--mysql-host=10.21.3.109 --mysql-port=3399 --mysql-user=zuo --mysql-password=123
run
    sysbench 0.5:  multi-threaded system evaluation benchmark
    Running the test with following options:
    Number of threads: 64
    Report intermediate results every 1 second(s)
    Initializing random number generator from timer.
    Random number generator seed is 0 and will be ignored
    Threads started!
    [   1s] threads: 64, tps: 84.58, reads/s: 1846.37, writes/s: 365.15, response time:
826.38ms (95%)
    [   2s] threads: 64, tps: 122.04, reads/s: 1594.50, writes/s: 474.15, response
time: 765.42ms (95%)
    [   3s] threads: 64, tps: 124.00, reads/s: 1841.01, writes/s: 515.00, response
time: 627.45ms (95%)
    ...
```

可以看到 CPU 负载在 1.5 和 18 时的 tps 差不多，所以结论为：负载到一定程度时，Percona Cluster 压测的结果比较稳定，而 MySQL、MariaDB 版本在负载十分高的情况下，tps 的下降峰值表现得不够稳定，因此这里在生产环境中选择了 Percona 集群。

4.3.4 Keepalived

Keepalived 是一种基于 VRRP 协议来实现的高可用方案，可以利用其来避免单点故障。通常有两台甚至多台服务器运行 Keepalived，一台为主服务器（MASTER），其他为备份服务器，但是对外表现为一个虚拟 IP，主服务器会发送特定的消息给备份服务器，当备份服务器收不到这个消息时，即认为主服务器宕机，备份服务器就会接管虚拟 IP，继续提供服务，从而保证了高可用性。其安装配置如下：

```
echo 'net.ipv4.ip_nonlocal_bind = 1'>>/etc/sysctl.conf
cd ~
wget  http://www.keepalived.org/software/keepalived-1.2.9.tar.gz
tar  -xf keepalived-1.2.9.tar.gz
mkdir /usr/local/keepalived
cd ~/ keepalived-1.2.9
./configure  --prefix=/usr/local/keepalived/
make && make install
cp /usr/local/keepalived/sbin/keepalived  /usr/sbin/
cp /usr/local/keepalived/etc/sysconfig/keepalived  /etc/sysconfig
cp /usr/local/keepalived/etc/rc.d/init.d/keepalived  /etc/init.d/
chmod +x /etc/init.d/keepalived
mkdir /etc/keepalived
cat > /etc/keepalived/check_haproxy.sh  <<OO
```

```
#!/bin/bash
A=`ps -C haproxy --no-header |wc -l`
if [ $A -eq 0 ];then
/etc/init.d/haproxyd restart
echo "Start haproxy" &> /dev/null
sleep 3
if [ `ps -C haproxy --no-header |wc -l` -eq 0 ];then
/etc/init.d/keepalived stop
echo "Stop keepalived" &> /dev/null
fi
fi
OO
chmod +x /etc/keepalived/check_haproxy.sh
cat > /etc/keepalived/keepalived.conf << GG
global_defs {
   notification_email {
   zuoyuezong@163.com
   }
   notification_email_from zuoyuezong@163.com
   smtp_server smtp.163.com   # 邮件服务器地址
   smtp_connect_timeout 30     # 连接超时时间
   router_id LVS_master
}
vrrp_script chk_http_port {
   script  "/etc/keepalived/check_haproxy.sh"    # HAProxy 运行检测脚本[HAProxy
宕掉重启 HAProxy 服务]
   interval       5                   # 脚本执行间隔
   weight        -5                     # 执行脚本后优先级变更：5 表示优先级+5；-5 则表示优
先级-5
}
vrrp_instance VI_A {
   state      BACKUP                # 主节点上此值为 MASTER，从节点上为 BACKUP
   interface   eth0
   virtual_router_id 50              # 此值主从必须一致
   priority 80                     # 此值在 MASTER 上比 BACKUP 大
   advert_int 1
   authentication {                  # authentication 两个参数值，主从也必须一致
      auth_type PASS
      auth_pass yiban
   }
track_script {
   chk_http_port
   }
   virtual_ipaddress {
      10.21.82.41               # HAProxy 提供的虚拟 IP 地址
   }
}
```

```
GG
```

以下为 backup 上 Keepalived 服务的配置：

```
cat keepalived.conf
global_defs {
   notification_email {
 zuoyuezong@163.com
   }
   notification_email_from zuoyuezong@163.com
   smtp_server smtp.163.com   # 邮件服务器地址
   smtp_connect_timeout 30     # 连接超时时间
   router_id LVS_master
}
vrrp_script chk_http_port {
   script  "/etc/keepalived/check_haproxy.sh"    # HAProxy 运行检测脚本[HAProxy
宕掉重启 HAProxy 服务]
   interval       5                      # 脚本执行间隔
   weight        -5                     # 执行脚本后优先级变更:5 表示优先级+5;-5 则表示优
先级-5
  }
  vrrp_instance VI_A {
    state MASTER                # 主节点上此值为 MASTER,从节点上为 BACKUP
    interface eth0
    virtual_router_id 50            # 此值主从必须一致
    priority 100
    advert_int 1
    authentication {   # authentication 两个参数值,主从也必须一致
       auth_type PASS
    auth_pass yiban
    }
  track_script {
    chk_http_port
    }
    virtual_ipaddress {
         10.21.82.41
    }
  }
```

4.4 MHA+Keepalived 集群搭建

MHA（master High Availability）是采用 Perl 语言编写的一个实现 MySQL 高可用方案的脚本管理工具。

MHA 高可用的实现方式是，当 MySQL-master 出现故障时，会挑选一个 MySQL-slave 作

为新的主节点（master），并构建新的主从架构关系。从 master 出现故障到构建成新的主从架构的时间是 10～30 秒。在 MySQL-master 出现故障时可能会出现 MySQL-slave 同步的数据不一致的现象，此工具可以自动应用差异的中继日志到其他 MySQL-slave 上来保证数据的一致性。

下面的架构图（见图 4-30）是为了把大部分负载压在 MYSQL.COM 与 SLAVE2.COM 上，因而设计为 MYSQL.COM 宕机后，SLAVE1.COM 一定会被提升为主节点。通过 Keepalived 服务配置客户端访问的 vip 会从 MYSQL.COM 切换到 SLAVE1.COM 上，基于负载的目的可以配合 Mycat 做读写分离等。

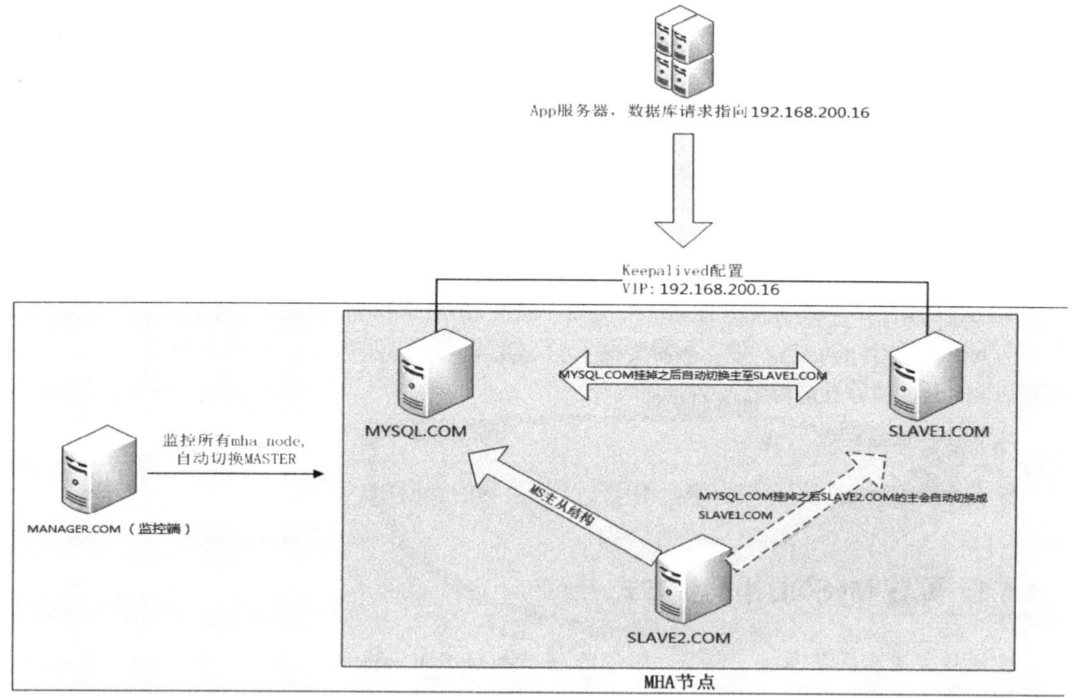

图 4-30

MHA 的优点如下。

（1）自动监控 master，故障转移的速度很快。

在 9～12 秒内可以检测到 master 故障；7～10 秒内可以关闭 master 机器以避免脑裂；在几秒内可以应用差异日志，并构建新的主从架构。整个过程可以在 10～30 秒内完成，可最大化地减少故障修复时间。

（2）master 宕机时可以最大化地减少数据的丢失。

当 master 宕机时，MHA 会自动检测和选择数据同步最全的 slave，并把差异日志应用到其他 slave 上，以保障数据的一致性。

（3）不需要修改现有的 MySQL 配置，支持 MySQL 5.0 及以上版本。

MHA 与最常见的双主多从、一主多从环境结合。MHA 项目既可以与 MySQL 同步，也可以与 MySQL 半同步复制结合。启动、停止、升级、降级、安装、卸载 MHA 可以不用改变（包括启动、停止）MySQL 原有的复制。

（4）不需要增加太多的服务器。

MHA 的更改升级配置等不影响线上正在运行的数据库。MHA 由 MHA Node 和 MHA Manager 组成。MHA Node 运行在 MySQL 服务器上，所以不会因为 MHA Node 增加新的服务器。MHA Manager 通常需要独立运行在一台服务器上，所以你需要增加一台服务器用于监控管理和运行 MHA Manager，但是一台服务器上的 MHA Manager 可以同时监控、管理多达百台 master，所以总的来说服务器的增加不会太多。MHA Manger 也可以运行在一台 slave 上，这样总的服务器数量也不会增加。

（5）整体性能较好。

MHA 工作在异步或半同步的主从架构上。当监控 master 时，MHA 会每隔几秒（默认为 3 秒）向 master 发出 ping 包，并且不需要很长的 SQL 语句用于监控 master 的健康状况。slave 需要开启 binlog，整体性能降低太多。

（6）适合任何存储引擎

它支持能主从复制的存储引擎，不限于支持事物的 InnoDB 引擎。

4.4.1 配置 MySQL 半同步方式

这里共有 4 台服务器：一台管理服务器、一台 MySQL-master 服务器、两台 MySQL-slave 服务器；操作系统为 Centos 6.4 64 bit；数据库版本为 MySQL-5.6.10；服务器的 IP 如下：

- 192.168.186.141 MYSQL.COM
- 192.168.186.142 SLAVE1.COM
- 192.168.186.146 SLAVE2.COM
- 192.168.186.144 MANAGER.COM

1. 首先在三台机器上编译安装 MySQL 5.6.10

关闭 selinux、iptables 服务以便后期的主从同步不出错，installmysql5.sh 脚本使用的是编译安装方式。为了弄清楚编译安装的过程，我们可以使用源代码编译、二进制包或 rpm 包等方式进行安装。

```
[root@MYSQL ~]# cd /usr/local/src/
[root@MYSQL src]# ls
installmysql5.sh
[root@MYSQL src]# wget
http://101.96.10.42/downloads.mysql.com/archives/get/file/mysql-5.6.10.tar.gz
[root@MYSQL src]# vim  installmysql5.sh
export
PATH=/usr/local/sbin:/usr/local/bin:/sbin:/bin:/usr/sbin:/usr/bin:/usr/local/mys
ql/bin:/root/bin:/usr/local/mysql/bin
DATE=`date "+%Y%m%d %H:%M:%S"`
MYSQL_DIR=/usr/local/mysql
#DATA_DIR=/home/mysql/data
CHECKINSTALL="is not installed"
RPMLIST="make gcc gcc-c++  autoconf automake  bison  ncurses-devel
libtool-ltdl-devel* cmake"
TAR=/usr/local/src
function install_sh()
{
if [ -f /etc/my.cnf ];then
mv /etc/my.cnf{,.bak}
fi
mysqlnum=`netstat -tualnp |grep mysql |wc -l`
if [ $mysqlnum -gt 0 ];then
echo  -e "$DATE \033[32m MYSQL-server already having runing \033[0m"
echo  -e "$DATE \033[32m MYSQL-server already having runing \033[0m" >>
/tmp/tarmysql.log
exit 0
fi
cat >>/etc/profile <<EOF
export PATH=$PATH:/usr/local/mysql/bin
EOF
source /etc/profile
cp -a /etc/profile  /etc/profile.bak
read -p "please enter you mysql version (eg:/mysql-5.5.34):" BANBEN
read -p "please enter you mysql datadir (eg:/data/mysql/data):" DATA_DIR
echo '执行完该脚本后启动你的 MYSQL，在/etc/profile 文件里写入 source 并执行，这样才能直接
使用 mysql 命令 export PATH=$PATH:/usr/local/mysql/bin'
sleep 1
echo  '装包部分开始'
sleep 1
```

```
rpm -qa |grep mysql  > /tmp/mysqlremove.txt
if [ $? -eq 0 ];then
        for i in $(cat /tmp/mysqlremove.txt); do yum -y remove $i ; done
        echo ""
        echo -e "$DATE \033[32m MYSQL already removed \033[0m" >> /tmp/tarmysql.log
        echo ""
else
        echo -e "$DATE \033[32m MYSQL does not exist \033[0m"  >> /tmp/tarmysql.log
fi
rpm -q --qf '%{NAME}-%{VERSION}-%{RELEASE} (%{ARCH})\n' gcc gcc-c++  autoconf
automake bison ncurses-devel libtool-ltdl-devel cmake > /tmp/rpmtoolinstall.log
        grep 'is not install' /tmp/rpmtoolinstall.log
        if [ $? -eq  0 ];then
         yum -y install $RPMLIST
else
        echo -e "$DATE \033[32m MYSQL tool already install \033[0m"  >>
/tmp/mysqltool.log
        fi
echo '开始创建 MySQL 相关的目录'
sleep 1
if [ ! -d /usr/local/mysql ];then
        mkdir /usr/local/mysql -p
else
        echo '/usr/local/mysql already having'   >> /tmp/tarmysql.log
fi
if [ ! -d $DATA_DIR ];then
        mkdir $DATA_DIR -p
else
        echo $DATA_DIR already having   >> /tmp/tarmysql.log
fi
echo '开始创建 MySQL 相关的用户和组'
sleep 1
grep mysql /etc/group &>/dev/null
if [ $? -eq 0 ];then
        echo "group:mysql is already exist"   >> /tmp/tarmysql.log
else
        groupadd mysql
fi
grep mysql /etc/passwd &>/dev/null
if [ $? -eq 0 ];then
        echo 'user:mysql is already exist'   >> /tmp/tarmysql.log
else
        useradd -g mysql mysql
fi
useradd mysql
chown mysql.mysql -R /usr/local/mysql/
echo '解压部分开始'
```

```
    sleep 1
    if [ ! -d $TAR/$BANBEN ];then
           tar -xf  $TAR/$BANBEN.tar.gz
    else
           echo  'tar -xf already ----> ok'
           echo 'tar -xf already ----> ok'     >> /tmp/tarmysql.log
    fi
    if [ -d $TAR/$BANBEN ];then
           cd $TAR/$BANBEN
    else
           echo  "没有数据库安装包,not having mysql-tar,,请把你下载的mysql的tar包放在
/usr/local/src目录下再执行此脚本"
           sleep 2
           exit 20
    fi
    echo '软件包安装开始'
    sleep 1
    if [ -f $TAR/$BANBEN/CMakeCache.txt ];
    then
    echo '你已经装好了一个数据库，最多没有执行如下代码，请先正常启动使用，如有问题请执行命令.'
/mysql_install_db --user=mysql --basedir=/usr/local/mysql --datadir=$DATA_DIR
--skip-grant-tables --skip-networking /usr/local/mysql/scripts/ '执行上一条命令不成
功时，请停止并删除你现有的数据库服务，然后执行该脚本'
    exit 1
    else
    cd $TAR/$BANBEN
    cmake -DCMAKE_INSTALL_PREFIX=/usr/local/mysql \
    -DMYSQL_UNIX_ADDR=/tmp/mysql.sock \
    -DDEFAULT_CHARSET=utf8 \
    -DDEFAULT_COLLATION=utf8_general_ci \
    -DWITH_EXTRA_CHARSETS:STRING=all \
    -DENABLED_LOCAL_INFILE=1 \
    -DWITH_ARCHIVE_STORAGE_ENGINE=1 \
    -DWITH_BLACKHOLE_STORAGE_ENGINE=1 \
    -DWITH_FEDERATED_STORAGE_ENGINE=1 \
    -DWITH_EXAMPLE_STORAGE_ENGINE=1 \
    -DMYSQL_DATADIR=$DATA_DIR \
    -DMYSQL_USER=mysql \
    -DMYSQL_TCP_PORT=3306
    sleep 1
    echo 'start make'
    sleep 1
    make
    sleep 2
    echo 'start make install'
    sleep 1
    make install
```

```
    fi
    /usr/local/mysql/scripts/mysql_install_db --user=mysql
--basedir=/usr/local/mysql --datadir=$DATA_DIR --skip-grant-tables
--skip-networking
    if [ -f /usr/local/mysql/support-files/my-small.cnf ]
    then
    cp /usr/local/mysql/support-files/my-small.cnf /etc/my.cnf
    else
    cp /usr/local/mysql/support-files/my-default.cnf /etc/my.cnf
    fi
    cp /usr/local/mysql/support-files/mysql.server /etc/init.d/mysqld
    export PATH=$PATH:/usr/local/mysql/bin
    chmod +x /etc/init.d/mysqld
    chkconfig --add mysqld
    chkconfig mysqld on
    chown mysql.mysql $DATA_DIR -R
    sed -i -e /server-id/d  -e
's/log_bin/\nlog_bin=mysql\n\server-id='$$'\nbinlog_format=row/g'   -e
's/log-bin=mysql-bin/\nlog_bin=mysql\n\server-id='$$'\nbinlog_format=row/g'
/etc/my.cnf
    service mysqld start
    source /etc/profile
    mysql -e "grant  USAGE,REPLICATION SLAVE,REPLICATION CLIENT on *.* to repl@'%'
identified by 'repl123';"
    }
    install_sh

    #以上为安装脚本
     [root@MYSQL src]# sh installmysql5.sh
    please enter you MySQL version (eg:/mysql-5.5.34):mysql-5.6.10
    please enter you MySQL datadir (eg:/data/mysql/data):/data/mysql/data
    [root@MYSQL src]#source /etc/profile
```

2. 配置 HOSTS 环境

```
[root@MYSQL etc]# vi /etc/hosts
127.0.0.1   localhost localhost.localdomain localhost4 localhost4.localdomain4
::1         localhost localhost.localdomain localhost6 localhost6.localdomain6
192.168.186.141 MYSQL.COM
192.168.186.142 SLAVE1.COM
192.168.186.146 SLAVE2.COM
192.168.186.144 MANAGER.COM
[root@MYSQL etc]# for i in  142 146 144;do scp /etc/hosts
192.168.186.$i:/etc/;
done
root@192.168.186.142's password:
Permission denied, please try again.
```

```
    root@192.168.186.142's password:
    Permission denied, please try again.
    root@192.168.186.142's password:
    hosts                                                100%  266
0.3KB/s   00:00
    root@192.168.186.146's password:
    hosts                                                100%  266
0.3KB/s   00:00
    root@192.168.186.144's password:
```

3. 安装 MySQL 主从半同步

```
#所有 MySQL 数据库服务器安装半同步插件（semisync_master.so,semisync_slave.so）
MySQL> install plugin rpl_semi_sync_master soname 'semisync_master.so';
MySQL> install plugin rpl_semi_sync_slave soname 'semisync_slave.so';
[root@MYSQL etc]vi /etc/my.cnf
[mysqld]
rpl_semi_sync_master_enabled=1
rpl_semi_sync_master_timeout=1000
rpl_semi_sync_slave_enabled=1
relay_log_purge=0
skip-name-resolve
#socket=/usr/mysql.sock
#auto_increment_offset = 2
#auto_increment_increment = 2
server-id = 1
log-bin=mysql-bin
read_only=1
slave-skip-errors=1396
#三台机器需要全部开启半同步功能，其中参数配置仅 server-id 处不同

MySQL> show variables like '%sync%';
# 查看半同步状态：
MySQL> show status like '%sync%';
# 有几个状态参数的值得关注：
rpl_semi_sync_master_status：显示主服务是异步复制模式还是半同步复制模式
rpl_semi_sync_master_clients：显示有多少个从服务器配置为半同步复制模式
rpl_semi_sync_master_yes_tx：显示从服务器确认成功提交的数量
rpl_semi_sync_master_no_tx：显示从服务器确认不成功提交的数量
rpl_semi_sync_master_tx_avg_wait_time：事务因开启 semi_sync 需要额外的平均等待时间
rpl_semi_sync_master_net_avg_wait_time：事务进入等待队列后到网络的平均等待时间
[root@MYSQL src]# service mysqld restart 每台机器重启
```

4. 每台机器配置 ssh 免密码登录

```
[root@MYSQL src]# cat /etc/hosts
```

```
127.0.0.1   localhost localhost.localdomain localhost4 localhost4.localdomain4
::1         localhost localhost.localdomain localhost6 localhost6.localdomain6
192.168.186.141 MYSQL.COM
192.168.186.142 SLAVE1.COM
192.168.186.146 SLAVE2.COM
192.168.186.144 MANAGER.COM
[root@MYSQL src]# ssh-keygen
[root@MYSQL src]# ssh-copy-id 192.168.186.142
[root@MYSQL src]# ssh-copy-id 192.168.186.144
[root@MYSQL src]# ssh-copy-id 192.168.186.146
```

为了保证每一台机器之间使用 ssh 命令登录不需要输入密码，配置完成后每一台服务器之间互登录一次，首次连接需要输入一次“YES”以确定，在 know-hosts 文件中进行记录。可以在每台服务器上执行如下脚本：

```
for  i  in  MANAGER.COM SLAVE1.COM SALVE2.COM MYSQL.COM;
do ssh $i date;done
[root@MYSQL ~]# ssh MANAGER.COM date
[root@MYSQL ~]# ssh SLAVE1.COM  date
[root@MYSQL ~]# ssh SALVE2.COM  date
```

5. 配置主从

编写主从脚本如下：

```
[root@MYSQL src]#vi masterslave.sh
#! /bin/bash

repl_user=repl
repl_pass=repl123
mysql_install_dir=/usr/local/mysql

usage() {
  echo usage: $filename [ -p slaveport -s slaveip -m masterip -o masterport ]
  echo ""
}
declare base_port
declare dbslave
declare dbmaster
declare master_port

while getopts "p:s:m:o:" arg
do
  case $arg in
    p)
      base_port="$OPTARG"
      ;;
```

```
        s)
          dbslave="$OPTARG"
         ;;
        m)
          dbmaster="$OPTARG"
         ;;
        o)
          master_port=""$OPTARG""
         ;;
        ?)
         { usage; exit 1; }
         ;;
      esac
    done

    if [ -z $master_port ] ; then
       master_port=3306 ;
    fi
    if [ -z $dbslave ] ; then
       dbslave="127.0.0.1" ;
    fi
    if [ -z $master_port ];then
         master_port="3306";
    fi
    if [ -z $dbmaster ] ; then
       echo this is master ;
    else
    pos=`$mysql_install_dir/bin/mysql  -P$base_port -u$repl_user  -p$repl_pass
-h$dbmaster  -e "show master   status;" |awk -F ' ' '{print $2}'|tail -1`

    log=`$mysql_install_dir/bin/mysql  -P$base_port -u$repl_user  -p$repl_pass
-h$dbmaster -e "show master   status;" |awk -F ' ' '{print $1}'|tail -1`

    intchange="stop slave;CHANGE MASTER TO
master_host="\"$dbmaster"\",master_port=$master_port,
master_user="\"$repl_user"\",
master_password="\"$repl_pass"\",master_log_file="\"$log"\",master_log_pos=$pos;
start slave;"
    (echo $intchange)|$mysql_install_dir/bin/mysql -uroot  -h127.0.0.1
-P$base_port

    $mysql_install_dir/bin/mysql -u$repl_user -p$repl_pass -h$dbmaster
-P$master_port -e "use test;create table if not exists a(id int);truncate table
a;insert into a select 1;"
    sleep 1
    ISE=`$mysql_install_dir/bin/mysql -uroot  -h127.0.0.1 -P$base_port -Ne "select
id from test.a;"`
```

```
echo $ISE
if [ $ISE -eq 1 ];then
echo "####[`date "+%F %T"`] [info] master-slave  on [$base_port] suceessful"
echo "[`date "+%F %T"`] [info] master-slave on [$base_port] suceessful" >
/tmp/zyz.log
echo "#################################################################"
else
echo "####[`date "+%F %T"`] [info] master-slave  on [$base_port] faild"
echo "[`date "+%F %T"`] [info] master-slave on [$base_port] faild" >> /tmp/zyz.log
echo "#################################################################"
fi
fi
echo "done"
exit 0
#本脚本的 mysql_install_dir 路径为 MYSQL-SLAVE 数据库软件的安装目录，请根据实际情况修改为
你的安装路径
```

在 MYSQL-SLAVE 上执行数据库主从安装脚本，如下所示：

```
[root@MYSQL src]# sh masterslave.sh  -p 3306 -s 192.168.223.129 -m
192.168.223.128 -o 3306
#-S 为 MYSQL-SLAVE 服务器 IP 地址一般为本机 IP 地址。
#-M 为已安装的 MYSQL-MASTER 服务器地址
#-O 为所需要同步的 MYSQL-MASTER 端口
```

至此 MySQL 安装主从半同步的配置完成。

4.4.2 安装配置 MHA

1. 安装

在安装时，对每台 MHA 节点服务器进行如下操作：

```
[root@SLAVE2data]#rpm -ivh
http://dl.fedoraproject.org/pub/epel/6/x86_64/epel-release-6-8.noarch.rpm
[root@MANAGER src]# yum clean all
Loaded plugins: fastestmirror, refresh-packagekit, security
Cleaning repos: epel name
Cleaning up Everything
Cleaning up list of fastest mirrors
[root@MANAGER src]# yum makecache
[root@MANAGER src]# rpm --import /etc/pki/rpm-gpg/*
[root@SLAVE2 data]# yum  -y install perl-DBD-mysql perl-Config-Tiny
perl-Log-Dispatch perl-Parallel-ForkManager perl-Config-IniFiles  ncftp
perl-Params-Validate  perl-CPAN perl-Test-Mock-LWP.noarch
perl-LWP-Authen-Negotiate.noarch perl-devel
[root@SLAVE2 data]#yum -y install perl-ExtUtils-CBuilder
```

```
perl-ExtUtils-MakeMaker
```

如果安装上面的软件包时，依赖关系有错，则请先安装 mysql-share-compat 包。

关于 MHA 包的下载，请参考 http://pan.baidu.com/s/1b4JxE2。

```
    #以下操作在 MHA 管理节点上需要安装 mha4mysql-node-0.53.tar.gz 与
mha4mysql-manager-0.53.tar.gz 软件包，其他 3 台数据库节点只需要安装
mha4mysql-node-0.53.tar.gz 软件包：
    [root@MANAGER src]# tar -xf mha4mysql-node-0.53.tar.gz
    [root@MANAGER src]# cd mha4mysql-node-0.53
    [root@MANAGER mha4mysql-node-0.53]# perl Makefile.PL
    [root@MANAGER mha4mysql-node-0.53]# make && make install
    [root@MANAGER src]# tar -xf  mha4mysql-manager-0.53.tar.gz
    [root@MANAGER src]# cd mha4mysql-manager-0.53
    [root@MANAGER mha4mysql-manager-0.53]# perl Makefile.PL
    [root@MANAGER mha4mysql-manager-0.53]# make && make install
    [root@MANAGER src]# mkdir /etc/masterha
    [root@MANAGER mha]# mkdir -p /master/app1
    [root@MANAGERmha]# mkdir -p /scripts
    [root@MANAGER mha]# cp samples/conf/* /etc/masterha/
    [root@MANAGERmha]# cp samples/scripts/*  /scripts
    [root@MANAGER mha4mysql-manager-0.53]#  cp samples/conf/*
    /etc/masterha/
```

2. 配置

```
[root@MANAGER masterha]# vi app1.cnf
```

内容如下：

```
[server default]
manager_workdir=/masterha/app1
manager_log=/masterha/app1/manager.log
user=mha_mon
password=123
ssh_user=root
repl_user=slave   做主从的用户  这个也是每一台都要授权的
repl_password=yunwei123  做主从的密码
ping_interval=1
shutdown_script=""
master_ip_online_change_script=""
report_script=""
[server1]
hostname=192.168.186.141
master_binlog_dir=/data/mysql/data
candidate_master=1
[server2]
hostname=192.168.186.142
```

```
master_binlog_dir=/data/mysql/data
candidate_master=1
[server3]
hostname=192.168.186.146
master_binlog_dir=/data/mysql/data
no_master=1
```

保存后退出。

```
[root@MANAGER masterha]# >masterha_default.cnf
```

3. 测试 ssh

```
[root@MANAGER masterha]# masterha_check_ssh --global_conf=
/etc/masterha/masterha_default.cnf  --conf=/etc/masterha/app1.cnf
Wed Jul  9 02:26:57 2014 - [info] Reading default configuratoins from
Wed Jul  9 02:26:57 2014 - [info] Starting SSH connection tests..
Wed Jul  9 02:26:58 2014 - [debug]
Wed Jul  9 02:26:57 2014 - [debug]  Connecting via SSH from
root@192.168.186.141(192.168.186.141:22) to
root@192.168.186.142(192.168.186.142:22)..
Wed Jul  9 02:26:57 2014 - [debug]   ok.
...
...
Wed Jul  9 02:26:58 2014 - [info] All SSH connection tests passed successfully.
```

登录每台数据库节点进行如下操作：

```
MySQL> grant all privileges on *.* to mha_mon@'%' identified by '123';
Query OK, 0 rows affected (1.00 sec)
MySQL> flush privileges;
Query OK, 0 rows affected (0.01 sec)
```

4. 注意事项

在每台 MYSQL-SLAVE 服务器上执行：

```
MySQL>set global read_only=1;  set  global  relay_log_purge=0;
```

或者写到数据库配置文件里面最好。

```
[root@SLAVE1 ~]# vi /etc/my.cnf
read_only=1
slave-skip-errors=1396
```

为什么要跳过 1396 这个错误呢？因为在主节点里面删除用户时，可能会报错说没有这个用户，从而跳过这个错误。

需要删除数据库中空域名的用户，否则 MHA 连接 MySQL 时可能会报错。

```
MySQL> select user,host from mysql.user;
+---------+---------------+
| user    | host          |
+---------+---------------+
| root    | 127.0.0.1     |
| mha_mon | 192.168.186.% |
| repl    | 192.168.186.% |
| slave   | 192.168.186.% |
| root    | ::1           |
|         | SLAVE2.COM    |
| root    | SLAVE2.COM    |
| root    | localhost     |
+---------+---------------+
8 rows in set (0.00 sec)
MySQL> drop user 'root'@SLAVE2.COM;
Query OK, 0 rows affected (0.00 sec)
```

5. 测试 MySQL

```
[root@MANAGER masterha]#  masterha_check_repl   --conf=/etc/masterha/app1.cnf
Wed Jul  9 04:23:16 2014 - [warning] Global configuration file
/etc/masterha_default.cnf not found. Skipping.
Wed Jul  9 04:23:16 2014 - [info] Reading application default configurations from
/etc/masterha/app1.cnf..
Wed Jul  9 04:23:17 2014 - [info] Alive Servers:
...
...
Wed Jul  9 04:23:21 2014 - [warning] shutdown_script is not defined.
Wed Jul  9 04:23:21 2014 - [info] Got exit code 0 (Not master dead).
MySQL Replication Health is OK.
```

至此说明你的 MHA 已经配置好了。

6. 启动

```
[root@MANAGER ~]#nohup masterha_manager --conf=/etc/mastermha/app1.cnf >
/tmp/mha_manager.log  </dev/null 2>&1 &启动 MHA
```

4.4.3　测试重构

1. 测试

将 MYSQL.COM 机器上的 MySQL 服务关闭，观察 manager.log 日志会发现：SLAVE1.COM 机器成为主节点，SLAVE2.COM 机器则成为 SLAVE1.com 的从节点。

[root@MANAGER app1]# tail -f manager.log 是启动后还没关闭主数据库的日志内容。

```
192.168.186.141 (current master)
 +--192.168.186.142
 +--SLAVE2.COM
Wed Jul  9 18:52:32 2014 - [info] Starting ping health check on
192.168.186.141(192.168.186.141:3306)..
Wed Jul  9 18:52:32 2014 - [info] Ping(SELECT) succeeded, waiting until MySQL
doesn't respond..
[root@MYSQL ~]# service mysqld stop
Shutting down MySQL..... SUCCESS!

[root@MANAGER app1]# tail -f manager.log
192.168.186.141 (current master)
 +--192.168.186.142
 +--SLAVE2.COM
...
Selected 192.168.186.142 as a new master.
192.168.186.142: OK: Applying all logs succeeded.
SLAVE2.COM: This host has the latest relay log events.
Generating relay diff files from the latest slave succeeded.
SLAVE2.COM: OK: Applying all logs succeeded. slave started, replicating from
192.168.186.142.
192.168.186.142: Resetting slave info succeeded.
master failover to 192.168.186.142(192.168.186.142:3306) completed successfully.
[root@slave2 ~]# mysql -e "show slave status\G"
*************************** 1. row ***************************
              slave_IO_State: Waiting for master to send event
                 master_Host: 192.168.186.142
                 master_User: repl
                 master_Port: 3306
               Connect_Retry: 60
             master_Log_File: MySQL-bin.000007
         Read_master_Log_Pos: 504
              Relay_Log_File: slave2-relay-bin.000002
               Relay_Log_Pos: 283
       Relay_master_Log_File: mysql-bin.000007
            slave_IO_Running: Yes
           slave_SQL_Running: Yes
```

看到已经切换为 192.168.186.142 同步了，此时的 SALVE1.COM（192.168.186.142）已经变成主节点了，说明切换成功。

2. 重构

切换后 SLAVE1.COM 由原先的数据库从角色变成了数据库主角色，如果需要在集群中加入新节点，则只需把新加入的服务器信息配置到主节点的 app1.conf 文件中。通过 rm -rf app1.failover.complete 删除该文件后，重新启动主节点，新节点将加入整个集群中。

4.4.4　扩展 Keepalived

1. 安装 Keepalived

```
[root@MYSQL src]# wget
http://www.keepalived.org/software/keepalived-1.2.12.tar.gz
[root@MYSQL src]# tar -xf keepalived-1.2.12.tar.gz
[root@MYSQL src]# cd keepalived-1.2.12
[root@MYSQL src]# yum -y install gcc gcc-c++ gcc-g77 ncurses-devel bison
libaio-devel cmake libnl* libpopt* popt-static openssl-devel
[root@MYSQL keepalived-1.2.12]# ./configure
[root@MYSQL keepalived-1.2.12]# make && make install
[root@MYSQL src]#mkdir /etc/keepalived/
[root@MYSQL src]# cp /usr/local/etc/keepalived/keepalived.conf
/etc/keepalived/
[root@MYSQL src]#  cp /usr/local/etc/rc.d/init.d/keepalived
/etc/init.d/
[root@MYSQL src]# cp /usr/local/etc/sysconfig/keepalived
/etc/sysconfig/
[root@MYSQL src]#cp /usr/local/sbin/keepalived /usr/sbin/
```

2. 配置 Keepalived

```
[root@MYSQL keepalived]# vi keepalived.conf
! Configuration File for keepalived
global_defs {
   notification_email {
     acassen@firewall.loc
     failover@firewall.loc
     sysadmin@firewall.loc
   }
   notification_email_from Alexandre.Cassen@firewall.loc
   smtp_server 192.168.200.1
   smtp_connect_timeout 30
   router_id LVS_DEVEL            ##配置是为了标识当前节点，两个节点的此项设置可相同，也
可不相同
}
vrrp_instance VI_1 {
    state MASTER                  #指定 A 节点为主节点，在备用节点上设置为 BACKUP 即可
    interface eth0
    virtual_router_id 51              #VRRP 组名，两个节点的设置必须一样，以指明各个节点
```

```
属于同一 VRRP
        priority 100      #主节点的优先级（1-254 之间），备用节点必须比主节点优先级低
             advert_int 1
        authentication {          #设置验证信息，两个节点必须一致
            auth_type PASS
            auth_pass 1111
        }
    virtual_ipaddress {
    192.168.200.16
        }
    }
```

3. 检测 MySQL 服务脚本

```
[root@MYSQL keepalived]# vi /root/check_mysql.sh
```

至于这里为何要写脚本，然后通过 crontab 去检测而不是放到 Keepalived 中，是为了当 Keepalived 发生问题并宕机后，可以通过脚本检测到并关闭本机的 MySQL 服务，保证集群的高可用性。

```
#!/bin/bash
MYSQL=/usr/local/mysql/bin/mysql
MYSQL_HOST=127.0.0.1
MYSQL_USER=root
MYSQL_PASSWORD=123
CHECK_TIME=3
#MySQL  is working MYSQL_OK is 1 , MySQL down MYSQL_OK is 0
MYSQL_OK=1
function check_MySQL_helth (){
$MYSQL -h $MYSQL_HOST -u $MYSQL_USER -e "show status;" >/dev/null 2>&1
if [ $? = 0 ] ;then
     MYSQL_OK=1
else
     MYSQL_OK=0
fi
     return $MYSQL_OK
}
while [ $CHECK_TIME -ne 0 ]
do
     let "CHECK_TIME -= 1"
     check_MySQL_helth
if [ $MYSQL_OK = 1 ] ; then
     CHECK_TIME=0
     exit 0
if
A=`ps -ef |grep keepalive |grep -v grep |grep -c keepalive`
if [ $A -ne 0 ];then
```

```
service mysqld stop
exit 0
fi
if [ $MYSQL_OK -eq 0 ] && [ $CHECK_TIME -eq 0 ]
then
    pkill keepalived
exit 1
fi
sleep 1
Done
```

该脚本使用以一分钟为频率的定时任务定时检查 MYSQL-SERVER，保证 client 端在 MYSQL-SERVER 宕机时，vip 自动从 MYSQL.COM 机器切换到 SLAVE1.COM 机器，此时 SLAVE1.COM 将成为主节点对外提供服务。

```
[root@MYSQL keepalived]# yum -y install cronie
[root@MYSQL ~]# crontab -l
*/1 * * * * bash /root/checkmysql.sh
```

4. 启动测试

MYSQL.COM 和 SLAVE1.COM 同时启动 Keepalived：

```
[root@MYSQL keepalived]# keepalived -f
/etc/keepalived/keepalived.conf
[root@MYSQL keepalived]# ps -ef |grep keep
root      3230     1  0 23:27 ?        00:00:00 keepalived -f
/etc/keepalived/keepalived.conf
root      3231  3230  0 23:27 ?        00:00:00 keepalived -f
/etc/keepalived/keepalived.conf
root      3232  3230  0 23:27 ?        00:00:00 keepalived -f
/etc/keepalived/keepalived.conf
root      3234  2538  0 23:27 pts/0    00:00:00 grep keep

[root@MYSQL keepalived]# ip a
1: lo: <LOOPBACK,UP,LOWER_UP> mtu 16436 qdisc noqueue state UNKNOWN
   link/loopback 00:00:00:00:00:00 brd 00:00:00:00:00:00
   inet 127.0.0.1/8 scope host lo
   inet6 ::1/128 scope host
     valid_lft forever preferred_lft forever
2: eth0: <BROADCAST,MULTICAST,UP,LOWER_UP> mtu 1500 qdisc pfifo_fast state UP
qlen 1000
    link/ether 00:0c:29:c9:85:ba brd ff:ff:ff:ff:ff:ff
    inet 192.168.186.141/24 brd 192.168.186.255 scope global eth0
    inet 192.168.200.16/32 scope global eth0
    inet6 fe80::20c:29ff:fec9:85ba/64 scope link
      valid_lft forever preferred_lft forever
```

在测试过程中关闭 MYSQL.COM 服务器上的 MySQL 服务，查看到 vip 成功切换到了 SLAVE1.COM 节点，MHA 中的 MySQL 主节点也切换到了 SLAVE1.COM 机器。

4.5 用 ZooKeeper 搭建 Mycat 高可用集群

随着并发量、数据量越来越大，单机处理能力已经出现瓶颈，需要多台机器组成一个分布式系统进行并行处理。目前大部分分布式系统采用 master/slave 架构，master 在整个集群中充当资源分配和任务调度角色，客户端的所有请求都提交到 master，master 负责把任务分配到 slave 执行，如果其中一台 slave 机器宕机，则 master 会把该机器的任务转移到其他 slave 机器去执行，确保任务能正常执行，但是如果 master 宕机，则整个集群将会不可用。高可用性 HA（High Availability）是目前企业防止主节点因故障而宕机的有效手段。目前 Mycat 高可用集群是基于 HAProxy 和 Keepalived 搭建的，具体实现请参考 4.3 节，基于 ZooKeeper 搭建高可用集群模块还处于开发阶段。

4.5.1 ZooKeeper 概述

ZooKeeper 是一个开源分布式应用协调服务，是 Google 的 Chubby 的一个开源实现，是 Hadoop 的一个子项目，现已成为 Apache 基金会的顶级项目。

ZooKeeper 角色有 Leader、Follower、ObServe 三类。Leader 负责进行投票的发起和决议，并更新系统状态；Follower 接收客户端的请求，向客户端返回结果，在选择 Leader 的过程中参与投票；ObServe 可以接收客户端的连接，把请求转发给 Leader 节点，但是选择 Leader 的过程不参与投票，只是同步 Leader 的状态。

ZooKeeper 的系统模型如图 4-31 所示。

ZooKeeper 的核心是原子广播，这种机制保证了各 Server 之间的同步，实现这种机制的协议是 Zab（ZooKeeper Atomic Broadcast）。Zab 有两种协议模式，分别是恢复模式（选主）和广播模式（同步）。当服务启动或者 Leader 崩溃后，Zab 将进入恢复模式，当 Leader 节点崩溃时，ZooKeeper 将进入新一轮选举，选出新的 Leader 节点。这时如果旧的 Leader 节点已经恢复，则可能还认为自己是 Leader 节点，集群中将会同时出现两个 Leader 节点，这种情况叫作“脑裂”。恢复模式用于确保集群只有一个 Leader，如果集群中有超过一半的节点与新的 Leader 服务器完成状态同步，则 Zab 协议退出恢复模式；如果集群中有超过一半的 Follower 节点与 Leader 服务器完成状态同步，则整个服务框架进入消息广播模式。

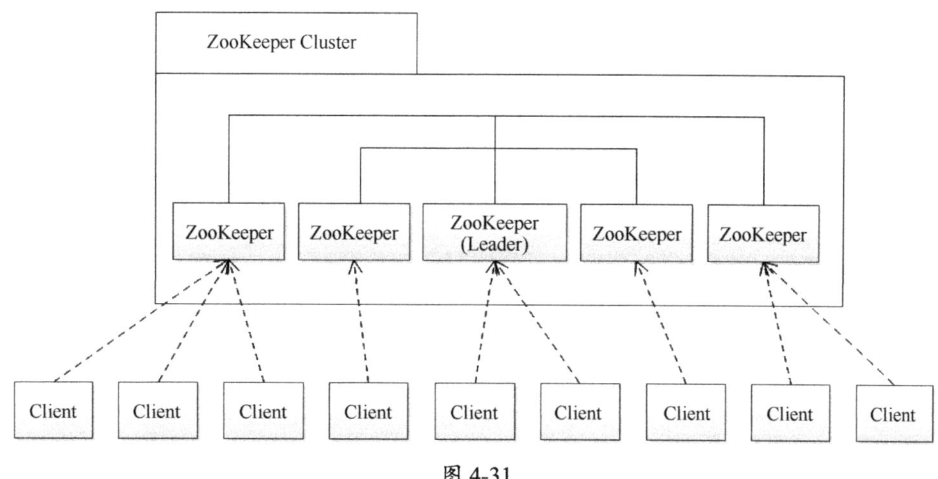

图 4-31

ZooKeeper 选举相关类图如图 4-32 所示。

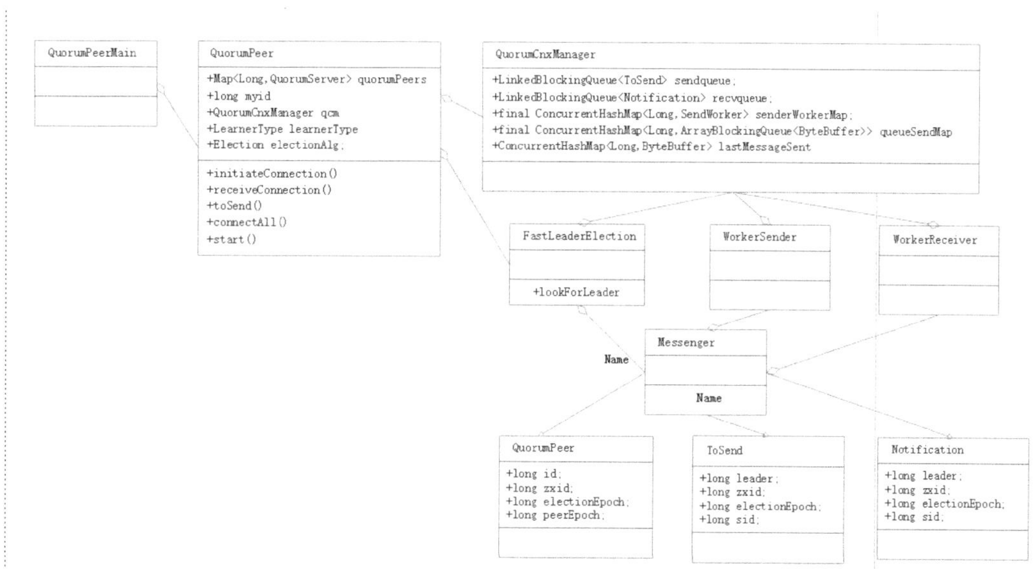

图 4-32

QuorumPeerMain 入口方法如下：

```
 public class QuorumPeerMain {
public static void main(String[] args) {
   QuorumPeerMain main = new QuorumPeerMain();
   try {
      main.initializeAndRun(args);
   } catch (IllegalArgumentException e) {
```

```
            LOG.fatal("Invalid arguments, exiting abnormally", e);
            LOG.info(USAGE);
            System.err.println(USAGE);
            System.exit(2);
        } catch (ConfigException e) {
            LOG.fatal("Invalid config, exiting abnormally", e);
            System.err.println("Invalid config, exiting abnormally");
            System.exit(2);
        } catch (Exception e) {
            LOG.fatal("Unexpected exception, exiting abnormally", e);
            System.exit(1);
        }
        LOG.info("Exiting normally");
        System.exit(0);
    }

    protected void initializeAndRun(String[] args)
        throws ConfigException, IOException
    {
        QuorumPeerConfig config = new QuorumPeerConfig();
        if (args.length == 1) {
            config.parse(args[0]);
        }

        if (args.length == 1 && config.servers.size() > 0) {
            runFromConfig(config);
        } else {
            LOG.warn("Either no config or no quorum defined in config, running "
                    + " in standalone mode");
            // there is only server in the quorum -- run as standalone
            ZooKeeperServerMain.main(args);
        }
    }

    public void runFromConfig(QuorumPeerConfig config) throws IOException {
      try {
          ManagedUtil.registerLog4jMBeans();
      } catch (JMException e) {
          LOG.warn("Unable to register log4j JMX control", e);
      }

      LOG.info("Starting quorum peer");
      try {
          NIOServerCnxn.Factory cnxnFactory =
              new NIOServerCnxn.Factory(config.getClientPortAddress(),
                      config.getMaxClientCnxns());
```

```
            quorumPeer = new QuorumPeer();
            quorumPeer.setClientPortAddress(config.getClientPortAddress());
            quorumPeer.setTxnFactory(new FileTxnSnapLog(
                    new File(config.getDataLogDir()),
                    new File(config.getDataDir())));
            quorumPeer.setQuorumPeers(config.getServers());
            quorumPeer.setElectionType(config.getElectionAlg());
            quorumPeer.setMyid(config.getServerId());
            quorumPeer.setTickTime(config.getTickTime());
            quorumPeer.setMinSessionTimeout(config.getMinSessionTimeout());
            quorumPeer.setMaxSessionTimeout(config.getMaxSessionTimeout());
            quorumPeer.setInitLimit(config.getInitLimit());
            quorumPeer.setSyncLimit(config.getSyncLimit());
            quorumPeer.setQuorumVerifier(config.getQuorumVerifier());
            quorumPeer.setCnxnFactory(cnxnFactory);
            quorumPeer.setZKDatabase(new ZKDatabase
(quorumPeer.getTxnFactory()));
            quorumPeer.setLearnerType(config.getPeerType());

            quorumPeer.start();
            quorumPeer.join();
        } catch (InterruptedException e) {
            // warn, but generally this is ok
            LOG.warn("Quorum Peer interrupted", e);
        }
      }
    }
    QuorumPeer 启动方法 Start
      public synchronized void start() {
       try {
             zkDb.loadDataBase();
          } catch(IOException ie) {
                LOG.fatal("Unable to load database on disk", ie);
                throw new RuntimeException("Unable to run quorum server ", ie);
          }
          cnxnFactory.start();
          startLeaderElection();
          super.start();
      }
```

4.5.2　ZooKeeper 的运用场景

1. ZooKeeper 在 Hadoop 中的使用

在 Hadoop 2.0 之前的版本中，两大核心模块 NameNode、JobTracker 都存在单节点问题，其中 NameNode 的单点问题更为严重，因为 NameNode 节点可以说是整个 Hadoop 集群的心脏部分，保存了整个 HDFS 的所有数据快照表，如果 NameNode 节点挂掉，则整个 HDFS 将无法访问，重启 NameNode 节点和恢复数据过程比较耗时，极大程度地限制了 Hadoop 的使用场景。在 Hadoop 2.0 中，NameNode 和 Yarn ResourceManager 的单点问题已经得到解决，NameNode 和 Yarn ResourceManager 的高可用（High Availability）方案类似。以下是 NameNode 的高可用架构图，从图中可以看到 Active NameNode 和 Standby NameNode 之间形成了主从关系：一台处于 Active 状态，对外提供服务；另一台处于 Standby 状态，不对外提供服务，定期检查 NFS 或者 JN 的最新 edits 文件，并把 edits 和 fsimage 文件合并成新的 fsimage 文件，合并之后通知 Active 节点获取最新的 fsimage 文件并替换原来的 fsimage 文件。如图 4-33 所示。

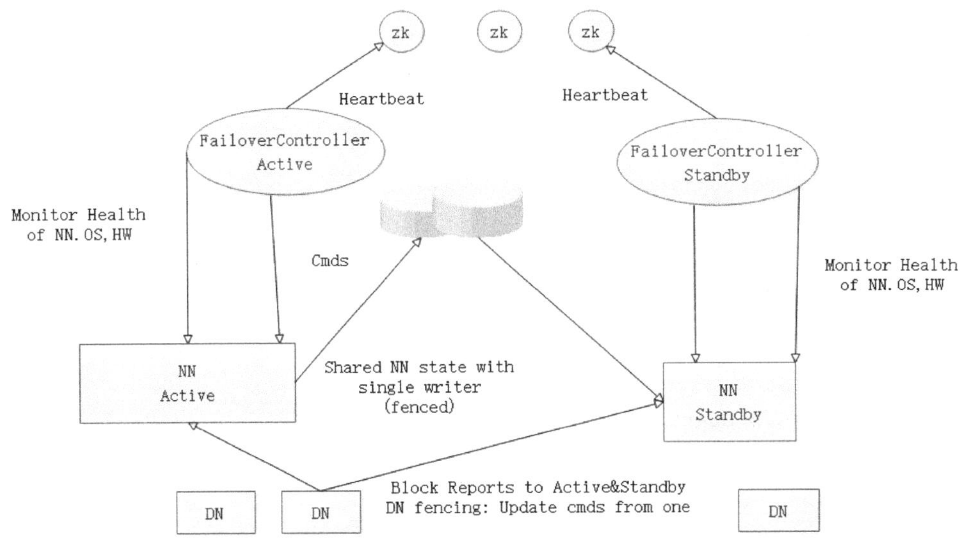

图 4-33

2. ZooKeeper 在 HBase 中的使用

ZooKeeper 为 HBase 提供稳定服务和 Failover 机制。在 HBase 中有两种特殊的表：.META.和-ROOT-。.META.表记录用户的 Region 信息，.META.可以有多个 Region。-ROOT-表记录.META.表的 Region 信息，该表中只有一个 Region。ZooKeeper 记录-ROOT-表的 Location。Client 在访问 HBase 之前先访问 ZooKeeper，然后访问-ROOT-表，接着访问.META.表，找到数据所在的位置。ZooKeeper Quorum 存储-ROOT-表的地址和 Hmaster 地址，HRegionServer 会以 Ephemeral 方式把自己注册到 ZooKeeper 中，使得 Hmaster 随时感知 HRegionServer 的健康状态。Hmaster 没有单点问题，在 HBase 中可以启动多个 Hmaster，通过 ZooKeeper 的 master Election 机制保证

一个 master 的运行。如图 4-34 所示。

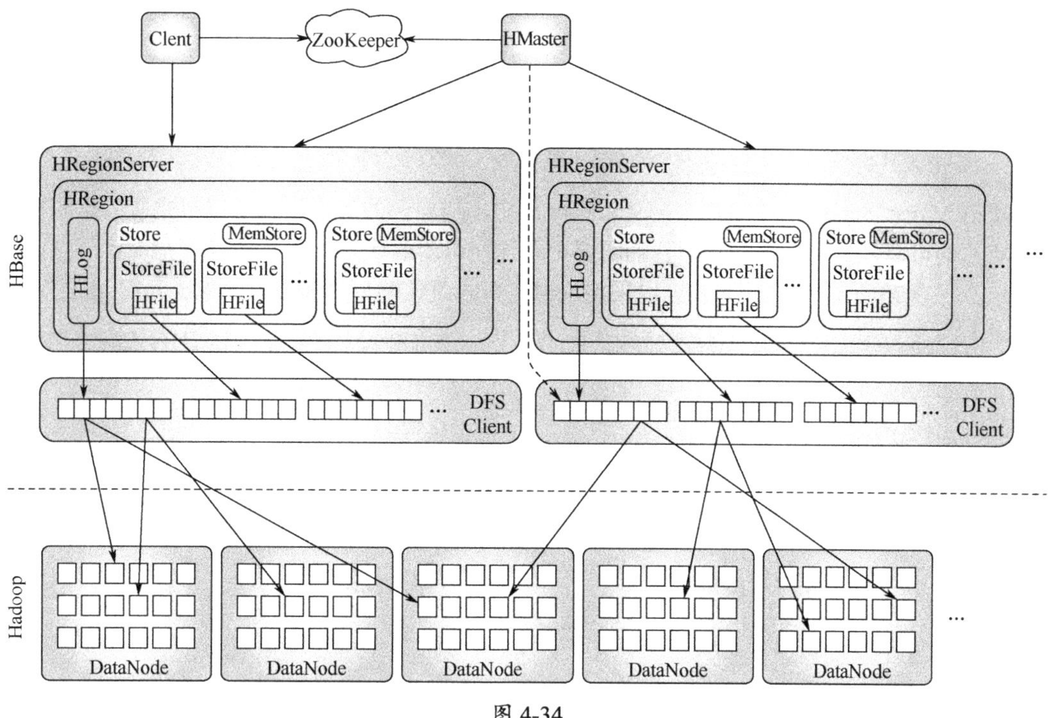

图 4-34

4.5.3　ZooKeeper 在 Mycat 中的使用

ZooKeeper 具有以下两大特点。

- 客户端如果对 ZooKeeper 的一个数据节点注册 Watcher 监听，那么该节点的内容或者其子节点列表发生变化时，ZooKeeper 服务器将会以订阅的方式通知客户端。
- 在 ZooKeeper 上创建临时节点，一旦客户端与服务器之间会话结束，临时节点将会自动删除。

利用 ZooKeeper 的这两大特点，就可以实现对 Mycat 集群机器存活性的监控，例如在 ZooKeeper 中添加 Mycat_Cluster 节点并注册一个 Watcher 监听，但凡启动 Mycat Server 时都向该节点下创建各自的节点：/Mycat_Cluster/Mycat(n)，这样一来所有加入的 Mycat 节点就都纳入 ZooKeeper 的监控范围了。如图 4-35 所示。

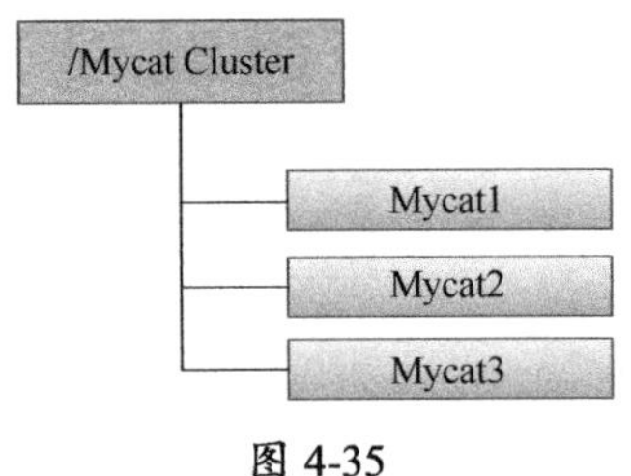

图 4-35

所有 Mycat Server 启动后都会向 ZooKeeper 中注册自己的节点，但是这些注册节点的机器都有宕机的可能，针对这个问题，我们需要记录这些机器的状态，需要在每个 Mycat 节点同时创建一个保存 Mycat 状态的节点，可以采用定时汇报心跳策略反馈各节点的健康状况。如图 4-36 所示。

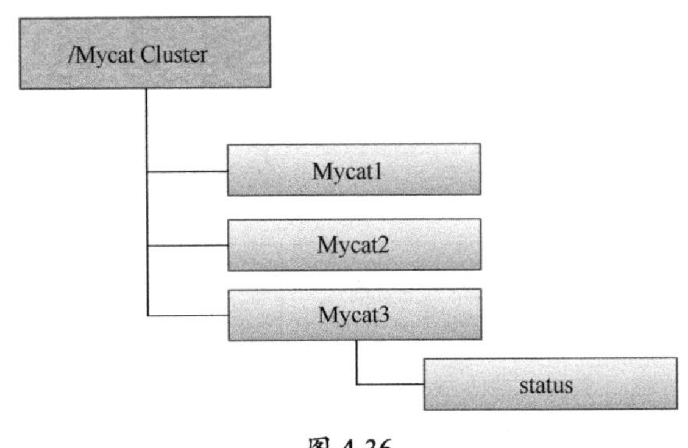

图 4-36

那么 ZooKeeper 集群如何选择一个节点作为 master 节点呢？针对这种情况，通常可以采用关系型数据库的主键唯一性来实现，集群中的所有机器同时向数据库中插入一条 ID 相同的记录，数据库会自动帮助解决主键冲突检查的问题，也就是说所有请求插入的节点中只有一个节点的插入是成功的，可以认为插入成功的节点成为集群的 master 节点。乍一看这种方案也是可行的，通过关系型数据库可以选出唯一的 master 节点，但是如果 master 节点挂掉，则关系型数据库没办法通知我们 master 节点已经挂掉，集群也就无法进入下一轮的新 master 的选择。使用 ZooKeeper 可以做到这一点吗？利用 ZooKeeper 的强一致性，能够很好地保证在分布式环境下创建唯一的数据节点，ZooKeeper 可以保证客户端无法重复创建一个已经存在的数据节点，也就是说，多个客户端同时请求创建相同的数据节点时，最终只有一个客户端的请求创建是成功的。利用这个特性，就很容易在分布式环境下进行 master 选举。

在 Mycat 中 master 的选择方案一目了然，所有 Mycat Server 启动时都将各自机器的信息向 ZooKeeper 中注册相同的数据节点，并同时注册 Watcher 监听，但是只有一个 Server 创建节点是成功的，成功的节点将成为 Mycat 集群中的 master 节点。如果 master 节点挂掉，则 ZooKeeper

通过订阅方式通知所有 Mycat Server 进入新一轮 master 选择即新一轮写入，客户端可以通过 ZooKeeper 找到 Mycat master 节点。整体的集群架构如图 4-37 所示。

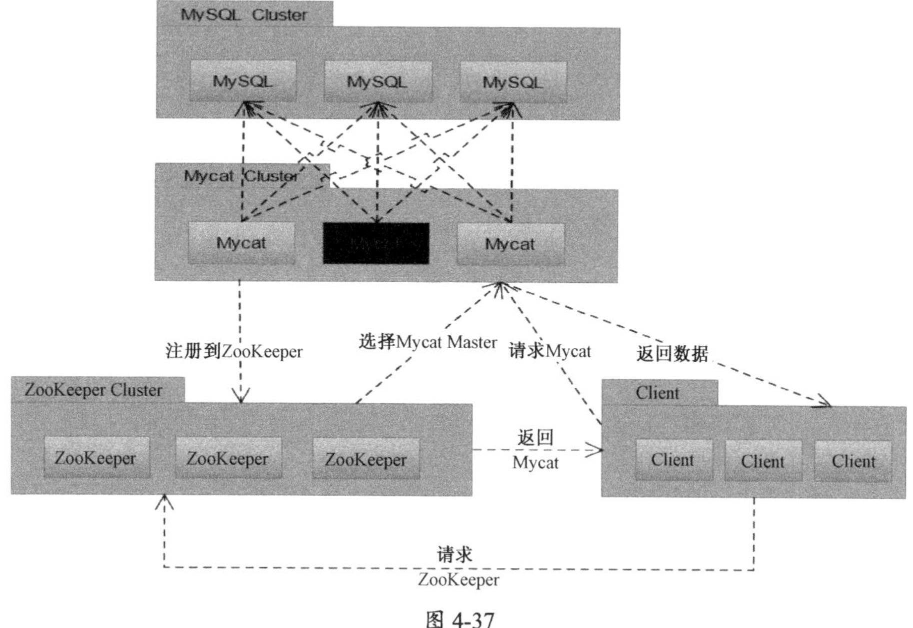

图 4-37

4.6　Mycat 高可用配置

HAProxy 相对于 LVS 使用起来要简单很多，在功能方面也很丰富。当前，HAProxy 支持两种主要的代理模式：4 层模式（TCP，大多用于邮件服务器、内部协议通信服务器等）和 7 层模式（HTTP）。在 4 层模式下，HAProxy 仅在客户端和服务器之间转发双向流量；在 7 层模式下，HAProxy 会分析协议，并且能通过允许、拒绝、交换、增加、修改、删除请求（request）或者在回应（response）里指定内容来控制协议，这种操作要基于特定的规则。

采用 HAProxy 主要有以下优点：免费、开源、稳定性较好，自带强大的监控服务器状态的页面，常用于 MySQL（读）负载均衡。

而 Mycat 是遵循 MySQL 的通信协议的，所以 HAProxy 也满足 Mycat 的高可用及负载均衡需求。

Keepalived 可提供 vrrp 及 health-check 功能，可以只用它提供双机浮动的 vip（vrrp 虚拟路由功能），这样就可以简单实现 HAProxy 的热备高可用功能。

下面我们从一个实际的例子来说明 HAProxy+Keepalived 如何满足 Mycat 的高可用需求。

服务器信息如表 4-5 所示。

表 4-5

	Keepalived	HAProxy1	HAProxy2	Mycat1	Mycat2	Pxc1	Pxc2	Pxc3
IP	10.230.4.210	10.230.4.130	10.230.4.131	10.230.4.131	10.230.4.131	10.230.3.194	10.230.3.195	10.230.3.196
port	7066			8088	8099	3306	3306	3306

其拓扑图如图 4-38 所示。

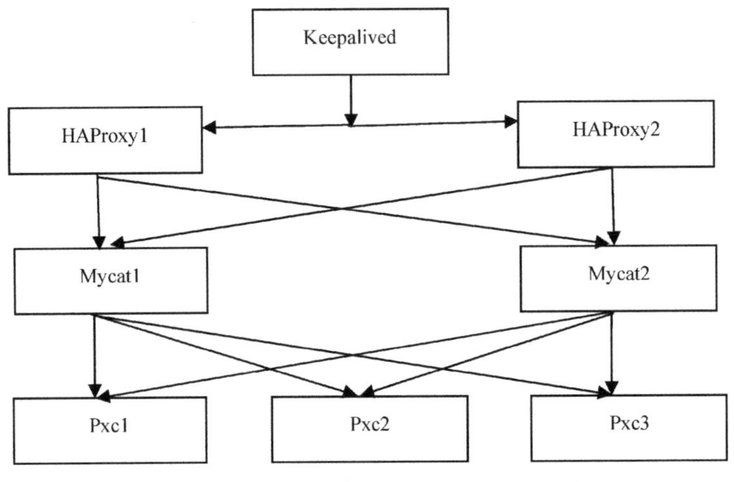

图 4-38

注意：Mycat1 和 Mycat2 在生产环境中应放在不同的服务器上。

10.230.4.130 的 Keepalived 配置如下：

```
global_defs {
notification_email {
yangfengb@xxxxx.com
}
notification_email_from yangfengb@xxxxx.com
smtp_server 192.168.210.1
smtp_connect_timeout 30
router_id LVS_DEVEL
}
vrrp_instance VI_1 {
state BACKUP # 10.230.4.130 上配置为 BACKUP
interface eth0
virtual_router_id 51
priority 30 # 10.230.4.130 上配置为 30，值越大权重越大
```

```
advert_int 1
authentication {
auth_type PASS
auth_pass 1111
}
virtual_ipaddress {
10.230.4.210 # vip 地址
}
10.230.4.131 的 Keepalived 的配置：
global_defs {
notification_email {
yangfengb@xxxxx.com
}
notification_email_from yangfengb@xxxxx.com
smtp_server 192.168.210.1
smtp_connect_timeout 30
router_id LVS_DEVEL
}
vrrp_instance VI_1 {
state MASTER # 10.230.4.131 上配置为 BACKUP
interface eth0
virtual_router_id 51
priority 100 # 10.230.4.131 上配置为 80，值越大权重越大
advert_int 1
authentication {
auth_type PASS
auth_pass 1111
}
virtual_ipaddress {
10.230.4.210 # vip 地址
}
}
```

10.230.4.130 上的 HAProxy 配置如下：

```
# this config needs haproxy-1.1.28 or haproxy-1.2.1
global
log 127.0.0.1 local0
log 127.0.0.1 local1 notice
#log loghost local0 info
maxconn 4096
daemon
nbproc 1
pidfile /usr/local/haproxy-1.5.8/haproxy.pid
#debug
#quiet
defaults
log global
```

```
mode http
option httplog
option dontlognull
retries 3
option redispatch
maxconn 2100
timeout connect 5000
timeout client 50000
timeout server 50000
listen admin_stats 10.230.4.130:8888
option httplog
stats enable
stats refresh 30s
stats uri /stats
stats realm Haproxy Manager
stats auth admin:admin
#stats hide-version
listen mycat_proxy 10.230.4.210:7066
mode tcp
balance roundrobin
option tcplog
option httpchk
server mycat_1 10.230.4.131:8088 weight 1
server mycat_2 10.230.4.131:8077 weight 1
```

10.230.4.131 上的 HAProxy 配置如下：

```
# this config needs haproxy-1.1.28 or haproxy-1.2.1
global
log 127.0.0.1 local0
log 127.0.0.1 local1 notice
#log loghost local0 info
maxconn 4096
daemon
nbproc 1
pidfile /usr/local/haproxy-1.5.8/haproxy.pid
#debug
#quiet
defaults
log global
mode http
option httplog
option dontlognull
retries 3
option redispatch
maxconn 2100
timeout connect 5000
timeout client 50000
```

```
timeout server 50000
listen admin_stats 10.230.4.131:8888
option httplog
stats enable
stats refresh 30s
stats uri /stats
stats realm Haproxy Manager
stats auth admin:admin
#stats hide-version
listen mycat_proxy 10.230.4.210:7066
mode tcp
balance roundrobin
option tcplog
option httpchk
server mycat_1 10.230.4.131:8088 weight 1
server mycat_2 10.230.4.131:8077 weight 1
```

为了说明 mycat1 和 mycat2 发生了切换，特意让 mycat1 和 mycat2 内的 schema 配置不同，这样就能直观地看到是哪个 Mycat 承载了前端的连接。

mycat1 的 schema 配置如下：

```
<schema name="mycat01" checkSQLschema="false" sqlMaxLimit="100" >
</schema>
```

mycat2 的 schema 配置如下：

```
<schema name=" order_io " checkSQLschema="false" sqlMaxLimit="100" >
 </schema>
<schema name=" bi " checkSQLschema="false" sqlMaxLimit="100" >
 </schema>
```

先停止 mycat1 的服务：

```
[MySQL@xjw-MySQL-01 ~]$ MySQL -h10.230.4.210 --default-character-set=utf8
-uxjw_dba  -p123 -P7066
MySQL> show databases;
+----------+
| DATABASE |
+----------+
| mycat01  |
+----------+
1 row in set (0.00 sec)
MySQL> show databases;
ERROR 2006 (HY000): MySQL server has gone away
No connection. Trying to reconnect...
Connection id:    132575
Current database: *** NONE ***
+----------+
| DATABASE |
```

```
+-----------+
| bi        |
| order_io |
+-----------+
2 rows in set (0.00 sec)
```

然后开启 mycat1 的服务，停止 mycat2 的服务：

```
MySQL> show databases;
ERROR 2006 (HY000): MySQL server has gone away
No connection. Trying to reconnect...
Connection id:    3
Current database: *** NONE ***
+-----------+
| DATABASE |
+-----------+
| mycat01  |
+-----------+
1 row in set (0.01 sec)
MySQL> show databases;
+-----------+
| DATABASE |
+-----------+
| mycat01  |
+-----------+
1 row in set (0.01 sec)
```

4.7 Mycat 注解技术

我们知道 MySQL 数据库有自己的 SQL 注解(hint)，比如 use index、force index、ignore index 等都是会经常用到的。Oracle 数据库的 hint 更加强大和复杂。Mycat 也一样实现了自己的 6 种注解技术。Mycat 作为一个数据库中间件，最重要的功能是 SQL 路由，所以 Mycat 中的注解基本上和路由功能相关。使用这些注解可以使我们更方便地为 SQL 指定路由信息。

4.7.1 balance 注解实战

引入 balance 注解的主要目的是实现负载均衡。在 Mycat 中每个 dataHost 可以有多个 writeHost，而一个 writeHost 又可以有多个 readHost。balance 注解主要是用于写分离时，实现 select、show 等查询语句的负载均衡，减轻 writeHost 的压力。至于在哪些 db 上进行负载均衡，则由 server.xml 中的 balance 属性值控制：

```
<dataHost name="localhost1" maxCon="1000" minCon="10" balance="0" …
```

- balance= “0” 表示关闭读写分离，所以 balance 注解也是无效的。
- balance=“1”表示全部的 readHost、stand by writeHost 都参与 select 语句的负载均衡（即 master writeHost 不参与 select 的执行）。
- balance= “2” 表示所有的 select 读操作都随机在所有 writeHost、readHost 上分发。
- balance= “3” 表示所有的 select 读操作都随机分发至 writeHost 对应的 readHost 上，writeHost 不负担读操作。

所以只有在 dataHost 元素的 balance 不为 0 时，balance 注解才会起作用。它的基本原理如下：

```
canRunInReadDB = (sqlType == ServerParse.SELECT
|| sqlType == ServerParse.SHOW);
hasBlanceFlag = (statement != null)
              && statement.startsWith("/*balance*/");
public boolean canRunnINReadDB(boolean autocommit) {
   return canRunInReadDB && autocommit && !hasBlanceFlag
      || canRunInReadDB && !autocommit && hasBlanceFlag;
}
```

在执行 SQL 语句时会调用 canRunnINReadDB()函数，如果返回 true，则会根据 dataHost 元素中的 balance 属性决定具体从哪一台机器获取数据。这里 autocommit 用来表示 SQL 语句是否在事务中，如果在事务中，则 autocommit=false；如果不在事务中，则 autocommit=true。根据上面的代码我们知道，/*balance*/注解实现对 select、show 语句的负载均衡时需要使用事务且带有/*balance*/前缀。在没有使用事物但是带有/*balance*/前缀时，会在 writeHost 上执行但不会进行负载均衡。

balance 的注解示例请参考如下 schema.xml 配置：

```
<dataHost name="localhost1" maxCon="1000" minCon="10" balance="1"
          writeType="0" dbType="MySQL" dbDriver="native"
switchType="1"  slaveThreshold="100">
   <heartbeat>select user()</heartbeat>
   <writeHost host="hostM1" url="192.168.1.3:3306"
 user="root" password="xxx">
      <readHost host="hostM1" url="192.168.1.200:3306"
user="root" password="xxx" />
   </writeHost>
</dataHost>
```

保证 balance 属性值不等于 0，然后为 SQL 加上/*balance*/前缀即可：

```
/*balance*/select * from company where id=100;
```

设置 log4j.xml 中的 level=debug，然后写段简单的 Java 代码。注意不能直接使用 MySQL client 连上去测试，因为 MySQL client 发送的 SQL 语句中的所有注解被删除了。

如果在日志文件中看到如下信息：

```
DEBUG [$_NIOREACTOR-3-RW] (PhysicalDBPool.java:452) -select read source hostS1
for dataHost:localhost1
```

或者看到类似下面的信息：

```
MySQLConnection [id=6 … fromslaveDB=true …
```

则表示在 readHost 上执行。

如果看到类似下面的信息：

```
MySQLConnection [id=6 … fromslaveDB=false …
```

则 SQL 显然是在 writeHost 上执行的。

4.7.2 master/slave 注解实战

master/slave 注解在 Mycat 1.6 中被正式引入。master/slave 注解的引入，主要是为了让一条 select 语句强制在 master db 或者 slave db 上执行。对于数据库读写分离架构的系统，slave db 可能会存在一定的复制延迟。在一些业务上不能接受任何延迟的 select 语句就可以使用 master 注解，强制其在 master db 上执行；而一些很消耗资源、对系统的性能影响较大，同时能接受一定延迟的 select 语句，则可以使用 slave 注解，强制其在 slave db 上执行，减轻其对 master db 的压力。

master 和 slave 注解的基本原理为，根据注解的值是 master 还是 slave，分别获得对应的 writeHost 上的 master db、readHost 的 slave db 的数据库连接，来执行 SQL 语句。其具体实现也不复杂，就是先解析得到被注解的 SQL 语句的路由结果，然后在该结果中增加一个 boolean 字段 runOnSave，来存放 master 或者 slave 的值。在执行 SQL 时，根据该字段的值，到相应的 master db 或者 slave db 获取数据库连接。

因为 Mycat 目前支持两种基本的注解语法格式，所以对应的 master/slave 的注解写法也分为如下两种。

（1）强制 SQL 在 master db 上执行，语法如下：

```
/*!mycat:db_type=master*/select * from company where id=100;
或者：
/*#mycat:db_type=master*/select * from company where id=100;
强制 SQL 在 slave db 上执行，语法如下：
/*!mycat:db_type=slave*/select * from company where id=100;
```

或者：

```
/*#mycat:db_type=slave*/select * from company where id=100;
```

（2）使用/*#mycat:xxx */ 形式的注解，这种形式的注解可能和 MySQL 自身的注解存在兼容性问题。

对 master/slave 注解测试的方法和 4.7.1 节中 balance 注解测试的方法类似，主要看日志文件中数据库连接的 from slaveDB 属性的值。

4.7.3　SQL 注解实战

SQL 注解是 Mycat 中最早引入的一种注解，功能强大，用途广泛。

SQL 注解的实现如下：注解部分是一条 SQL 语句，称其为注解 SQL，被注解的部分也是一条 SQL 语句，称其为原始 SQL。其原理就是用解析注解 SQL 的路由结果，来执行被注解的 SQL 语句。其处理过程是先对注解中的 SQL 语句进行解析，得到路由结果，然后将被注解的 SQL 语句在该路由结果表示的数据库分片上执行。因为注解 SQL 仅仅是用来获得路由结果的，所以一般尽量使其为一条简单的 select 语句，使用分片键所在的字段即可。而原始 SQL 语句可以是 select、delete、update、insert、call 等。原始 SQL 会发送到注解 SQL 解析得到的路由结果集上执行。

SQL 注解使用示例如下。

（1）使用 SQL 注解来实现在指定的 dataNode 上创建存储过程。

如果需要在 dn1 的 dataNode 上创建一个存储过程，则可以通过下面的语句进行：

```
MySQL> explain select id from hotnews where id=201;
+-----------+-------------------------------------+
| DATA_NODE | SQL                                 |
+-----------+-------------------------------------+
| dn1       | select id from hotnews where id=201 |
+-----------+-------------------------------------+
1 row in set (0.00 sec)
```

可以知道分片表 hotnews 中 id=201 的记录在 dn1 上，可以通过使用如下所示的 SQL 注解来达到在 dn1 节点上创建一个存储过程的目的：

```
MySQL> set autocommit=0;
Query OK, 0 rows affected (0.04 sec)
MySQL> delimiter //
MySQL> /*!mycat:sql=select id from hotnews where id=201*/
    -> create procedure test_sql_hint_proc()
    -> begin
    -> select 1;
```

```
    -> end;
    -> //
Query OK, 0 rows affected (0.10 sec)
```

注解 sql: select id from hotnews where id=201 得到路由结果后，在该路由结果对应的分片上创建存储过程。可以直连 MySQL-server，查看 dn1 节点对应的 db1 数据库中是否有该存储过程：

```
MySQL> select name from MySQL.proc where db = 'db1' and type = 'PROCEDURE';
+--------------------+
| name               |
+--------------------+
| test_sql_hint_proc |
+--------------------+
1 row in set (0.00 sec)
```

可以看到存储过程在 dn1 节点对应的 db1 数据库中。也许你会感到奇怪，为什么要在开始之前执行 set autocommit=0。如果掌握了 balance 注解的原理，这里就很容易理解了。因为 MySQL client 默认是自动提交的，所以注解语句 select id from hotnews where id=201 返回的路由结果可能会是 readHost（配置文件 schema.xml 中 dataHost 元素的 balance 属性不等于 0 时），而我们希望存储过程是创建在 writeHost 上的。这是 SQL 注解存在的一个小小的缺陷，将会在 Mycat 的后续版本中修复。

（2）使用 SQL 注解来调用指定的 dataNode 上的存储过程。

前面使用 SQL 注解在 dn1 节点上创建了一个存储过程，那么如何调用它呢？简单地执行 call test_sql_hint_proc()是无法正确地进行路由的，所以需要通过使用 SQL 注解来实现调用 dn1 节点上的存储过程：

```
MySQL> /*!mycat:sql=select id from hotnews where id=204*/
      -> call test_sql_hint_proc();
+---+
| 1 |
+---+
| 1 |
+---+
1 row in set (0.01 sec)
Query OK, 0 rows affected (0.01 sec)
```

如果直接执行 call test_sql_hint_proc()，则会报错：

```
MySQL> call test_sql_hint_proc();
ERROR 1305 (42000): PROCEDURE db3.test_sql_hint_proc does not exist
MySQL> call test_sql_hint_proc();
ERROR 1305 (42000): PROCEDURE db2.test_sql_hint_proc does not exist
```

可以看到无法对存储过程的调用正确地进行路由。

（3）使用 SQL 注解来支持 insert、select 语句。

Mycat 一直是不支持 insert、select 语句的，原因可能是该语句需要在所有分片上执行，显然对事务不好控制。如果一个分片执行成功，一个分片执行失败，那么该怎么办呢？通过 SQL 注解可以在某一个分片上运行 insert、select 语句。

首先，连接 MySQL-server，在 dn1 节点对应的 db1 上为 customer 表创建一个结构相同的 customer2 表：

```
MySQL> create table customer2 like customer;
```

在 server.xml 配置文件中加入 customer2 的配置，重启 Mycat 或者重新加载配置文件，连接 Mycat，执行下面的命令：

```
MySQL> set autocommit=0;
Query OK, 0 rows affected (0.03 sec)
MySQL> /*!mycat:sql=select id from hotnews where id=201*/
    -> insert into customer2 select * from customer;
Query OK, 3 rows affected (0.40 sec)
Records: 3  Duplicates: 0  Warnings: 0
MySQL> /*!mycat:sql=select id from hotnews where id=201*/
    -> select * from customer2;
+----+------------+----------+
| id | company_id | name     |
+----+------------+----------+
|  1 |          1 | yuanfang |
|  2 |          2 | 2        |
|  3 |          3 | 3        |
+----+------------+----------+
3 rows in set (0.03 sec)
```

连接 MySQL-server，查看 db1 上 customer2 的结果集：

```
MySQL> use db1;
Database changed
MySQL> select * from customer2;
+----+------------+----------+
| id | company_id | name     |
+----+------------+----------+
|  1 |          1 | yuanfang |
|  2 |          2 | 2        |
|  3 |          3 | 3        |
+----+------------+----------+
3 rows in set (0.00 sec)
```

这里 set autocommit=0 的作用是防止注解 sql: select id from hotnews where id=201;返回 readHost 的路由结果，导致 insert、select 语句在 readHost 上执行，而不是在 writeHost 上执行。

4.7.4 schema 注解实战

schema 注解可以很方便地使 SQL 仅在某一个 schema 中执行，适用于使用 Mycat 架构的多租户系统。在多租户系统中，一个租户对应一个 schema，所以我们可以维护一个租户和租户 schema 的一个映射关系。某个租户登录时，就获取其对应的 schema 的名字，使用 schema 注解可以将所有的 SQL 语句仅发送到该租户的 schema 中。注意这里的 schema 是 Mycat 配置文件 schema.xml 配置的逻辑库的名称，不是 MySQL-server 中的某个数据库的名称。

schema 注解的实现原理十分简单，即根据注解中的 schema 名称或从系统维护的配置信息中读取其对应的 schema 的配置信息，就可以获取对应的路由结果。

schema 注解的使用方法为：将 select * from customer;语句仅仅在 TESTDB 这个 schema 中执行：

```
MySQL> /*!mycat:schema=TESTDB*/select * from customer;
+----+------------+----------+
| id | company_id | name     |
+----+------------+----------+
|  1 |          1 | yuanfang |
|  2 |          2 | digdeep  |
+----+------------+----------+
2 rows in set (0.09 sec)
```

使用 use TESTDB 查询语句进行验证：

```
MySQL> use TESTDB;
Database changed
MySQL> select * from customer;
+----+------------+----------+
| id | company_id | name     |
+----+------------+----------+
|  1 |          1 | yuanfang |
|  2 |          2 | digdeep  |
+----+------------+----------+
6 rows in set (0.01 sec)
```

4.7.5 dataNode 注解实战

dataNode 注解和 schema 注解类似。schema 注解可以让 SQL 语句仅仅在注解指定的 schema 中执行，而 dataNode 注解可以让 SQL 语句仅仅在注解指定的 dataNode 中执行。

dataNode 注解的实现原理为：根据注解中的 dataNode 的名称或从系统维护的配置信息中获取其对应的节点信息，将 SQL 语句发往该节点。本质上，路由结果就是一些节点信息，表示

SQL 语句应该路由到哪些节点执行。

dataNode 注解的使用方法为将 select * from customer;语句在注解指定的节点中执行：

```
MySQL> /*!mycat:dataNode=dn1*/select * from customer;
+----+------------+----------+
| id | company_id | name     |
+----+------------+----------+
|  1 |          1 | yuanfang |
+----+------------+----------+
1 rows in set (0.01 sec)

MySQL> /*!mycat:dataNode=dn2*/select * from customer;
+----+------------+---------+
| id | company_id | name    |
+----+------------+---------+
|  2 |          2 | digdeep |
+----+------------+---------+
1 rows in set (0.00 sec)
```

4.7.6　catlet 注解实战

catlet 注解的引入是为了实现跨分片 Join，在 Mycat 中称其为 shareJoin，用在一些比较特殊的场景下。目前 shareJoin 可以支持两个表的跨分片 Join，不仅支持 joinKey 字段为整型的 join，也支持 joinKey 字段为字符型的 Join。joinKey 字段一般为整型。

目前只实现了 catlet 注解中的 shareJoin 方式。shareJoin 的基本原理是，将一条跨分片的两个表的 Join，分为两条 select 语句来执行。比如下面的 catlet 注解的 sharejoin：

```
/!mycat:catlet=demo.catlets.ShareJoin/
select a.id as aid, b.id as bid, b.name as name
from customer a, company b
where a.company_id=b.id and a.id=1;
```

执行时，其内部原理是以 Join 的子表、依赖表（也就是有外键的表）为驱动表，从原始 SQL 语句中解析得到内部的 SQL 语句：

```
select a.id as aid, a.company_id as company_id from customer a where a.id=1;
```

执行后得到中间结果。显然我们获得了符合条件的所有 company 的 id 值，然后通过这些 id 值，使用 in 关键字构造下面的 SQL 语句：

```
/!mycat:catlet=demo.catlets.ShareJoin/
select a.id as aid, b.id as bid, b.name as name
from customer a, company b
where a.company_id=b.id and a.id=1;
```

执行并聚合就可以获得我们想要的结果。在查询 company_id 记录条数大于 1000 时，则取 1000 条记录。

catlet 注解示例如下：

```
select b.id as bid, b.name as name from company b where b.id in(….)
```

第 5 章 Mycat 企业运维

数据库性能优化的重要性已经不仅仅是 DBA、运维人员所关心的了，更是广大开发人员所关注的。在数据量少且并发量不高的情况下，很多问题体现不出来，隐藏的问题也难以被发现，通过监控能为数据库性能优化提供许多参考依据。

5.1 Mycat 性能监控——Mycat-web 详解

5.1.1 Mycat-web 简介

Mycat-web 是 Mycat-server 可视化运维的管理和监控平台，弥补了 Mycat-server 在监控上的空白，是由 rainbow 和冰风影等人主导的一个开源项目，主要目的是为 Mycat-server 分担统计任务和配置管理任务等，使得 Mycat-server 更专注于数据服务。Mycat-web 引入了 ZooKeeper 作为配置中心，可以管理多个节点。Mycat-web 主要管理和监控 Mycat 的流量、连接、活动线程和内存等，具备 IP 白名单、邮件告警等模块，还可以统计 SQL 并分析慢 SQL 和高频 SQL 等，为 SQL 优化提供了重要的参考依据。

Mycat-web 支持 Windows 和 Linux 版本，用户可以通过 GitHub 的 Mycat-Download 下载 Mycat-web 安装文件，也可以下载源码自己编译。Mycat-web 目前已更新至较为成熟的 1.0 版本，下面以其 1.0 版本为基础进行介绍。

在安装 Mycat-web 前需要先安装 ZooKeeper，Mycat-web 的安装步骤如下。

1. 安装 ZooKeeper

（1）下载并解压 ZooKeeper 安装包。

（2）将 zookeeper conf 目录下的 zoo_sample.cfg 重命名为 zoo.cfg。

（3）启动 ZooKeeper。在 Windows 环境下的命令为 bin\zkServer.bat，在 Linux 环境下的命令为 bin\zkServer.sh start。

2. 安装 Mycat-web

（1）下载并安装 Mycat-web。

（2）启动 Mycat-web，之前需要确保 ZooKeeper 已经启动，启动命令为 sh start.sh。启动完成后即可访问 Mycat-web，默认地址是 http://localhost:8082/mycat/。

5.1.2 Mycat-web 的配置和使用

1. 连接 ZooKeeper

初次使用 Mycat-web 时需要配置 ZooKeeper 的地址，有以下两种方法。

（1）访问 http://localhost:8082/mycat/，在注册中心填写 ZooKeeper 的 IP 地址和端口即可，第一次访问时才需要进行配置。

（2）修改 mycat.properties 文件，填写格式为 zookeeper=IP:2181

2. 连接 Mycat

访问管理界面 http://localhost:8082/mycat/，选择 Mycat 配置→Mycat 服务管理菜单，单击“新增”按钮，依次填写 Mycat 的连接信息，如图 5-1 所示。

Mycat 的名称可以自由定义，管理端口默认为 9066，服务端口默认为 8066，数据库的名称为 Mycat 的 schema 的名称。

3. Mycat-VM 管理

Mycat-VM 管理是一个基于 JMX 的图形监控工具，用于连接正在运行的 JVM，以图表化的形式显示各种数据，并可通过远程连接监视远程服务器的 VM 情况，可以较直观地观察各种变化，如图 5-2 所示。

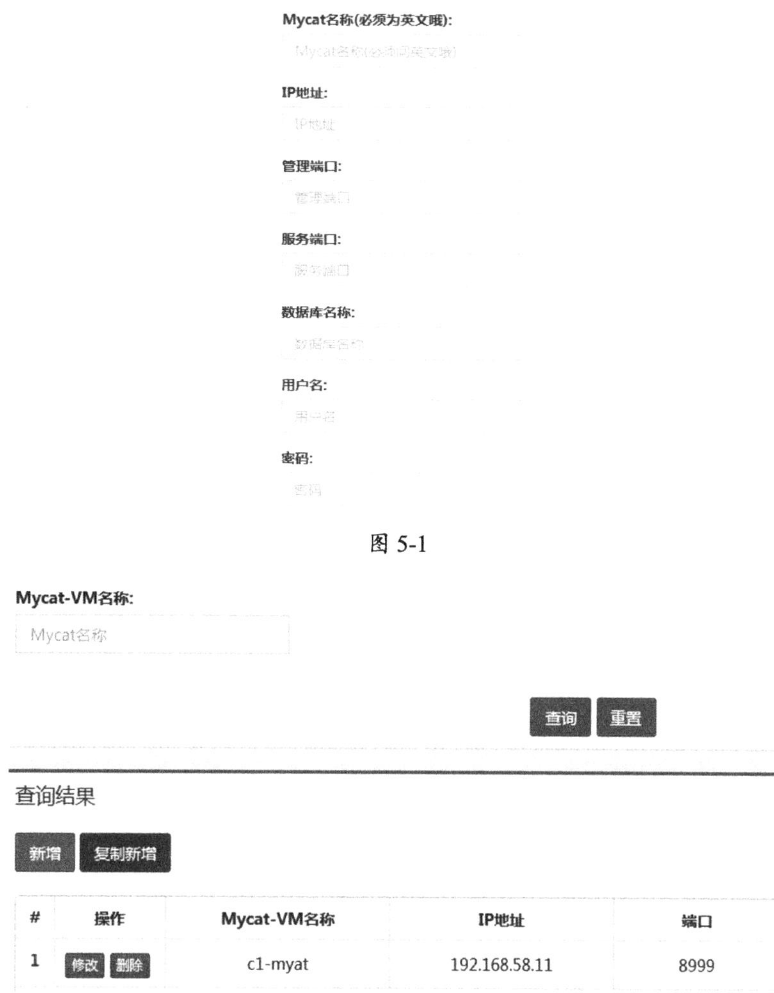

图 5-1

图 5-2

5.1.3　Mycat 性能监控指标

Mycat 性能监控涉及对 Mycat 的流量、连接、活动线程、缓冲队列、MycatTPS 和内存的分析和监控等，如图 5-3 所示。

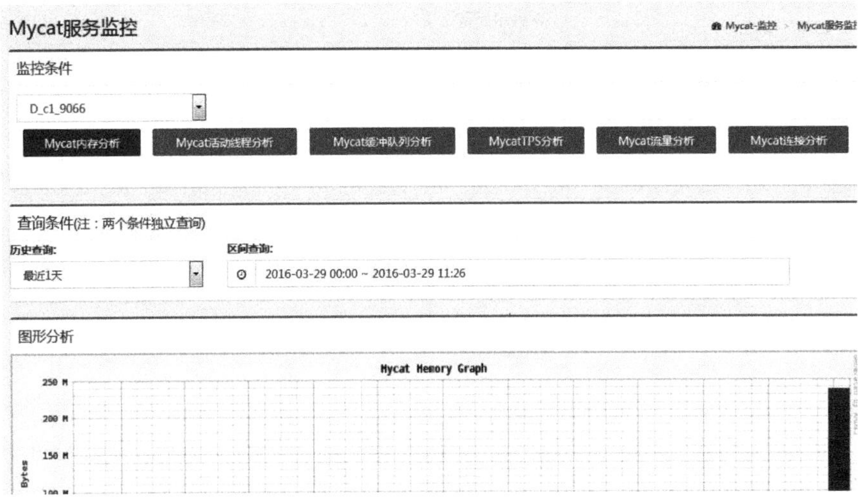

图 5-3

（1）Mycat 内存分析反映了当前的内存使用情况及历史时间段的峰值、平均值。

（2）Mycat 活动线程分析反映了 Mycat 线程的活动情况。

（3）Mycat 流量分析统计了历史时间段的流量峰值、当前值、平均值，是 Mycat 数据传输的重要指标。

（4）Mycat 连接分析反映了 Mycat 的连接数。

（5）MycatTPS 是 LoadRunner 中重要的性能参数指标，为系统在每秒内能够处理的交易或事务的数量。MycatTPS 分析统计了 Mycat 在单位时间内处理的交易或事务的数量，是衡量 Mycat 处理能力的重要指标。

Mycat-web 的 SQL 监控以用户为单位统计读写次数、读占比、最大并发数、分段请求数和耗时请求数，以时间为维度分析 SQL 在不同时间段内的执行、响应时间分布，最后统计高频 SQL、慢 SQL、表读写占比，是 SQL 优化不可或缺的采集数据。

SQL 统计界面按用户统计 SQL 的执行时间分布、响应时间分布和 SQL 读写比例，也可以通过 show @@sql.sum 命令获取该数据。

SQL 表统计以数据表为维度统计了表的读写次数、读占比、关联表、关联和次数，运维人员可以直观地看到数据表被访问的情况，也可以通过 show @@sql.sum.table 命令获取该数据。

SQL 语句监控业务系统执行的 SQL 语句耗时多少毫秒，也可以通过 show @@sql 命令获取该数据。

高频 SQL 统计执行频率高的 SQL 语句，记录高频语句被执行的最大耗时、最小耗时，也可以通过 show @@sql.high 命令获取该数据。

慢 SQL 统计通过设置阈值来定义慢 SQL，默认查询耗时 1000 毫秒的 SQL 语句，记录了慢 SQL 的执行耗时、执行时间和执行节点，也可以通过 show @@sql.slow 命令获取该数据。

5.2　Mycat 性能优化

Mycat 性能优化的第 1 步是 JVM、操作系统、MySQL 和 Mycat 本身的调优。

1. JVM 调优

内存占用分为两部分：Java 堆内存和直接内存映射（DirectBuffer 占用），建议堆内存大小适度，直接映射的内存尽可能大，总计占用操作系统 50%～67%的内存。下面以 16GB 内存的服务器为例，Mycat 堆内存为 4GB，直接内存映射为 6GB，JVM 参数如下：

```
-server -Xms4G - Xmx4G XX:MaxPermSize=64M -XX:MaxDirectMemorySize=6G
```

Mycat 中 JVM 参数的配置在 conf\wrapper.con 配置文件中，下面是一段实例：

```
# Java Additional Parameters
wrapper.java.additional.5=-XX:MaxDirectMemorySize=2G
wrapper.java.additional.6=-Dcom.sun.management.jmxremote
# Initial Java Heap Size (in MB)
wrapper.java.initmemory=2048
# Maximum Java Heap Size (in MB)
wrapper.java.maxmemory=2048
```

2. 操作系统调优

分别把 Mycat Server 和 MySQL 数据库机器的最大文件句柄数量设置为 5000～10000。Linux 操作系统对每个进程打开的文件句柄数量是有限制的（包含打开的 Socket 数量，影响 MySQL 的并发连接数量）。这个值可通过 ulimit 命令来修改，但 ulimit 命令的修改只对当前登录的用户有效，系统重启或者用户退出后就会失效。

3. Mycat 调优

conf/schema.xml 调优如下：

```
<system>
    <!-- CPU 核心数越多，可以越大，当发现系统 CPU 压力很小的情况下，可以适当调大此参数，如
4 核心的 4CPU，可以设置为 16，24 核心的可以最大设置为 128-->
    <property name="processors">1</property>
```

```
下面这个参数为每个 processor 的线程池大小，建议可以是 16-64，根据系统能力来测试和确定。
<property name="processorExecutor">16</property>
</system>
```

- processorBufferPool：每个 processor 分配的 Socket Direct Buffer，用于网络通信。
- Processor：每个 processor 上管理的所有连接共享。
- processorBufferChunk：为 Pool 的最小分配单元，每个 Pool 的容量为 processorBufferPool / processorBufferChunk，processorBufferPool 的默认值为 16MB，processorBufferChunk 的默认值为 4096 字节。对 processorBufferPool 参数的调整需要通过观察 show@@processor 的结果来确定。
- BU_PERCENT：已使用的百分比。
- BU_WARNS：当 Socket Buffer Pool 不够时，临时创建新 Buffer 的百分比如果经常超过 90%并且 BU_WARNS>0，则表明 Buffer 不够，需要增大 processorBufferPool。基本上连接数越多，并发越高，需要的 Pool 越大，建议 BU_PERCENT 最大为 40%～80%。

conf/schema.xml 调优如下：

```
<schema name="TESTDB" checkSQLschema="true">
```

checkSQLschema 属性建议设置为 false，不能在 SQL 中添加数据库的名称，这样可以优化 SQL 解析。

```
<dataHost name="localhost1" maxCon="500" minCon="10" balance="0"
      dbType="MySQL" dbDriver="native" banlance="0">
```

最大连接池 maxCon 的值可以改为 1000～2000，同一个 MySQL 实例上的所有 dataNode 节点共享本 dataHost 上的所有物理连接。

性能测试时，建议 minCon、maxCon、MySQL max_connections 的值相等，设为 2000 左右。另外，读写分离是否开启根据环境的配置来决定。

4. 缓存优化调整

show @@cache 命令展示了缓存的使用情况，需要经常观察其结果，并在需要时进行调整。若 CUR 接近 MAX，而 PUT 比 MAX 大很多，则表明 MAX 需要增大。HIT/ACCESS 为缓存命中率，这个值越高越好。在重新调整缓存的最大值以后，观测指标都会跟着发生变化，如果想知道调整是否有效，则需要观察缓存命中率是否在提升，PUT 是否在下降。

目前缓存服务的配置文件为 cacheservice.properties，主要使用的缓存为 enhache，在 enhache.xml 里设定了 enhance 缓存的全局属性。下面定义了几个缓存：

```
#used for mycat cache service conf
```

```
factory.encache=org.opencloudb.cache.impl.EnchachePooFactory
#key is pool name ,value is type,max size, expire seconds
pool.SQLRouteCache=encache,10000,1800
pool.ER_SQL2PARENTID=encache,1000,1800
layedpool.TableID2DataNodeCache=encache,10000,18000
layedpool.TableID2DataNodeCache.TESTDB_ORDERS=50000,18000
```

- SQLRouteCache：SQL 解析和路由选择的缓存，其大小基本上相对固定，就是所有 Select 语句的数量。
- ER_SQL2PARENTID：在 ER 分片时根据关联 SQL 查询父表的节点时用到，如果没有使用到 ER 分片，则用不到这个缓存。
- TableID2DataNodeCache：当某个表的分片字段不是主键时，缓存主键到分片 ID 的关系，这里命名为 schema_tableName（tablename 要大写）如“TEST_ORDERS”；当根据主键查询比较多时，这个缓存往往需要设置得比较大才能更好地提升性能。

5. Mycat 大数据量查询调优

如果返回的结果比较多，则建议调整 frontWriteQueueSize 的大小，即将默认值乘以 3，原因是返回数据太多。这里做了一个改进，就是超过 Pool 以后，仍然创建临时的 Buffer 以供使用，但对这些缓冲区不进行回收。在这样的情况下，需要增大 Buffer 参数 processorBufferPool。

6. Buffer Pool 调优

所有 NIOProcessor 共享一个 Buffer Pool。Buffer Pool 的总长度为 bufferPool 与 bufferChunk 的比值。

我们可以连接到 Mycat 管理端口，使用 show @@processor 命令列出所有 processor 的状态。

查看列 FREE_BUFFER、TOTAL_BUFFER、BU_PERCENT，如果 FREE_BUFFER 的数值过小，则说明配置的 Buffer Pool 的大小可能不够，这时就要根据公式手动配置这个属性了，bufferPool 的大小最好是 bufferChunk 的整数倍。例如配置 Buffer Pool 的大小为 5000，在 server.xml 文件中定义：

```
<property name="processorBufferPool">20480000</property>
```

另一个 Buffer Pool 是线程内的 Buffer Pool，这个值可以根据 processors 的数值计算出来。具体看 server.xml 配置详解。

7. Mycat I/O 调优

NIOProcessor 类持有所有的前后端连接，定期进行空闲检查和写队列检查。Mycat 是通过遍历 NIOProcessor 持有的所有连接来完成这个动作的。可以适当地根据系统性能调整 NIOProcessor 的数量，使得前、后端连接可以均匀地分布在每个 NIOProcessor 上，这样就可以加快每次的空闲检查和写队列检查，快速地将空闲的连接关闭，减少服务器的内存使用量。

NIOReactor 是 NIO 中具体执行 selector 的类，当该类事件发生时，就通知上层逻辑进行具体处理。NIOReactor 的数量与具体事件处理器的数量相等，如果系统配置允许，则应该尽可能地增加 NIOReactor 的数量，默认值是 CPU 的核心数。

AsynchronousChannelGroup 是 AIO 中必须提供的一个组成部分，根据 processors 的数值确定实例数和 channelGroup 组内线程池的大小。后端 AIO 连接循环读取 AsynchronousChannelGroup 数组中的实例。如果在 AIO 模式下使用 Mycat，则调整这个参数也是有必要的，默认值是 CPU 的核心数。

5.3 MySQL 优化技术

5.3.1 数据库建表设计规范

1. MySQL 字符集

MySQL 的字符集支持（Character Set Support）涉及两个方面：字符集（Character set）和排序方式（Collation）。

对于字符集的支持可细化到四个层次：服务器（server）、数据库（database）、数据表（table）、连接（connection）。

连接 MySQL 服务并通过如下命令查看字符集的详情：

```
SHOW VARIABLES LIKE 'character_set%';show variables like '%collation_%';
```

客户端字符集的设置如下：

```
character_set_client= utf8 ;
```

连接层字符集的设置如下：

```
character_set_connection= utf8
```

数据库端字符集的默认设置如下：

```
character_set_database= utf8
```

服务端字符集的设置如下：

```
character_set_server= utf8
```

系统元数据字符集的设置如下：

```
character_set_system= utf8
```

查询结果字符集的设置如下：

```
character_set_results= utf8
```

排序方式如下：

```
 collation_connection    utf8_general_ci
 collation_database      utf8_general_ci
 collation_server        utf8_general_ci
```

修改 MySQL-SERVER 配置文件：

```
default-character-set  =utf8  #在[client]下设置，如此设置影响 mysqldump 还原，在还原时可以注释无须重启服务。
character_set_server  = utf8 #在[mysqld]下设置。
```

以上设置可以在编译安装时修改配置实现，也可以通过命令行来设置。

如果需要存储 emoj 表情，则需要使用 UTF8mb4。

连接 MySQL-SERVER 的设置方式如下：

```
set character_set_client = utf8;
set character_set_connection = utf8;
set character_set_database = utf8;
set character_set_results = utf8;
set character_set_server = utf8;
set collation_connection = utf8_general_ci;
set collation_database = utf8_general_ci;
set collation_server=utf8_general_ci;
```

简单设置方式如下：

```
SET NAMES 'utf8';等同于
set character_set_client = utf8;
set character_set_connection = utf8;
set character_set_results = utf8;
set collation_connection = utf8_general_ci;
```

对导入的文本字符集的转换可以通过系统命令 iconv 进行。

2. 命名规则

对数据库的命名（例如 gzyz_user、mycat、Mycat_epl）必须遵循如下规则。

（1）为字母、数字或下画线的组合，尽量避免仅使用数字。

（2）禁止使用关键字。

（3）字母遵循英文简称或简写模式。

（4）名称尽量与业务或企业文化相关。

对表名称的命名（示例：User、UserInfo、UserCon、user_list,user,user_info）必须遵循易懂、简单、无二义性原则。

（1）为字母或字母与数字的组合，总字符数不得超过 64 个（建议不超过 32 个）。

（2）禁止使用关键字。

（3）分区表允许使用下画线。

（4）表名遵循驼峰或者下画线命名规则，例如 Mbs、User、UserInfo、UseMycat，mycat_user。

（5）表名称应该与业务关联，相同的业务表应该带有相同的表头标识。

对字段的命名必须遵循易懂、简单的原则（所有新建表）。

（1）为字母或字母与数字的组合，总字符数不得超过 64 个（建议不超过 25 个）。

（2）禁止使用关键字（MySQL 系统保留字）。

（3）字母遵循英文简称、简写或简写+数字的格式。

（4）下画线方式：新建的表字段与其他已存在的表字段如果没有关联关系，则命令的名称应须能够基本表达该新建字段属性的含义，例如 age、sex、content、title；如果新建的表字段为组合含义字段，则命令规则为从第 2 个单词开始采用下画线分割或驼峰的方式，例如对于 user_name，如果存在 user 表，则需要注释为“非关联”；如果为驼峰方式，则新建表字段从第 2 个单词开始采用单词首字母大写的格式，例如 userName、subTitle、subTitleTag（user_name 与 userName 不同）。

（5）引用字段必须采用“被引用表名”+“被引用字段”的格式。例如表 user 的字段 id 在表 user_info 中被引用，则采用 userinfo 表字段可命名为 user_id。如果与上述第 4 条中 user_name 类似的起名方式发生冲突，则可以通过采用驼峰与下画线交互的方式区分。例如，可将无业务关联的字段命名为 user_name，将有业务关联其他表的字段命名为 userName。如果因为表名本身带有下画线、较长且未按照引用规则，则必须注释该列与哪张表的哪一列关联。

（6）当一个表中存在多次引用相同表的同一字段的情况时，按照业务标识+被引用表名+被引用字段名称的规则命名。例如 Message 表在业务上为短信表，该表的发送者 id 及收信人 id 均引用 User 表的 id 字段，则可以命名为 from_user_id、to_user_id。

（7）建议引用字段的字符数不超过 35 个。

3. 字段类型的选择

字段类型的选择遵循能占一个字节绝不占两个字节的原则，因此在设计表结构时需要预估字段值的范围。

1）数字类型

（1）整数

- TINYINT：1 字节，(-128,127) (0,255)，小整数值。
- SMALLINT：2 字节，(-32 768,32 767) (0,65 535)，大整数值。
- MEDIUMINT：3 字节，(-8 388 608,8 388 607) (0,16777 215)，大整数值。
- INT 或 INTEGER：4 字节，(-2 147 483 648,2 147 483 647) (0,4 294 967 295)，大整数值。
- BIGINT：8 字节，(-9 233 372 036 854 775 808,9 223 372 036 854 775 807) (0,18 446 744 073 709 551 615)，极大整数值。

对于自增字段，如果记录经常做物理删除（delete）或记录数可能会超过 21 亿个，则必须用 BIGINT，默认使用 UNSIGNED 类型。表示状态、类型、种类时一律用 TINYINT。依据域范围合理选择 SMALLINT、MEDIUMINT。

（2）小数

FLOAT：4 字节单精度浮点数值，格式为 FLOAT(*M*,*N*)，其中 *M*<=*N*，*M* 表示显示 *M* 位整数，*N* 表示小数点后面最多有 *N* 位。需要注意四舍五入可能带来的问题。

DOUBLE：8 字节双精度浮点数值，在代码中允许有近似值存在，一律使用 FLOAT、DOUBLE 类型。对于货币、金额等不允许四舍五入的情况，一律用 DECIMAL。

2）字符串类型

（1）CHAR(*N*)

占 *N* 个字节，1<=*N*<=255，适用于值的范围较为固定时，例如 ip、url、phonenum 等。

（2）VARCHAR(*M*)

VARCHAR(*M*)占 *L* 个字节，1<= *L* <=65535，其中最大能存储 *N* 个字符（utf8 编码），1<=*N*<=21845。如果字符串列的最大长度比平均长度大很多且更新不频繁，则建议使用 VARCHAR(*M*)。

例如：新闻、名称、标题、内容、commentContent、content、description、news 等。

```
CREATE TABLE zyz(content VARCHAR(N)) CHARSET=utf8;
```

此处 N 的最大值为(65535-1-2)/3= 21845，减 1 是因为实际的行存储从第 2 个字节开始；减

2 是因为 VARCHAR 头部的两个字节表示长度，除 3 是因为字符编码为 utf8。

VARCHAR(10)与 VARCHAR(100)的区别为：对于 VARCHAR 数据类型来说，硬盘上的存储空间是根据实际字符长度来分配的，对于内存来说，是使用固定大小的内存块来保存值的，因而 VARCHAR(100)会占用更大的内存。

CHAR(1)与 VARCHAR(1)的区别为：虽然它们都只能用来保存单个字符，但是 VARCHAR 要比 CHAR 多占用一个存储位置，这主要是因为在使用 VARCHAR 数据类型时，会多用 1 个字节来存储长度信息。

3）ENUM

ENUM 占 1、2 个字节，具体取决于枚举值的数量，最大为 65535。ENUM 适合取值较少且不经常变更的字段，例如 sex、state、type 等。

4）TEXT 和 BLOB 类型

TEXT 和 BLOB 都是为了存储较大的数据而设计的字符串类型，分别采用字符和二进制方式存储。下面主要介绍几种不同的类型。

（1）TINYBLOB、TINYTEXT：允许最大存储 255 个字符。

（2）BLOB、TEXT：能存储 64 000 个字节。

（3）MEDIUMBLOB、MEDIUMTEXT：限制在 16MB 内。

（4）LONGBLOB、LONGTEXT：存储容量可超过 4GB。

需要注意 BLOB 类型存储的是二进制数据，没有排序规则和字符集，而 TEXT 类型有字符集和排序规则。

5）DATETIME 和 TIMESTAMP

DATETIME 占 8 个字节，范围为 1001—9999 年（和时区无关）。

TIMESTAMP 占 4 个字节，范围为 1970 年 1 月 1 日（格林尼治标准时间）—2028 年（和时区有关）。如果在多个时区存储或访问数据，则会出现不同的结果。

对于非定义表（表内容经常需要变更）必须有一个时间字段，以方便 DBA 查找问题。

4. 默认设置

1）主键

采用 InnoDB 引擎创建的任何一张表必须有主键，默认需要设置成自增主键（ID）。

2）默认值

字段只要有可能用到或者未来有可能用到，则 WHERE 条件的字段必须不为空（NOT NULL），同时有默认值（可以选择空字符或 0 等）。

5. 存储引擎

默认使用 InnoDB 存储引擎，有特殊需要时可使用其他存储引擎。需要注意，MyISAM 支持全文检索，在 MySQL 5.5 中 InnoDB 不支持全文检索。

6. MySQL 关键字

MySQL 关键字如表 5-1 所示。

表 5-1

ADD	ALL	ALTER
ANALYZE	AND	AS
ASC	ASENSITIVE	BEFORE
BETWEEN	BIGINT	BINARY
BLOB	BOTH	BY
CALL	CASCADE	CASE
CHANGE	CHAR	CHARACTER
CHECK	COLLATE	COLUMN
CONDITION	CONNECTION	CONSTRAINT
CONTINUE	CONVERT	CREATE
CROSS	CURRENT_DATE	CURRENT_TIME
CURRENT_TIMESTAMP	CURRENT_USER	CURSOR
DATABASE	DATABASES	DAY_HOUR
DAY_MICROSECOND	DAY_MINUTE	DAY_SECOND
DEC	DECIMAL	DECLARE
DEFAULT	DELAYED	DELETE
DESC	DESCRIBE	DETERMINISTIC
DISTINCT	DISTINCTROW	DIV
DOUBLE	DROP	DUAL
EACH	ELSE	ELSEIF
ENCLOSED	ESCAPED	EXISTS

续表

ADD	ALL	ALTER
EXIT	EXPLAIN	FALSE
FETCH	FLOAT	FLOAT4
FLOAT8	FOR	FORCE
FOREIGN	FROM	FULLTEXT
GOTO	GRANT	GROUP
HAVING	HIGH_PRIORITY	HOUR_MICROSECOND
HOUR_MINUTE	HOUR_SECOND	IF
IGNORE	IN	INDEX
INFILE	INNER	INOUT
INSENSITIVE	INSERT	INT
INT1	INT2	INT3
INT4	INT8	INTEGER
INTERVAL	INTO	IS
ITERATE	JOIN	KEY
KEYS	KILL	LABEL
LEADING	LEAVE	LEFT
LIKE	LIMIT	LINEAR
LINES	LOAD	LOCALTIME
LOCALTIMESTAMP	LOCK	LONG
LONGBLOB	LONGTEXT	LOOP
LOW_PRIORITY	MATCH	MEDIUMBLOB
MEDIUMINT	MEDIUMTEXT	MIDDLEINT
MINUTE_MICROSECOND	MINUTE_SECOND	MOD
MODIFIES	NATURAL	NOT
NO_WRITE_TO_BINLOG	NULL	NUMERIC
ON	OPTIMIZE	OPTION
OPTIONALLY	OR	ORDER
OUT	OUTER	OUTFILE
PRECISION	PRIMARY	PROCEDURE
PURGE	RAID0	RANGE
READ	READS	REAL
REFERENCES	REGEXP	RELEASE

续表

ADD	ALL	ALTER
RENAME	REPEAT	REPLACE
REQUIRE	RESTRICT	RETURN
REVOKE	RIGHT	RLIKE
SCHEMA	SCHEMAS	SECOND_MICROSECOND
SELECT	SENSITIVE	SEPARATOR
SET	SHOW	SMALLINT
SPATIAL	SPECIFIC	SQL
SQLEXCEPTION	SQLSTATE	SQLWARNING
SQL_BIG_RESULT	SQL_CALC_FOUND_ROWS	SQL_SMALL_RESULT
SSL	STARTING	STRAIGHT_JOIN
TABLE	TERMINATED	THEN
TINYBLOB	TINYINT	TINYTEXT
TO	TRAILING	TRIGGER
TRUE	UNDO	UNION
UNIQUE	UNLOCK	UNSIGNED
UPDATE	USAGE	USE
USING	UTC_DATE	UTC_TIME
UTC_TIMESTAMP	VALUES	VARBINARY
VARCHAR	VARCHARACTER	VARYING
WHEN	WHERE	WHILE
WITH	WRITE	X509
XOR	YEAR_MONTH	ZEROFILL

7. 尽量遵循三范式

（1）第一范式：属性原子化，一个栏位不要包含多个属性。

所谓第一范式（1NF）是指在关系模型中对域添加的一个规范要求，所有的域都应该是原子性的，即数据库表的每一列都是不可分割的原子数据项，而不能是集合、数组、记录等非原子数据项。

（2）第二范式：主键依赖，可以根据主键的唯一性确定其他属性。

在 1NF 的基础上，非码属性必须完全依赖于候选码（在 1NF 基础上消除非主属性对主码的部分函数依赖）。在数据库层可以这样理解：数据库表中的非关键字段对任一候选关键字段都不存在部分函数依赖（当一个表是复合主键时，非主键的字段不依赖于部分主键，即必须依赖于全部的主键字段）。第二范式可以说是消除部分依赖，可以减少插入异常、删除异常和修改异常。

例如学生某次的上课安排如下：

学生	课程	教师	教师职称	教材	教室	上课时间
小鲁	高级 Java	亚瑟	高级讲师	《Java 架构分析》	101	9:00
小优	mycat 实战	小巧	Java 讲师	《分布式数据库》	201	14:00

其中，组合主键为学生、课程。这里通过(学生,课程)可以确定教师、教师职称、教材、教室，所以可以把(学生,课程)作为主键。但是，通过课程可以知道教材，出现了部分依赖，因此不满足第二范式。可以将教材字段删除，新建一张课程与教材字段表。

（3）第三范式：属性不依赖于其他非主属性，也就是在满足 2NF 的基础上，任何非主属性不得传递依赖于主属性，也不得依赖于其他非主属性（在 2NF 基础上消除传递依赖）。

学生	课程	教师	教师职称	教室	上课时间
小鲁	高级 Java	亚瑟	高级讲师	101	9:00
小优	mycat 实战	小巧	Java 讲师	201	14:00

在满足第二范式的基础上，在上例中修改后的选课表中，一个教师能确定一个教师职称。这样，教师依赖于(学生,课程)，而教师职称又依赖于教师这个其他非主属性，这便叫作传递依赖。第三范式就是要消除传递依赖。最终应修改为三张表。

选课表：

学生	课程	教师	教室	上课时间
小鲁	高级 Java	亚瑟	101	9:00
小优	mycat 实战	小巧	201	14:00

课程表：

课程	教材
高级 Java	《Java 架构分析》
mycat 实战	《分布式数据库》

教师表：

教师	教师职称

亚瑟　　高级讲师

小巧　　Java 讲师

5.3.2　SQL 语句与索引

在应用系统开发早期，数据库中的数据量并不大，但是系统提交到实际应用场景之后，数据量和并发量会急剧增长，这时如何缩短系统的响应时间成为最迫切解决的问题之一。系统优化中很重要的一部分是对 SQL 语句的优化，当数据量到达一定程度时，优劣 SQL 语句之间的处理速度差别可以达到上百倍甚至上千倍。由此可见，对于应用系统来说不是简单地实现功能就可以了，而需要有高质量的 SQL 语句来提升系统的性能。

1. 如何定位执行慢 SQL 语句

可以通过 MySQL 慢日志去定位慢 SQL 语句，但是慢日志记录的 SQL 语句也会消耗一定的 I/O 资源，开启慢语句记录功能后可以通过分析 SQL 语句找到优化 SQL 语句的方案。可以通在 my.cnf 配置文件中加入以下参数来开启慢语句的功能：

```
slow_query_log=1
slow_query_log_file=/usr/local/mysql/slow_sql_query.log
long_query_time=10(超过 10 秒的 SQL 会记录下来)
```

Percona 官方提供了 Percona-toolkit 工具集，其中的 pt-query-digest 命令为分析慢日志，Percona 下载地址为 https://www.percona.com/software/database-tools/percona-toolkit，文档地址为 https://www.percona.com/doc/percona-toolkit/2.2/pt-query-digest.html。整体输出结果分为以下三部分。

（1）第 1 部分：

```
# 524.1s user time, 27.1s system time, 24.82M rss, 176.29M vsz
# Current date: Mon Aug  8 22:40:29 2016
# Hostname: zyz
# Files: /usr/local/mysql/data/mysql-slow.log
# Overall: 20.31k total, 42 unique, 0.75 QPS, 5.36x concurrency ___________
# Time range: 2016-10-05 16:00:06 to 09:32:34
# Attribute          total     min     max     avg     95%  stddev  median
# ============     ======= ======= ======= ======= ======= ======= =======
# Exec time        145438s   100ms   3161s      7s     35s     38s   189ms
# Lock time        127593s       0    267s      6s     33s     30s    49us
# Rows sent          2.06M       0    1000   54.85  174.84   58.51   49.17
# Rows examine      33.14M       0   8.16M   1.17k  363.48  61.46k   49.17
# Query size         9.57M       6     379  235.96  363.48   64.22  212.52
```

该部分是一个大致的概要信息，通过它可以对当前 MySQL 的查询性能做一个初步的评估，比如各个指标的最大值(max)、平均值(min)、95%分布值、中位数(median)及标准偏差(stddev)。这些指标有查询的执行时间（Exec time）、锁占用的时间（Lock time）、MySQL 执行器需要检查的行数（Rows examine）、最后返回给客户端的行数（Rows sent）及查询的大小。

（2）第 2 部分：

```
# Profile
# Rank Query ID           Response time    Calls R/Call    V/M   Item
# ==== ================== ================ ===== ========= ===== =========
#    1 0x83CFBBF241DC77E2 97667.5895 67.2%  2466   39.6057 11...
SELECT  ?mycat_hot
#    2 0x68661F3BFDF2EE7B 30777.9161 21.2%   415   74.1637 63.51 SELECT
d?mycat_?_hot?
#    3 0x92BB3C8B7FDD4727  3801.0833  2.6%  6751    0.5630 13.30 SELECT
d?mycatserver
#    4 0xCBBBD7282212BB0D  3161.0466  2.2%     1 3161.0466  0.00 DELETE
d?mycat_log_in?day
# MISC 0xMISC             5132.6567   3.5% 10581    0.4851   0.0 <33 ITEMS>
```

在 Rank 的整个分析中此类“SQL”的排名一般为响应占总比差的性能排名。

- Response time：“SQL”的响应时间及整体占比情况。
- Calls：此类“SQL”的执行次数。
- R/Call：每次执行的平均响应时间。
- V/M：响应时间的差异平均对比率。

最后一行输出显示了其他 33 个占比较低而不值得单独显示的查询的统计数据。

（3）第 3 部分：

```
# Query 1: 0.09 QPS, 3.60x concurrency, ID 0x83CFBBF241DC77E2 at byte 1597952728
# Scores: V/M = 118.68
# Time range: 2016-08-05 16:00:06 to 23:31:51
# Attribute    pct   total     min     max     avg     95%  stddev  median
# ============ === ======= ======= ======= ======= ======= ======= =======
# Count         12    2466
# Exec time     67  97668s   100ms    268s     40s    246s     69s      1s
# Lock time     75  95791s    47us    267s     39s    246s     68s   733us
# Rows sent      0   2.24k       0       1    0.93    0.99    0.26    0.99
# Rows examine   0   4.47k       0       2    1.86    1.96    0.51    1.96
# Query size    12 580.86k     238     243  241.20  234.30       0  234.30
# String:
# Hosts
# Users        stats
# Query_time distribution
```

```
#   1us
#  10us
# 100us
#   1ms
#  10ms
# 100ms  #################################################################
#    1s  #
#  10s+  #################################################################
# Tables
#    SHOW TABLE STATUS LIKE 'm3mycat_hot'\G
# EXPLAIN /*!50100 PARTITIONS*/
select * from 2016_mycat_hot;
```

以上结果是查询响应时间的分布情况，通过慢查询定义慢 SQL 之后将慢 SQL 提取，通过 EXPLAIN 进行分析。

在 MySQL 中提供了 EXPLAIN 语句用来分析查询语句。

（1）id 为 select 识别符，是 select 的查询序列号。

（2）select_type 有如下几种。

- SIMPLE：简单的 select 语句（不使用 UNION 或子查询）。
- PRIMARY：最外面的 select 语句。
- UNION：UNION 中的第 2 个或后面的 select 语句，这是 UNION 语句中的一个 SQL 元素。
- DEPENDENT UNION：UNION 中的第 2 个或后面的 select 语句，取决于外面的查询。
- UNION RESULT：UNION 的结果。
- SUBQUERY：子查询中的第 1 个 select。
- DEPENDENT SUBQUERY：子查询中的第 1 个 select，取决于外面的查询。
- DERIVED：导出表的 select（FROM 子句的子查询）。

（3）table 为输出的行所引用的表。

（4）type 有如下几种。

- System：表仅有一行，这是 CONST 类型的特例。
- Const：表最多只有一个匹配行，它将在查询开始时被读取，在余下的查询优化中被作为常量对待。CONST 表的查询速度很快，因为它们只读取一次。CONST 用于常数值比较 PRIMARY KEY 或 UNIQUE 索引所在的场合。

在下面的查询中，tbl_name 可以用于 CONST 的表：

```
SELECT * from tbl_name WHERE primary_key=1;
SELECT * from tbl_name WHERE primary_key_part1=1 和 primary_key_part2=2;
```

如下所示的 uid 为普通索引，因为没有用到指定的索引，所以不会使用 CONST。

```
[05:08:31]root@127.0.0.1:3306-test>  explain select * from lis3 where uid=9;
+----+-------------+-------+------+---------------+------+---------+-------+------+-------+
| id | select_type | table | type | possible_keys | key  | key_len | ref   | rows | Extra |
+----+-------------+-------+------+---------------+------+---------+-------+------+-------+
|  1 | SIMPLE      | lis3  | ref  | ll            | ll   | 5       | const |    1 | NULL  |
+----+-------------+-------+------+---------------+------+---------+-------+------+-------+
1 row in set (0.00 sec)
```

如下所示的 CONST id 是主键，所以使用了 CONST，可以理解为 CONST 是最优化的。

```
[05:07:08]root@127.0.0.1:3306-test>  explain select * from lis3 where id=1;
+----+-------------+-------+-------+---------------+---------+---------+-------+------+-------+
| id | select_type | table | type  | possible_keys | key     | key_len | ref   | rows | Extra |
+----+-------------+-------+-------+---------------+---------+---------+-------+------+-------+
|  1 | SIMPLE      | lis3  | const | PRIMARY       | PRIMARY | 4       | const |    1 | NULL  |
+----+-------------+-------+-------+---------------+---------+---------+-------+------+-------+
1 row in set (0.00 sec)
```

注意：不要混淆 ref 中的 CONST 与 type 中的 CONST。

（5）Eq_ref：对于每个来自于前面的表的行组合，从该表中读取一行。这可能是除 CONST 类型外最好的联接类型。它在一个索引的所有部分被联接使用，并且索引类型是 UNIQUE 或 PRIMARY KEY。EQ_REF 可以用等号比较带索引的列，比较的值可以为常量或该表的列的表达式。

在下面的例子中，MySQL 可以使用 EQ_REF 联接来处理 ref_tables：

```
 SELECT * FROM ref_table,other_table WHERE
ref_table.key_column=other_table.column;
 SELECT * FROM ref_table,other_table WHERE
ref_table.key_column_part1=other_table.column
 AND ref_table.key_column_part2=1;
```

如下所示，MySQL 使用 EQ_REF 联接来处理 lis2 表。

```
[03:25:31]root@127.0.0.1:3306-test> explain select * from lis,lis2 where lis.id=lis2.uid;
+----+-------------+-------+--------+---------------+---------+---------+---------------+------+-------------+
| id | select_type | table | type   | possible_keys | key     | key_len | ref           | rows | Extra       |
+----+-------------+-------+--------+---------------+---------+---------+---------------+------+-------------+
|  1 | SIMPLE      | lis2  | ALL    | NULL          | NULL    | NULL    | NULL          |    1 | Using where |
|  1 | SIMPLE      | lis   | eq_ref | PRIMARY       | PRIMARY | 4       | test.lis2.uid |    1 | NULL        |
+----+-------------+-------+--------+---------------+---------+---------+---------------+------+-------------+
2 rows in set (0.30 sec)
```

下面给出各种联接类型，按照从最佳类型到最坏类型进行排序。

- Ref：对于每个来自于前面的表的行组合，所有有匹配索引值的行将从这张表中读取。如果联接只使用键的最左边的前缀或者键不是 UNIQUE 或 PRIMARY KEY（即如果联接不能基于关键字选择单个行），则使用 REF。如果使用的键仅仅匹配少量的行，则该联接类型比较适合。

REF 可以用于使用“=”或“<=>”操作符的带索引的列。在下面的例子中，MySQL 可以使用 REF 联接来处理 ref_tables：

```
    SELECT * FROM ref_table WHERE key_column=expr;
    SELECT * FROM ref_table,other_tableWHERE
ref_table.key_column=other_table.column;
        SELECT * FROM ref_table,other_tableWHERE
ref_table.key_column_part1=other_table.columnAND ref_table.key_column_part2=1;
```

```
[03:37:25]root@127.0.0.1:3306-test> explain select * from lis3 where uid=2;
+----+-------------+-------+------+---------------+------+---------+-------+--------+-------+
| id | select_type | table | type | possible_keys | key  | key_len | ref   | rows   | Extra |
+----+-------------+-------+------+---------------+------+---------+-------+--------+-------+
|  1 | SIMPLE      | lis3  | ref  | ll            | ll   | 5       | const | 302820 | NULL  |
+----+-------------+-------+------+---------------+------+---------+-------+--------+-------+
1 row in set (0.00 sec)
```

- Ref_or_null1：该联接类型如同 REF，但是在 MySQL 中可以专门搜索包含 NULL 值的行。在解决子查询时经常使用该联接类型的优化。在下面的例子中，MySQL 可以使用 REF_OR_NULL 联接来处理 REF_TABLES：

```
SELECT * FROM ref_table WHERE key_column=expr OR key_column IS NULL;
```

- Index_merge：该联接类型表示使用了索引合并优化方法。在这种情况下，key 列包含了所使用索引的清单，key_len 包含了所使用索引的最长的关键元素。
- Unique_subquery：为一个索引查找类型，可以完全替换子查询，效率更高。该类型替换了如下所示的 IN 子查询的 ref：

```
value IN (SELECT primary_key FROM single_table WHERE some_expr)
```

- Index_subquery：该联接类型类似于 UNIQUE_SUBQUERY，不过索引类型不需要是唯一索引，可以替换 IN 子查询，但只适合下列形式的子查询中的非唯一索引：

```
value IN (SELECT key_column FROM single_table WHERE some_expr)
```

- Range：只检索给定范围的行，使用一个索引来检索行数据。key 列显示使用了哪个索引，key_len 显示了所使用索引的长度。在该类型中 ref 列为 NULL。当使用=、<>、>、>=、<、<=、IS NULL、<=>、BETWEEN 或者 IN 操作符用常量比较关键字列时，类型为 range。下面介绍几种检索指定行数据的情况。

```
SELECT * FROM tbl_nameWHERE key_column = 10;
SELECT * FROM tbl_nameWHERE key_column BETWEEN 10 and 20;
SELECT * FROM tbl_nameWHERE key_column IN (10,20,30);
 SELECT * FROM tbl_nameWHERE key_part1= 10 AND key_part2 IN (10,20,30);
```

- Index：该联接类型与 ALL 相同，除了只有索引树被扫描。它通常比 ALL 快，因为索引文件通常比数据文件小。当查询只使用作为单索引的一部分的列时，MySQL 可以使用该联接类型。

- All：对于每个来自于先前的表的行组合，进行完整的表扫描。如果是第 1 个没标记 CONST 的表，则使用它的效果通常会很差，可以增加更多的索引而不是使用 ALL，使得行能基于前面表中的常数值或列值被检索出。
- Possible_keys：列指出 MySQL 能使用哪个索引在该表中找到行。注意，该列完全独立于 EXPLAIN 输出所示的表的次序。这意味着 POSSIBLE_KEYS 中的某些键实际上不能按生成的表的次序使用。如果该列是 NULL，则没有相关的索引。在这种情况下，可以通过检查 WHERE 语句的查询条件，来判断是否存在适合建立索引的列来提高查询性能。如果是，则创建一个适当的索引并且再次用 EXPLAIN 检查查询。为了弄明白一张表有什么索引，可使用 SHOW INDEX FROM tbl_name。

（6）Key：列显示 MySQL 实际决定使用的键（索引）。如果没有选择索引，则键是 NULL。要想强制 MySQL 使用或忽视 POSSIBLE_KEYS 列中的索引，则在查询中使用 FORCE INDEX、USE INDEX 或者 IGNORE INDEX。

（7）Key_len：列显示 MySQL 决定使用的键长度。如果键是 NULL，则长度为 NULL。注意我们可以通过 key_len 值确定 MySQL 将实际使用一个多部关键字的几个部分。

（8）Ref：列显示使用哪个列或常数与 key 一起从表中选择行。

（9）Row：列显示 MySQL 认为它执行查询时必须检查的行数。

（10）Extra：该列包含 MySQL 解决查询的详细信息。下面解释了该列可以显示的不同的文本字符串。

- Distinct：MySQL 发现第 1 个匹配行后，停止为当前的行组合搜索更多的行。
- Not exists：MySQL 能够对查询进行 LEFT JOIN 优化，发现 1 个匹配 LEFT JOIN 标准的行后，不再为前面的行组合在该表内检查更多的行。
- Range checked for each record：没有找到合适的索引。
- Using filesort：这可能是一个 CPU 密集型的过程，MySQL 需要额外的一次传递，用于找出如何按照排序检索行。根据联接类型浏览所有的行，并为所有匹配 WHERE 子句的行保存排序的关键字和行的指针来完成排序。然后关键字被排序，并按照排序检索行。

```
[04:24:09]root@127.0.0.1:3306-test> explain select * from lis3  order by uid ;
+----+-------------+-------+------+---------------+------+---------+------+--------+----------------+
| id | select_type | table | type | possible_keys | key  | key_len | ref  | rows   | Extra          |
+----+-------------+-------+------+---------------+------+---------+------+--------+----------------+
|  1 | SIMPLE      | lis3  | ALL  | NULL          | NULL | NULL    | NULL | 605641 | Using filesort |
+----+-------------+-------+------+---------------+------+---------+------+--------+----------------+
1 row in set (0.00 sec)
```

- Using index：只使用索引树中的信息来检索表中的列信息。当查询只使用作为单一索引的一部分的列时，可以使用该策略。如下所示的 WHERE 条件为非数字类型。

```
[04:22:15]root@127.0.0.1:3306-test> explain select * from lis3 where heh='d';
+----+-------------+-------+------+---------------+------+---------+-------+------+-----------------------+
| id | select_type | table | type | possible_keys | key  | key_len | ref   | rows | Extra                 |
+----+-------------+-------+------+---------------+------+---------+-------+------+-----------------------+
|  1 | SIMPLE      | lis3  | ref  | lld           | lld  | 33      | const |    1 | Using index condition |
+----+-------------+-------+------+---------------+------+---------+-------+------+-----------------------+
1 row in set (0.00 sec)
```

- Using temporary：表示使用了内部临时（基于内存的）表，一个查询可能用到多个临时表。有很多原因都会导致 MySQL 在执行查询期间创建临时表，例如查询包含 GROUP BY 和 ORDER BY 子句。
- Using where：where 子句用于限制哪些行匹配下一个表或发送到客户端。除非你专门从表中索取或检查所有行，如果 Extra 值不为 Using WHERE 并且表联接类型为 ALL 或 index，则查询可能会有一些错误。
- Using index for group-by：类似于访问表的 Using index 方式，Using index for group-by 表示 MySQL 发现了一个索引，可以用来查询 GROUP BY 或 DISTINCT 查询所涉及的所有列，而不要额外搜索硬盘访问实际的表，并且按最有效的方式使用索引，以便对于每个组只读取少量的索引条目。

2. 使用索引

索引是一项重要的优化查询技术，利用好索引可以达到事半功倍的效果。如果没有利用好索引，则反而会降低 MySQL 的性能。如果在一张数据量较大的表中没有设置主键或者索引，则仅仅查询几条记录的效率是比较低的。如表 5-2 所示是 ads 表的结构。

表 5-2

company_id	ad_id	fee
14	48	0.01
15	49	0.02
16	50	0.01
13	51	0.03
17	52	0.04

如果在 ads 表中以 company_id 的值进行分类，那么在 company_id 列上加上索引，则索引包含了每一个数据行的项，加了索引之后进行的查询不需要把全部数据都扫描一遍。因为索引是经过分类的，假如要搜索 company_id 为 13 的所有数据，则当编号达到 14 的数据行时，这个值会高于要搜索的值，我们就知道不会有与 13 相匹配的内容了，因此不需要继续往下扫描。由此可见，索引是通过数据匹配到某行结束来提高查询效率的。另外也可以使用算法，使得不需

要从索引的开始位置按顺序搜索就能直接找到第 1 个匹配项，可以节省大量的检索时间。

在单表中使用索引可以使检索不需要将全部数据都扫描一遍，加快了搜索的速度。在关联表查询时索引的重要性更能体现出来，例如 table1、table2、table3 都包含数据列 c1、c2、c3，假如每一张表的数据量都是 1000 条记录，则可以通过如下关联查询语句进行查询：

```
select table1.*,table2.*,table3.* from t1,t2,t3 where t1.c1 = t2.c2 and
t2.c2= t3.c3;
```

假如上面三张表都没有添加索引，则查询出来的可能的组合有 1000 1000 1000 种，处理这些表关联需要更多的时间，导致 MySQL 的整体性能下降。如果三张表的 c1、c2、c3 列都使用索引，则会在很大程度上提高查询的效率。利用索引查询的流程如下。

（1）从 table1 表中选择第 1 行数据。

（2）直接通过索引找到与表 table1 的值相匹配的数据行，table3 使用索引直接找到与数据表 table2 的值相匹配的数据行。

（3）table1 表重复以上过程，直到 table1 表中的所有数据都扫描完毕。

索引有很多种类型，在 MySQL 中索引是在存储引擎层而不是服务器层实现的，所以没有统一的索引标准，不同存储引擎的索引的工作方式并不一样。

3. 索引结构

在讨论索引时，目前大部分数据库系统及文件系统都采用 B-Tree 或者 B+Tree。

1）B-Tree

B-Tree 通常意味着所有的值都是按照顺序存储的，图 5-4 是 B-Tree 的抽象表示，大致反映了 MySQL 中的 InnoDB 索引是如何工作的。

B-Tree 索引之所以能够快速检索数据，是因为存储引擎不需要进行全表扫描就可以获取需要的数据。检索是从索引的根节点开始的，根节点存放了指向子节点的指针，存储引擎根据指针向下查询。通过比较节点页的值和要查找的值，可以找到合适的指针进入下层节点，这些指针实际上定义了子节点页的上下限，最终检索的结果是要么找到相应的值，要么查询记录不存在。B-Tree 对索引是顺序存储的，很适合范围查询，例如在一个基于文本域的索引树上，根据字母顺序传递的值进行查找非常合适，所以像“找出所有以 I 到 *K* 开头的名字”这样查询的效率非常高。

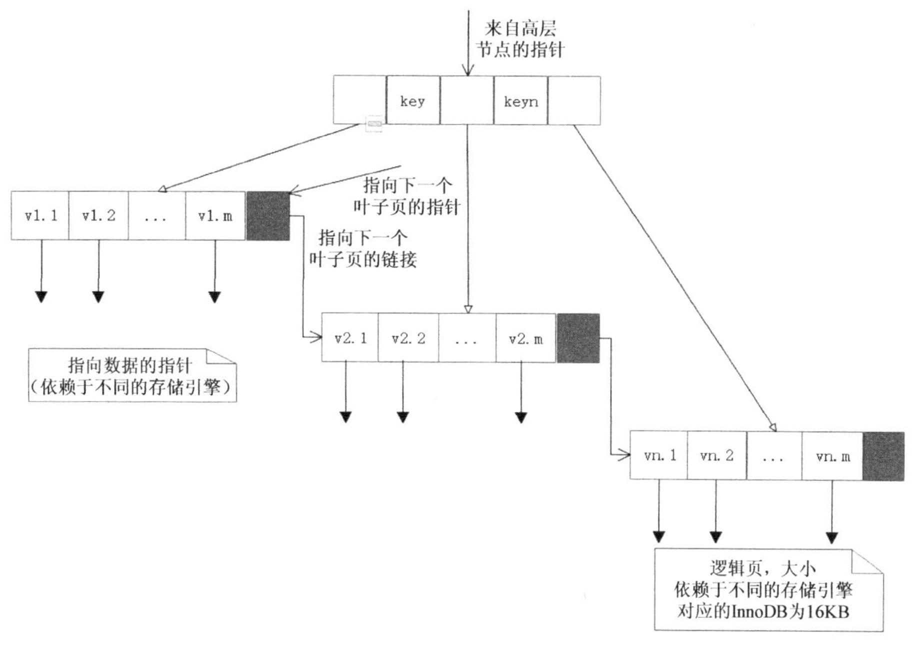

图 5-4

因为索引树中的节点是有序的，所以索引还可以用于查询中的 order by 排序操作。下面是一些关于 B-Tree 索引的限制。

（1）如果不是按照索引列的最左原则开始查找，则索引无效。

（2）不能跳过索引中的列。

（3）如果查询中有某个列的范围查询，则其右边的所有列都无法使用索引优化查找。

2）B+Tree

不同的存储引擎可能使用不同的数据结构存储，InnoDB 使用的是 B+Tree，B+Tree 是为了满足文件系统的需求而形成的一种 B-Tree 的变型树。一棵 *m* 阶的 B+树和 *m* 阶的 B-树的差异如下。

（1）有 *n* 棵子树的结点中含有 *n* 个关键字。

（2）在所有的叶子结点中包含了全部关键字的信息，以及指向含这些关键字的记录的指针，且叶子结点本身根据关键字自小到大链接。

（3）所有的非终端结点可以看作索引部分，结点中仅包含其子树（根结点）中的最大（或最小）关键字。

B-树的关键字和记录是放在一起的，叶子节点可以看作外部节点，不包含任何信息。B+树的非叶子节点中只有关键字和指向下一个节点的索引，记录只放在叶子节点中。

在 B-树中，越靠近根节点的记录的查找时间越快，只要找到关键字即可确定记录的存在；而 B+树中每个记录的查找时间基本上一样，都需要从根节点走到叶子节点，而且在叶子节点中还要再比较关键字。从这个角度来看，B-树的性能好像要比 B+树好，而在实际应用中却是 B+树的性能要好些。因为 B+树的非叶子节点不存放实际的数据，这样每个节点可容纳的元素的数量比 B-树多，树高比 B-树小，这样带来的好处是减少了磁盘的访问次数。尽管 B+树找到一个记录所需的比较次数要比 B-树多，但是一次磁盘访问的时间相当于成百上千次内存比较的时间，因此 B+树的性能实际上可能会好些，而且 B+树的叶子节点使用指针连接在一起，方便顺序遍历（例如查看一个目录下的所有文件及一个表中的所有记录等），这也是很多数据库和文件系统使用 B+树的原因。

3）全文检索

全文检索是一种特殊类型的索引，查找文本中的关键字，而不是直接比较索引中的值。全文检索类似于搜索引擎，而不是简单地进行 where 条件查询；使用 match against 操作，而不是普通的 where 条件查询。

4. 索引的优缺点

索引可以使存储引擎快速定位到表指定的位置，但是这并不是索引唯一的作用，索引也有其他作用，因为索引是按照顺序存储数据的，所以 MySQL 可以用它来做 order by 和 group by 操作。

索引的优点如下。

（1）可以大大减少服务器需要扫描的数据量。

（2）可以帮助服务器避免排序和临时表。

（3）可以将随机 I/O 变成顺序 I/O。

索引的缺点如下。

（1）在绝大多数情况下使用索引可以提高查询效率，但是在有索引存在的表中进行 insert、delete、update 操作会导致速度降低，故索引不宜过多。

（2）如果涉及数据的修改操作，则不仅仅要修改数据库中的数据行，还要修改相应的索引。

（3）一张表中的索引越多，相应的改变也越多，会影响数据库的整体性能。

5. 避免使用 NULL

NULL 对于大多数数据库都需要进行特殊处理，MySQL 也不例外，它需要更多的代码、检查和特殊的索引逻辑。有些开发人员完全没有意识到，创建表时 NULL 是默认值，但大多数时候应该使用一个特殊值，例如将 0 或者-1 作为默认值。

6. 优化 count()查询

count()是一个特殊的函数，有两种不同的作用：统计某个列的数量；统计行的数量。在统计列值时要求列值非空（不统计 null）。最简单的就是我们使用 count(*)时，通配符*并不会像我们想象中的那样扩展到所有列，实际上它忽略了所有的列值并统计了所有的行数。如果在括号中指定一个列并希望统计列的行数，则在统计列的行数时是不统计列的非空值（null）的；如果要想知道所有结果集的行数，则最好使用 count(*)。

7. 优化子查询

遇到子查询时，MySQL 查询优化引擎并不总是最有效的，这就是为什么经常将子查询转换为连接查询。优化器已经能够正确处理连接查询了，当然要注意的一点是，确保连接表（第 2 个表）的连接列是有索引的，如果第 2 张表没有添加索引，那么第 1 个表上的每一行记录通常会相对于第 2 个表的查询子集进行一次全表扫描。

8. 优化 union 查询

MySQL 总是通过创建并填充临时表的方式执行 union 查询的，因此很多优化策略在 union 查询中都没法很好地使用。我们经常需要手工地将 where、limit、order by 等子句“下推”到 union 的各个子查询，以便可以利用这些条件优化。除非确实需要服务器消除重复的行，否则就一定要使用 union all，这一点很重要。如果没有 all 关键字，则 MySQL 会给临时表加上 distinct 选项，这会导致对整个临时表的数据进行唯一性检查，这样做的代价非常高。即使用 ALL 关键字，MySQL 仍然会使用临时表存储结果。实际上 MySQL 总是将结果放在临时表中，然后读出来，再返回给客户端。

9. 优化 limit 分页

在系统中需要分页操作时，我们通常使用 limit 加上偏移量的办法实现，同时适用于 order by 语句，否则 MySQL 需要做大量的文件排序操作。优化此类分页查询的一种最简单的办法是尽量使用索引覆盖扫描，而不是查询所有的列。

10. select 优化

很多没有经验的开发者在做查询时都是直接用 select *，在数据量和表字段比较少时一般没什么问题，但是，如果面对的是一张宽表而且数据量比较大，则这种查询将返回很多没必要的数据，会消耗大量的 MySQL I/O 性能。

11. 避免使用 order by rand()

如果需要查询随机打乱的数据，则有很多种方式可以实现这种目的，但是尽量避免使用 rand()函数，这会比较消耗 CPU 的性能，就算加上 limit 也无济于事。

12. 避免索引无效的查询

以下操作都会导致索引无效。

- 尽量避免在 where 子句中使用!=、<>操作符。
- 尽量避免在 where 子句中对字段进行 null 值判断。
- 尽量避免在 where 子句中使用 or 来连接条件。
- like 前面有百分号（%）。
- in 和 not in 也要慎用，例如 delete 时不建议使用 in，建议使用自连接或者 exist 替代。
- 在 where 子句中使用参数。
- 尽量避免在 where 子句中对字段进行表达式操作。例如 select id from t where age/4 = 20 应改为 select id from t where age = 20*4。
- 尽量避免在 where 子句中对字段进行函数操作。例如 select id from t where substring (name,1,3) = 'ABB';应改为 select id from t where name like 'ABB%'。
- 不要在 where 子句中的“=”左边进行函数、算法运算等。
- 如果使用复合索引，那么必须将该索引的第 1 个字段作为条件，才能保证系统使用该索引，否则该索引将不会被使用，并且尽可能让作为查询条件的字段的查询顺序和索引顺序相一致。

5.3.3 配置文件

MySQL 自身的优化主要是针对其配置文件 my.cnf 中的各项参数。下面通过介绍一些对性能影响较大的参数来了解 MySQL 配置。

```
[mysqld]
port = 3306
serverid = 1
```

```
socket = /tmp/mysql.sock
skip-locking
#避免 MySQL 的外部锁定，减少出错几率，增强稳定性。
skip-name-resolve
#禁止 MySQL 对外部连接进行 DNS 解析，可以消除 MySQL 进行 DNS 解析的时间。如果开启该选项，则所有的远程主机授权都要使用 IP 方式连接。
back_log = 384
#参数的值表示在 MySQL 暂时停止响应新请求之前的这段时间内，有多少个请求可以被存在堆栈中
key_buffer_size = 256M
#key_buffer_size 指定用于索引的缓冲区大小，增加它可得到更好的索引处理性能。
max_allowed_packet = 4M
thread_stack = 256K
table_cache = 128K
sort_buffer_size = 6M
#查询排序时所能使用的缓冲区大小。该参数对应的分配内存是每个连接独占的，如果有 10 个连接，那么实际分配的总排序缓冲区大小为 10 × 6 = 60MB。
read_buffer_size = 4M
#读查询操作所能使用的缓冲区大小，和 sort_buffer_size 一样。
join_buffer_size = 8M
#联合查询操作所能使用的缓冲区大小，和 sort_buffer_size 一样。
myisam_sort_buffer_size = 64M
table_cache = 512
thread_cache_size = 64
query_cache_size = 64M
#指定 MySQL 查询缓冲区的大小。
tmp_table_size = 256MB
max_connections = 768
#指定 MySQL 允许的最大连接进程数。
max_connect_errors = 10000000
wait_timeout = 10
#指定一个请求的最大连接时间。
thread_concurrency = 8
#该参数取值为服务器逻辑 CPU 数量*2。
skip-networking
#开启该选项可以彻底关闭 MySQL 的 TCP/IP 连接方式。
table_cache=1024
#物理内存越大，设置就越大，默认为 2402。
InnoDB_additional_mem_pool_size=4MB
#默认为 2MB
InnoDB_flush_log_at_trx_commit=1
#设置为 0 就是等到 InnoDB_log_buffer_size 列队满后再统一储存，默认为 1。
InnoDB_log_buffer_size=2MB
#默认为 1MB
InnoDB_thread_concurrency=8
#服务器 CPU 有几个就设置为几，建议默认为 8
key_buffer_size=256MB
#默认为 218，调到 128 最佳
```

```
tmp_table_size=64MB
#默认为 16MB，调到 64-256 最挂
read_buffer_size=4MB
#默认为 64KB
read_rnd_buffer_size=16MB
#默认为 256KB
sort_buffer_size=32MB
#默认为 256KB
thread_cache_size=120
#默认为 60
query_cache_size=32MB
```

1. 与 I/O 相关的一些参数

（1）InnoDB_buffer_pool_size：InnoDB 用来高速缓冲数据和索引内存缓冲的大小。更大的设置可以使访问数据时减少磁盘 I/O。在一个专用的数据库（InnoDB 专用）服务器上可以将它设置为物理内存的 80%。

（2）InnoDB_additional_mem_pool_size：InnoDB 用来存储数据字典（data dictionary）信息和其他内部数据结构（internal data structures）的存储器组合（memory pool）大小。理想的值为 2MB，如果有更多的表，则需要在这里重新分配。如果 InnoDB 用尽这个池中的所有内存，则它将从操作系统中分配内存，并将错误信息写入 MySQL 的错误日志中。

（3）InnoDB_file_io_threads：InnoDB 中的文件 I/O 线程，通常设置为 4。

（4）InnoDB_flush_method：控制 InnoDB 数据文件和 redo log 的打开、刷写模式。有 fdatasync（默认）、O_DSYNC、O_DIRECT 这三个值。

- fdatasync 模式：写数据时，write 这一步并不需要真正写到磁盘中才算完成（可能写入操作系统的 Buffer 中就会返回），真正完成的是 flush 操作，Buffer 交给操作系统去 flush，并且文件的元数据信息也都需要更新到磁盘中。
- O_DSYNC 模式：写日志操作是在 write 这一步完成的，而数据文件的写入是在 flush 这一步通过 fsync 完成的。
- O_DIRECT 模式：数据文件的写入操作是直接从 MySQL InnoDB Buffer 到磁盘的，并不用通过操作系统的缓冲，而真正的完成也是在 flush 这一步，日志还是要经过 OS 缓冲。在大量随机写的环境中，O_DIRECT 要比 fdatasync 的效率更高些，如果顺序写入磁盘的情况较多，则还是默认的 fdatasync 更高效。常规 SAS 设置为 O_DIRECT 较多。

（5）InnoDB_max_dirty_pages_pct：最大的脏页百分数，该值如果过大，那么内存很大或者服务器的压力也很大，效率降低；如果值过小，那么硬盘的压力会增加，建议为 75～80。InnoDB plugin 引进了 InnoDB_adaptive_flushng（自适应地刷新），该值影响每秒刷新脏页的数量。

（6）InnoDB_adaptive_flushing：设置为 ON（使刷新脏页更智能）。

（7）InnoDB_io_capacity：从缓冲区刷新脏页时一次刷新脏页的数量。根据磁盘 IOPS 的能力，一般建议设置 SAS 为 200（如果是 RAID 盘，则可根据实际的 IOPS 能力进行调整）。

（8）InnoDB_write_io_threads、InnoDB_read_io_threads：异步 I/O 线程数。

（9）InnoDB_flush_log_at_trx_commit：这是控制 MySQL 磁盘写入策略的关键参数。如果将 InnoDB_flush_log_at_trx_commit 设置为 0，则 log buffer 每秒进行一次写缓存（write cache）和刷写磁盘（flush disk）；设置为 1，则 log buffer 每次提交都写缓存和刷写磁盘；设置为 2，则 log buffer 每次提交都写缓存，然后根据 InnoDB_flush_log_at_timeout（默认为 1 秒）刷写磁盘。从这三个配置来看，InnoDB_flush_log_at_trx_commit=1 显然更安全，因为每次提交都保证在线日志（redo）写入了磁盘，但是其性能相对于 DML 来说较低，在我们的测试中发现，如果设置为 2，则 DML 的性能要比设置为 1 高 10 倍左右。

（10）InnoDB_log_file_size：该参数用于设置 InnoDB REDO 日志文件的大小。设置为较大的值时，可以减少刷新缓冲池的次数，从而减少磁盘 I/O。但是大的日志文件意味着在崩溃时需要更长的时间来恢复数据。

（11）InnoDB_log_buffer_size：InnoDB 将日志写入日志磁盘文件前的缓冲大小。

（12）sync_binlog：二进制日志（binary log）同步到磁盘的频率，binary log 写入 sync_binlog 次后，再刷写到磁盘。如果 autocommit 开启，则每个语句都写一次 binary log，否则对每次事务都写一次。默认值为 0，不主动同步，而依赖操作系统本身不定期把文件内容刷写到磁盘。设为 1 最安全，在每个语句或事务后同步一次 binary log，即使在崩溃时也最多丢失一个语句或事务的日志，但因此也最慢。在大多数情况下，对数据的一致性并没有很严格的要求，所以并不会把 sync_binlog 配置为 1。为了追求高并发及提升性能，可以设置为 100 或直接用 0。

2. 与缓存相关的一些参数

（1）query_cache_size、query_cache_type

Query Cache 作用于整个 MySQL Instance，主要用来缓存 MySQL 中的 ResultSet，也就是一条 SQL 语句执行的结果集，所以只能针对 select 语句。当我们打开了 Query Cache 功能时，MySQL 在接收到一条 select 语句的请求后，如果该语句满足 Query Cache 的要求（未显式申明不允许使用 Query Cache，或者已经显式申明需要使用 Query Cache），则 MySQL 会直接根据预先设定好的 hash 算法将接收到的 select 语句以字符串的方式进行 hash，然后到 Query Cache 中直接查找是否已经缓存。也就是说，如果已经在缓存中，则该 select 请求会直接将数据返回，从而省略了后面的所有步骤（例如 SQL 语句的解析、优化器优化及向存储引擎请求数据等），极大地提高性能。

当然，Query Cache 也有一个致命的缺陷，那就是当某个表的数据有任何变化时，都会导致所有引用了该表的 select 语句在 Query Cache 中的缓存数据失效。Query Cache 的使用需要多个参数配合，其中最为关键的是 query_cache_size 和 query_cache_type，前者设置用于缓存 ResultSet 的内存大小，后者设置在任何场景下使用 Query Cache。从以往的经验来看，如果不是用来缓存基本不变的数据的 MySQL 数据库，则 256MB 一般是 query_cache_size 的比较合适的大小。当然，这可以通过计算 Query Cache 的命中率（Qcache_hits/(Qcache_hits+Qcache_inserts)*100)）来进行调整。query_cache_type 可以设置为 0（OFF）、1（ON）或者 2（DEMOND），分别表示完全不使用 query cache、除显式要求不使用 query cache（使用 sql_no_cache）外的所有的 select 都使用 query cache、只有显示要求才使用 query cache（使用 sql_cache）。

查询缓存碎片率= Qcache_free_blocks / Qcache_total_blocks * 100%，如果查询缓存碎片率超过 20%，则可以用 FLUSH QUERY CACHE 整理缓存碎片，或者如果你的查询都是小数据量的，则可以试试减小 query_cache_min_res_unit。

查询缓存利用率= (query_cache_size - Qcache_free_memory) / query_cache_size * 100%，查询缓存利用率在 25%以下时，说明 query_cache_size 设置得过大，可适当减小；查询缓存利用率在 80%以上而且 Qcache_lowmem_prunes > 50 时，说明 query_cache_size 可能有点小，要不就是碎片太多。

查询缓存命中率= (Qcache_hits - Qcache_inserts) / Qcache_hits * 100%。例如服务器查询缓存碎片率＝20.46%，查询缓存利用率＝62.26%，则查询缓存命中率＝1.94%，命中率很差，可能写操作比较频繁，而且可能存在不少碎片。

（2）record_buffer_size

进行一个顺序扫描的每个线程为其扫描的每张表分配这个大小的一个缓冲区。如果你做很多顺序扫描，则你可能想要增加该值。默认数值是 131 072（128KB），可改为 16 773 120（16MB）。

（3）read_rnd_buffer_size

随机读缓冲区的大小，按任意顺序读取行时（例如，按照排序顺序）将分配一个随机读缓存区。进行排序查询时，MySQL 会首先扫描一遍该缓冲，以避免磁盘搜索，提高查询的速度；如果需要排序大量的数据，则可适当调高该值。但 MySQL 会为每个客户连接发放该缓冲空间，所以应尽量适当设置该值，以避免内存开销过大，一般可设置为 16MB。

（4）sort_buffer_size

为每个需要进行排序的线程分配该大小的一个缓冲区。增加这值可以加速 ORDER BY 或

GROUP BY 操作。默认值为 2 097 144（2MB），可改为 16 777 208（16MB）。

（5）join_buffer_size

联合查询操作所能使用的缓冲区大小。

（6）record_buffer_size，read_rnd_buffer_size，sort_buffer_size，join_buffer_size

为每个线程独占，也就是说如果有 100 个线程连接，则占用为 16MB×100。

（7）table_cache

表高速缓存的大小。每当 MySQL 访问一个表时，如果在表缓冲区中还有空间，则该表被打开并放入其中，这样可以更快地访问表的内容。通过检查峰值时间的状态值 Open_tables 和 Opened_tables，可以决定是否需要增加 table_cache 的值。如果你发现 open_tables 等于 table_cache，并且 opened_tables 不断增长，那么你就需要增加 table_cache 的值了（上述状态值可以使用 SHOW STATUS LIKE 'Open%tables'获得）。注意，不能盲目地把 table_cache 设置成很大的值。如果设置得太大，则可能会造成文件描述符不足，从而造成性能不稳定或者连接失败。对于 1GB 内存的机器，推荐值是 128～256。内存在 4GB 左右的服务器，该参数可设置为 256MB 或 384MB。

（8）max_heap_table_size

用户可以创建的内存表（memory table）的大小。这个值用来计算内存表的最大行数值。这个变量支持动态改变，即 set @max_heap_table_size=#。

这个变量和 tmp_table_size 一起限制了内部内存表的大小。如果某个内部 heap（堆积）表的大小超过了 tmp_table_size，则 MySQL 可以根据需要自动将内存中的 heap 表改为基于硬盘的 MyISAM 表。

（9）tmp_table_size

通过设置 tmp_table_size 选项来增加一张临时表的大小，例如做高级 GROUP BY 操作生成的临时表。如果调高该值，则 MySQL 将同时增加 heap 表的大小，可达到加快联接查询速度的效果，建议尽量优化查询，要确保在查询过程中生成的临时表在内存中，避免临时表过大而导致生成基于硬盘的 MyISAM 表。

每次创建临时表时 Created_tmp_tables 都会增加，如果临时表的大小超过 tmp_table_size，则在磁盘上创建临时表，Created_tmp_disk_tables 也增加，Created_tmp_files 表示 MySQL 服务创建的临时文件文件数，比较理想的配置是：Created_tmp_disk_tables / Created_tmp_tables * 100% <= 25%，比如上面的服务器 Created_tmp_disk_tables / Created_tmp_tables * 100% ＝ 1.20%比较好。默认为 16MB，可调到 64～256MB，线程独占，太大则可能内存不够，导致 I/O 堵塞。

（10）thread_cache_size

可以复用的保存在内存中的线程的数量。如果有，则新的线程从缓存中取得，当断开连接时如果有空间，则用户的线置将保存在缓存中。如果有很多新的线程，则为了提高性能，可以增加这个变量值。通过比较 Connections 和 Threads_created 状态的变量，可以看到这个变量的作用。

（11）InnoDB_buffer_pool_size

对于 InnoDB 表和 MyISAM 表来说，InnoDB_buffer_pool_size 的作用就相当于 key_buffer_size。InnoDB 使用该参数指定大小的内存来缓冲数据和索引。对于单独的 MySQL 数据库服务器，最大可以把该值设置成物理内存的 80%。

（12）InnoDB_log_buffer_size

log 缓存的大小，一般为 1～8MB，默认为 1MB。对于较大的事务，可以增大缓存的大小，可设置为 4MB 或 8MB。

（13）InnoDB_additional_mem_pool_size

该参数指定 InnoDB 用来存储数据字典和其他内部数据结构的内存池大小，默认值是 1MB。通常不用太大，只要够用就行，应该与表结构的复杂度有关系。如果不够用，则 MySQL 会在错误日志中写入一条警告信息。

（14）bulk_insert_buffer_size

和 key_buffer_size 一样，这个参数也仅用于使用 MyISAM 存储引擎，用来缓存批量插入的数据。

（15）binlog_cache_size

Binlog Cache 用于打开了二进制日志（binlog）记录功能的环境中，是 MySQL 用来提高 binlog 的记录效率而设计的一个用于短时间内临时缓存 binlog 数据的内存区域。建议设置为 2～4MB。

（16）key_buffer_size

key_buffer_size 参数可用于设置缓存在 MyISAM 存储引擎中的索引文件的内存区域大小。如果我们有足够的内存，则这个缓存区域最好能够存放下所有 MyISAM 引擎表的所有索引，以尽可能提高性能。

5.3.4 InnoDB 选择文件系统

MySQL 官方建议我们使用 XFS 文件系统，如表 5-3 所示为 MySQL 官方给出的一份对不同的文件系统进行测试的结果。

表 5-3

result	unit	ext3	reiserfs	jfs	xfs	average
restore time	secs	1241.2	868.1	27.6	785.8	955.7
create 2mln myisam	secs	40.2	35.4	36.4	34.9	36.7
create 1mln myisam	secs	19.2	17.2	17.2	17.0	17.6
create 1mln InnoDB	secs	93.9	86.9	95.3	95.2	92.8
create 1mln bdb	secs	20.0	18.0	17.7	17.7	18.3
overall write	score	79.3%	109.3%	102.5%	117.9%	
myisam tran/s	n/s	145.9	206.1	207.4	203.5	190.7
InnoDB tran/s	n/s	81.5	85.4	154.9	76.3	99.5
bdb tran/s	n/s	142.9	203.5	204.7	199.5	187.6
sysbench rw/s	n/s	2029.1	3672.6	3720.0	3612.9	3258.6
myisam tran/s	n/s	112.7	204.0	206.7	200.7	181.0
overall trans	score	64.1%	111.6%	114.7%	109.6%	

可以明显看出，XFS 各方面的性能优于其他文件系统，所以我们选择 XFS 作为 MySQL 文件系统。

5.3.5 系统架构

优秀的系统架构设计可以提高系统查询的效率，在高并发、大数据量的情况下这种作用尤其明显。生产上的业务环境大多是读多写少的，针对这样的情况，我们设计的架构通常在数据库 DB（MySQL）层之前搭建一套高并发数据缓存层（采用 Redis、Memcache 的集群），以减少对数据库的查询和访问，提高系统的效率。如果缓存 key 设置得合理，那么在命中率比较高时几乎都在查询高速缓存，可大大缓解数据库的压力。

在千万级别数据量的情况下，我们可以对 MySQL 层做分表的设计方式，采用如下架构可以满足这个级别的数据量，如图 5-5 所示。

如果数据量大到上亿或者几十亿级别，则再怎么分表都很难解决大数据量导致数据库性能降低的问题，因为主热备份方式的瓶颈在 MySQL I/O 上，数据量达到这种程度之后必须采用分片才可以解决这个问题。把数据根据某种算法分散到不同的 MySQL Server 中，以减少每一台 MySQL Server 的压力，加快整个系统的响应速度。实现这种分片的中间件有 Cobar、Mycat 等，但是个人建议使用 Mycat。如图 5-6 所示是使用 Mycat 中间件搭建 MySQL 分片的架构。

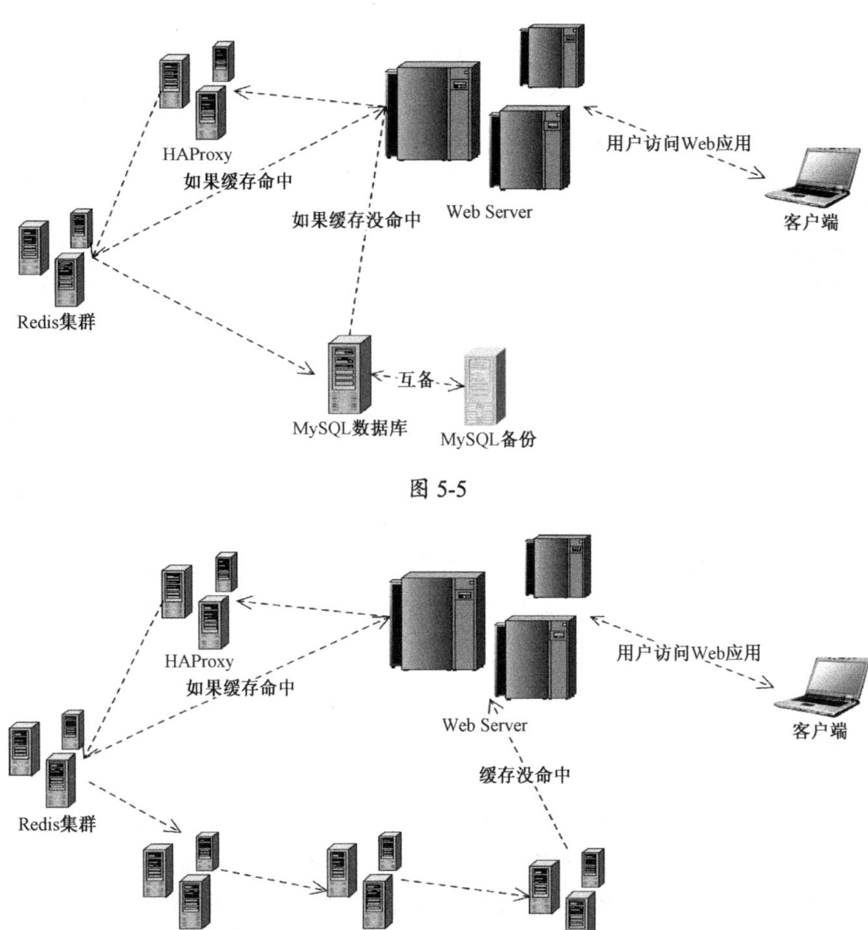
HAProxy
如果缓存命中
如果缓存没命中
Web Server
用户访问Web应用
客户端
Redis集群
互备
MySQL数据库
MySQL备份
HAProxy
如果缓存命中
Web Server
用户访问Web应用
客户端
缓存没命中
Redis集群
ZooKeeper集群
Mycat集群
MySQL集群

图 5-5

图 5-6

第 6 章

Mycat 架构剖析

本章主要介绍 Mycat 的总体实现架构。了解 Mycat 的架构对于 Mycat 优化及二次开发都至关重要，通过学习 Mycat 的架构也可以提升自身的架构设计水平。

6.1 Mycat 总体架构介绍

Mycat 在逻辑上由几个模块组成：通信协议、路由解析、结果集处理、数据库连接、监控等模块。模块之间的交互如图 6-1 所示。

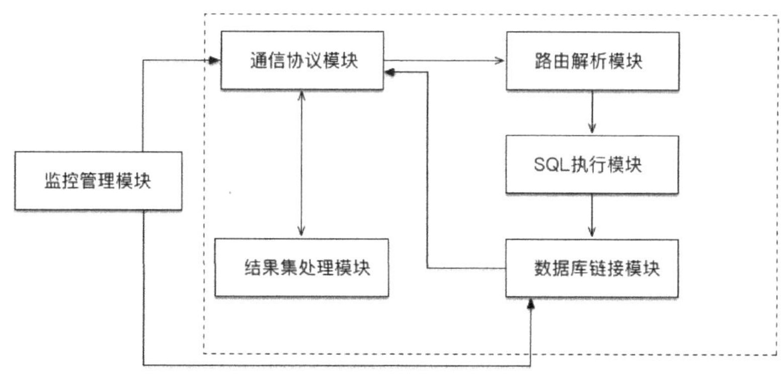

图 6-1

1. 通信协议模块

通信协议模块承担底层的收发数据、线程回调处理工作，主要采用 Reactor、Proactor 模式来提高效率。目前，Mycat 通信模块默认采用 Reactor 模式，在协议层采用 MySQL 协议。

2. 路由解析模块

路由解析模块负责对传入的 SQL 语句进行语法解析，解析从 MySQL 协议中解析出来并进入该模块的 SQL 语句的条件、语句类型、携带的关键字等，对符合要求的 SQL 语句进行相关优化，最后根据这些路由计算单元进行路由计算。

3. 结果集处理模块

结果集处理模块负责对跨分片的结果进行汇聚、排序、截取等。由于数据存储在不同的数据库中，所以对跨分片的数据需要进行汇聚。

4. 数据库连接模块

数据库连接模块负责创建、管理、维护后端的连接池。为了减少每次建立数据库连接的开销，数据库使用连接池机制对连接生命周期进行管理。

5. 监控管理模块

监控管理模块负责对 Mycat 中的连接、内存等资源进行监控和管理。监控主要是通过管理命令实时地展现一些监控数据，例如连接数、缓存命中数等；管理则主要通过轮询事件来检测和释放不使用的资源。

6. SQL 执行模块

SQL 执行模块负责从连接池中获取相应的目标连接，对目标连接进行信息同步后，再根据路由解析的结果，把 SQL 语句分发到相应的节点执行。

总体执行流程如下所述。

主要流程由通信协议模块的读写事件通知发起。读写事件通知具体的回调代码进行这次读写事件的处理。管理模块的执行流程由定时器事件进行资源检查和资源释放时发起（见图 6-2）。

由客户端发送过来的数据通过协议解析、路由解析等流程进入执行组件。通过执行组件把数据发送到通信协议模块，最终数据被写入目标数据库（见图 6-3）。

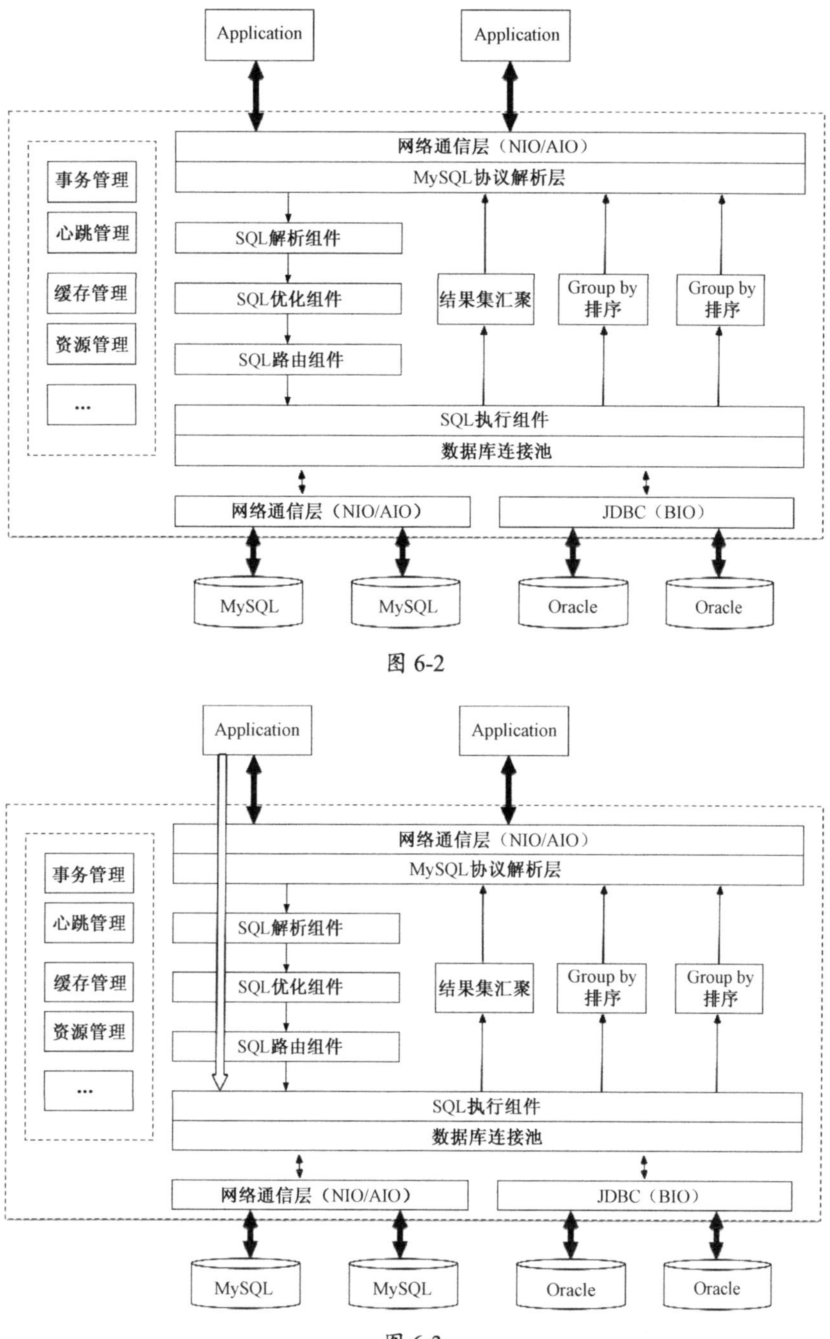
Application
Application
网络通信层（NIO/AIO）
MySQL协议解析层
事务管理
心跳管理
缓存管理
资源管理
...
SQL解析组件
SQL优化组件
SQL路由组件
结果集汇聚
Group by 排序
Group by 排序
SQL执行组件
数据库连接池
网络通信层（NIO/AIO）
JDBC（BIO）
MySQL
MySQL
Oracle
Oracle
Application
Application
网络通信层（NIO/AIO）
MySQL协议解析层
事务管理
心跳管理
缓存管理
资源管理
...
SQL解析组件
SQL优化组件
SQL路由组件
结果集汇聚
Group by 排序
Group by 排序
SQL执行组件
数据库连接池
网络通信层（NIO/AIO）
JDBC（BIO）
MySQL
MySQL
Oracle
Oracle

图 6-2

图 6-3

由后端数据库返回数据，通过协议解析后发送至回调模块。如果是涉及多节点的数据，则执行流程将会先进入结果集汇聚、排序等模块中，然后将处理后的数据通过通信协议模块返回到客户端（见图 6-4)。

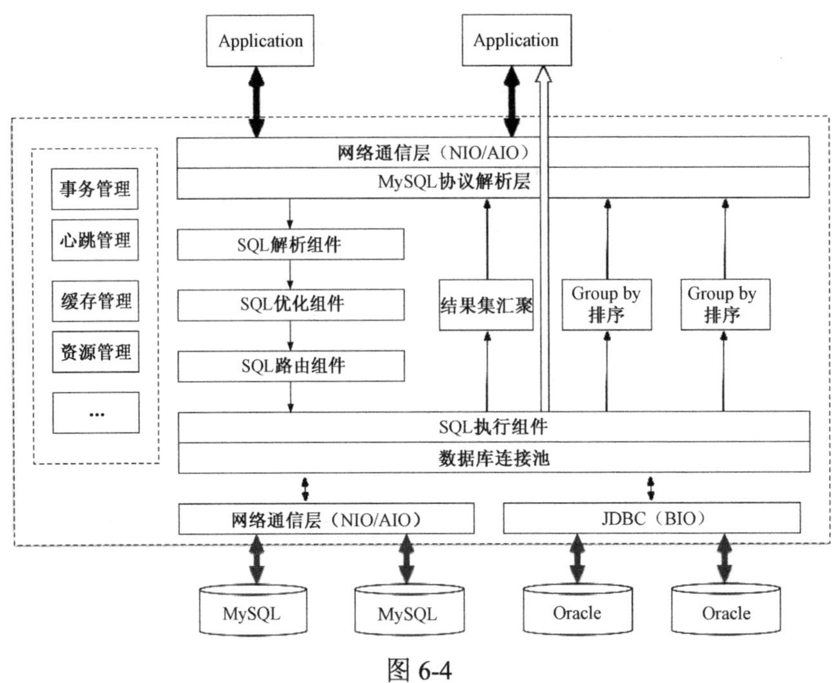

图 6-4

6.2 Mycat 网络 I/O 架构与实现

6.2.1 Mycat I/O 架构概述

在讨论 Mycat 的 I/O 架构前，我们先回顾一下网络通信中的 BIO、NIO 和 AIO。

BIO（阻塞 I/O）通常由一个独立的 Acceptor 线程负责监听客户端的连接，它在接收到客户端的连接请求后，会为每个客户端创建一个新的线程进行链路处理，处理完成之后，再通过输出流返回给客户端，完成后线程销毁，这就是典型的请求-应答通信模型。每个客户端连接过来后，服务端都会启动一个线程去处理该客户端的请求，阻塞 I/O 的通信模型如图 6-5 所示。

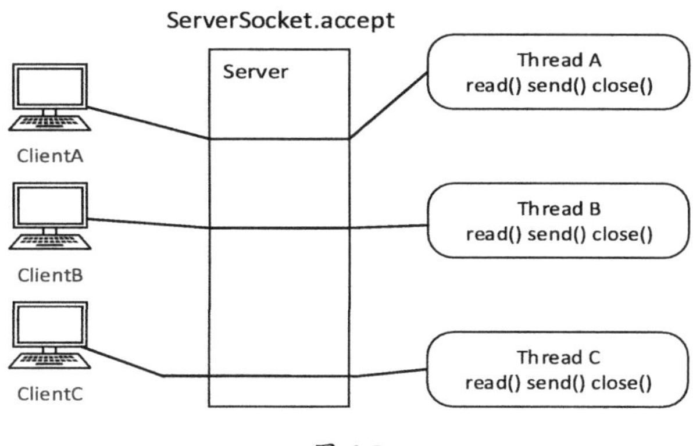

图 6-5

BIO 的主要缺点在于每当有一个新的客户端请求接入时，服务端必须创建一个新的线程处理新接入的客户端连接，一个线程只能处理一个客户端连接。当客户端变多时，就会创建大量的处理线程，且每个线程都要占用栈空间和一些 CPU；与此同时阻塞可能带来频繁的上下文切换，大部分上下文切换可能是无意义的。在高性能服务器应用领域，往往需要面对成千上万个客户端的并发连接，而这种模式显然无法满足高性能、高并发接入的要求，在这种情况下，非阻塞式 I/O 就有了它的应用场景。

NIO 基于 Reactor，当 Socket 有流可读或可写入 Socket 时，操作系统会通知相应的应用程序进行处理，应用程序再将流读取到缓冲区或写入操作系统。这时已经不是一个连接对应一个处理线程了，而是一个有效的请求对应一个线程，当连接没有数据时，就没有工作线程来处理。如图 6-6 所示。

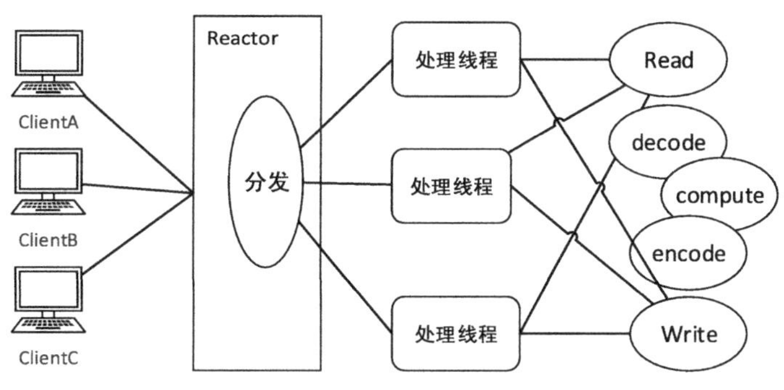

图 6-6

NIO 采用了双向通道（channel）进行数据传输，而不是单向的流（stream），在通道中我们可以注册以下四种事件中我们感兴趣的事件，如表 6-1 所示。

表 6-1

事　件　名	对 应 的 值
服务端接收客户端连接事件	SelectionKey.OP_ACCEPT(16)
客户端连接服务端事件	SelectionKey.OP_CONNECT(8)
读事件	SelectionKey.OP_READ(1)
写事件	SelectionKey.OP_WRITE(4)

NIO 的实现主要使用 Reactor 模式作为底层的通信模型，Reactor 模式可以将事件驱动的应用进行事件分派，将客户端发送过来的服务请求分派给合适的处理类（handler）。服务端和客户端各自维护一个管理通道的对象，我们称之为 Selector，该对象能检测一个或多个通道（channel）上的事件。NIO 的最大优点体现在线程轮询地访问 Selector，当 read 或 write 到达时则处理，未到达时则继续轮询。

AIO 与 NIO 的主要区别在于回调与轮询，客户端不需要关注服务处理事件是否完成，也不需要轮询，只需要关注自己的回调函数。熟悉 jQuery 的 Ajax 方法的程序员都知道，在提交一个 Ajax 请求后，无论成功与否，都有一个回调函数去处理。

Proactor 读取消息时会先检查队列是否为空，如果为空则发起 read 异步调用，并注册 CompletionHandler，然后返回。操作系统负责将消息写入并返回结果（写入的字节数）给 Proactor，Proactor 派发给 CompletionHandler。由此可见，写入的工作是操作系统处理的，无须用户线程参与，此时 AsynchronousChannelGroup 就扮演了 Proactor 的角色。CompletionHandler 有三个方法，分别对应处理成功、失败、被取消（通过返回的 Future）情况下的回调处理，如图 6-7 所示。

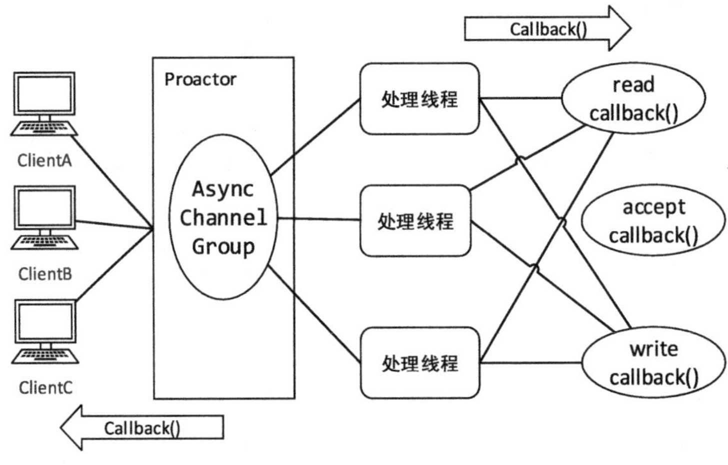

图 6-7

借助 Java 1.7 的 API，我们来看一下实现 AIO 的主要类。

- java.nio.channels.AsynchronousChannel 标记一个 channel，支持异步 I/O 操作。
- java.nio.channels.AsynchronousServerSocketChannel 是 ServerSocket 的 AIO 版本，创建 TCP 服务端、绑定地址、监听端口等。
- java.nio.channels.AsynchronousSocketChannel 面向流的异步 socket、channel，表示一个连接。
- java.nio.channels.AsynchronousChannelGroup 是对异步 channel 的分组管理，目的是资源共享。
- java.nio.channels.CompletionHandler 是异步 I/O 操作结果的回调接口，用于定义在 I/O 操作完成后所做的回调工作。

AIO 的 API 允许采用两种方式来处理异步操作的结果：返回 Future 模式或者注册 CompletionHandler。

在 Mycat 中实现了 NIO 与 AIO 这两种 I/O 模式，下面进行讲解。

6.2.2　前端通信框架

在 Mycat 中 I/O 是对外窗口，是与外部应用进行交互的基础，这部分的设计关系到系统的性能和稳定，那么 Mycat 的 I/O 是如何设计的呢？

Mycat 前端支持两种 I/O 架构，在 server.xml 加入配置：

```
<property name="usingAIO">1</property>
```

通过配置能够选择 AIO 或者 NIO 模式。我们先探讨 Mycat 的 AIO 架构。在启动过程中还涉及连接池的管理线程架构等，将在之后的章节中进行讲解。

1. Mycat 的 AIO 架构

MycatStartup 是程序的入口，也是 I/O 的入口。进入程序后，主要任务交给了 MycatServer，在获取 Mycat Home 的路径和初始化日志后，就进入了 Startup。AIO 架构如图 6-8 所示。

初始化系统配置并获取用户的 usingAIO 配置、进程数，记录日志后，新建两个重要的连接：一个是管理后台的连接，一个是 Server 的连接。管理后台的默认端口是 9066，Server 的默认端口是 8066。

程序根据 usingAIO 选择进入 AIO 还是 NIO，我们在这里先讨论进入 AIO 分支的情况。

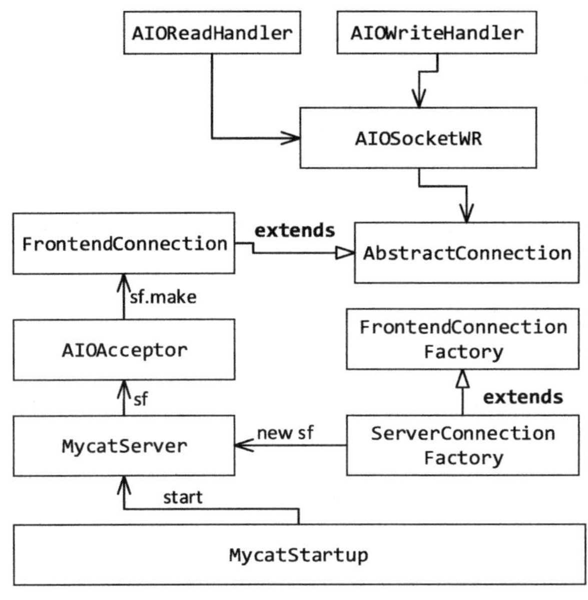

图 6-8

进入 AIO 分支后，主要由 AIOAcceptor 接收客户端的请求，绑定端口，创建服务端的异步 Socket，通过 AsynchronousServerSocketChannel.accept()进行写 accept 事件的注册，在 accept 中完成两件重要的大事：一是 FrontendConnection 的创建，这是前端框架连接的关键；二是 register 注册事件，MySQL 协议的握手包就在此时发送，一次连接只发送一次握手包，因为 AsynchronousServerSocketChannel 的 accept 方法注册的 completionHandler 只能被一次连接接入事件调用。

FrontendConnection 继承了 AbstractConnection，AbstractConnection 中根据配置启动 NIO 或 AIO 的读写通道。如果是 AIO，则启用 AIOSocketWR，AIOSocketWR 包含了两个内部类，分别是 AIOReadHandler、AIOWriteHandler。

AIOReadHandler 实现了 CompletionHandler<V,A>，用作 read 事件的用户句柄回调。根据 AsynchronousSocketChannel.read()的参数定义，对 AIOReadHandler 而言，V 已经固定为 Integer 类型表示读的字节数，A 可以自定义。

AIOReadHandler 的 completed 方法主要做两件事。

- 读 buffer 中的内容。
- 继续注册下一次读的回调句柄。

AIOWriteHandler 负责写操作，主要由 onWriteFinished 完成，业务线程发起写请求操作，当显式调用 AbstactConnection 时，若空闲则直接写，否则放入写队列等待。

2. Mycat 的 NIO 架构

在 Mycat 中是如何实现兼容两种 I/O 架构的呢？有两个关键之处。在 MycatServer 中有个 if 分支的判断，如图 6-9 所示。

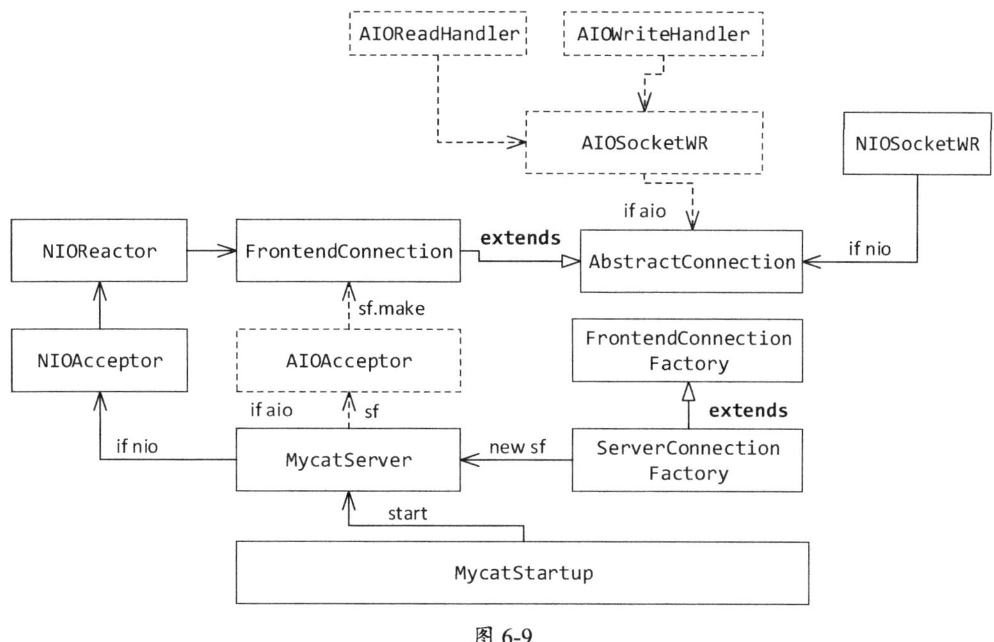

图 6-9

如果用户的设置是 usingAio 等于 0，那么将走 NIOAcceptor 通道，代码如下：

```
MycatServer.java
if (aio) {
aio...
}
else{
nio...
}
```

在 AbstractConnection 类中如果不是 AIO，那么将创建 NIO 的读写通道，代码如下：

```
if (isAIO) {
        socketWR = new AIOSocketWR(this);
     } else {
        socketWR = new NIOSocketWR(this);
     }
```

在上面的架构中 NIOAcceptor 负责处理 Accept 事件，服务端接收客户端的连接事件，就是 Mycat 作为服务端去处理前端业务程序发过来的连接请求，在连接建立后，NIOAcceptor 调用

NIOReactor.postRegister 进行注册。但是 NIOReactor.postRegister 并没有直接注册，而是把 AbstractConnection 对象加入缓冲队列，避免了加锁的竞争。

```
NIOReactor reactor = reactorPool.getNextReactor();
reactor.postRegister(c);
......
    final void postRegister(AbstractConnection c) {
        reactorR.registerQueue.offer(c);
        reactorR.selector.wakeup();
    }
```

NIOAcceptor 有三个重要的成员：Selector 是事件的选择器，它可以通过 open 方法创建选择器，然后利用 SelectionKey 对象注册选择通道到选择器，具体的事件有四种，见表 6-1；serverChannel 负责监听新进来的 TCP 连接的通道；reactorPool 负责分配 NIOReactor，当连接建立后，在 reactorPool 中分配一个 NIOReactor 用来处理 Read 和 Write，真正的内容读写是由 NIOSocketWR 和 NIOReactor 共同完成的。

6.3 Mycat 线程架构与实现

现在多线程的使用越来越广泛，各种分布式计算框架都毫无例外地使用了多线程、并行计算以提高程序执行的效率；同时多线程技术是一把双刃剑，设计得不好往往会出现死锁、线程泄露问题，并可能会冒出一些不知原因的怪异 Bug，所以设计好多线程的架构就成为系统性能的关键。

6.3.1 多线程基础

使用多线程架构往往是为了提高系统的性能或者因为业务的需要，那么究竟提高了多少效率呢，计算机界中的两个重要经验法则回答了这个问题。

1. Amdahl（阿姆达尔）定律

Amdahl 定律定义了串行系统并行化后的加速比计算公式：

加速比=优化前的任务时间/优化后的任务时间=1/((1-p)+p/n)

- p 就是代码中可以完全并行执行的部分。
- n 就是可用的执行单元的数量（处理器或者物理核心）。
- 加速比越高，表示优化效果越明显，很明显，随着核的增加，加速比会提高。但实际上，这只是一个理论加速比，因为不同核上执行的线程可能存在同步开销，操作系统也会因多核而产生额外的开销。

假设串行系统耗时如下：

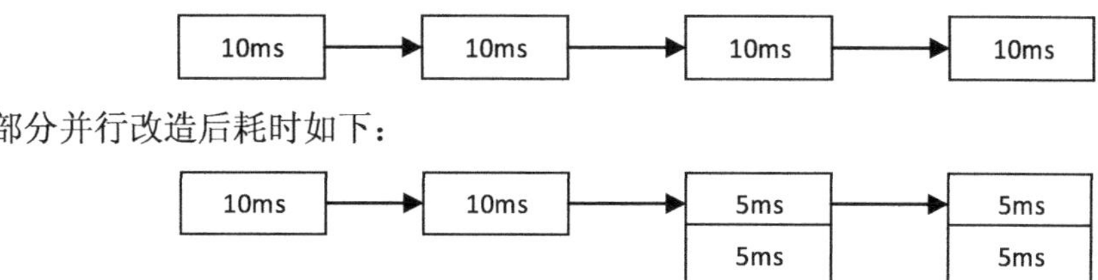

部分并行改造后耗时如下：

那么加速比为 40/30，即 1.33。

2. Gustafson 定律

系统优化某部件所获得的系统性能的改善程度，取决于该部件被使用的频率或所占总执行时间的比例。

Gustafson 定律定义了加速比：

$$加速比=采用改进措施前性能/采用改进措施后的性能=n-F(n-1)$$

其中，n 是处理器的个数，F 是程序中只能串行执行的比例。

假设串行比例比并行比例小得多，那么增加处理器的个数会使性能得到显著提高，如此看来是不是和 Amdahl 定律矛盾呢？其实不然，它们只是从不同的角度去描述问题。

Amdahl 定律强调串行比例一定时，加速比是有上限的，不管增加几个 CPU 都没有效果。

而 Gustafson 定律关注的是假设并行化的程序比例足够多，那么只要增加 CPU，性能就会呈线性增长。

线程池是一种多线程处理形式，在处理过程中将任务添加到队列，然后在创建线程后自动启动这些任务，解决了处理器单元内多个线程执行的问题，它可以显著减少处理器单元的闲置时间，增加处理器单元的吞吐能力。

当线程的创建时间和销毁时间远大于执行时间时，可以考虑使用线程池，一个线程池有四个基本组成部分。

- 线程池管理器（ThreadPool）：用于创建并管理线程池，包括创建线程池、销毁线程池、线程泄漏的控制、添加新任务。
- 工作线程：线程池中用于工作的线程。
- 任务接口：每个任务必须实现的接口，以供工作线程调度任务的执行。

- 任务队列：用于存放没有处理的任务，提供一种缓冲机制。

当从线程池中取出一个线程执行任务，在任务完成后该线程没有返回池中时，我们称之为线程泄露，随着泄露数量的增多，当线程池中没有可用的线程时，系统也就停止了动作。

当一个线程 A 持有对象 X 的独占锁，在等待对象 Y 的锁，而另一线程 B 持有对象 Y 的独占锁，却在等待对象 X 的锁时，它们将永远等待下去，此时我们称之为死锁。线程设计时应该考虑到死锁的情况，并有机制打破死锁。

线程池大小的设置一般与处理器的数量相等，当线程的数量小于处理器的数量时，增加额外的线程会提高系统的整体处理能力；当线程的数量大于处理器的数量时，增加额外的线程对提高性能不起作用。

在 Mycat 中，线程池 NameableExecutor 通过继承 JDK 的 ThreadPoolExecutor 类实现，它是 java.uitl.concurrent.ThreadPoolExecutor 类，是线程池中最核心的一个类，对外提供管理任务执行、线程调度、线程池管理等服务。在该类中提供了 4 个构造方法，在 Mycat 的线程池中主要用了以下构造方法：

```
public ThreadPoolExecutor(int corePoolSize,int maximumPoolSize,long
 keepAliveTime,TimeUnit unit,BlockingQueue<Runnable>
 workQueue,ThreadFactory threadFactory)
```

构造器中各个参数的含义如下。

- corePoolSize：核心池的大小。
- maximumPoolSize：线程池的最大线程数，它表示在线程池中最多能创建多少个线程。
- keepAliveTime：表示线程没有任务执行时最多保持多久会终止。
- TimeUnit：线程池维护线程所允许的空闲时间的单位。
- workQueue：阻塞任务队列。
- threadFactory：线程工厂，主要用来创建线程。

6.3.2 Mycat 线程架构

在 Java 中有两类线程，一类是用户线程，一类是守护线程。守护线程是指在程序运行时在后台提供一种通用服务的线程，比如垃圾回收线程就是一个很称职的守护者；用户线程一般指运行在前台的由用户创建并为应用程序服务的线程，用户也可以在编写程序时设置一个守护线程。

这里不讨论 JVM 系统的守护线程，仅讨论 Mycat 的线程。在 Mycat 中 NIO 与 AIO 线程结构略有不同，下面主要以 NIO 的线程结构进行讲解。Mycat 的线程架构如图 6-10 所示。

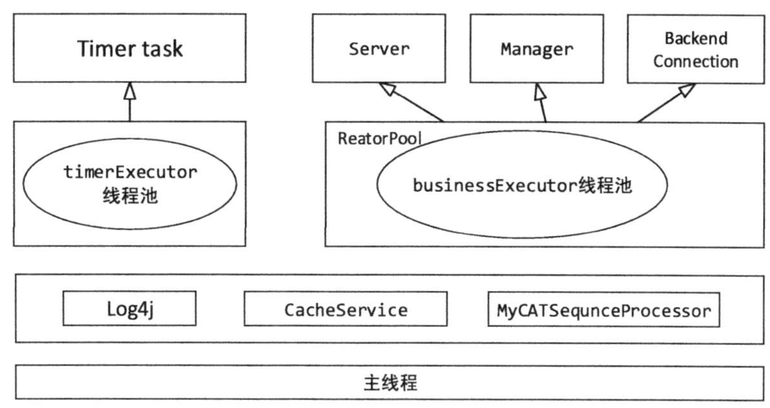

图 6-10

在 Mycat 中主要有两大线程池：timerExecutort 和 businessExecutor。

（1）timerExecutort 完成系统时间定时更新任务、处理器定时检查任务、数据节点定时连接空闲超时检查任务、数据节点定时进行心跳任务，用通俗的话来讲这个线程承担了后勤保障工作。

（2）businessExecutor 是 Mycat 最重要的线程资源池，该资源池的线程的使用范围非常广，具体如下。

- 后端用原生协议连接数据。
- JDBC 执行 SQL 语句。
- SQL 拦截。
- 数据合并服务。
- 批量 SQL 作业。
- 查询结果的异步分发。
- 基于 guava 实现异步回调。

通过以上列表我们可以看到，businessExecutor 线程池确实很重要，Mycat 的运作离不开它，那么该线程池是如何创建的呢？

从本质上看 Mycat 利用了 JDK 自带的 ThreadPoolExecutor 创建线程，创建的过程比较简单，这里再简单地介绍一下它的工作原理。ThreadPoolExecutor 将根据 corePoolSize 和 maximumPoolSize 设置的参数自动调整线程池的大小，当新任务在 execute(java.lang.Runnable)方法中提交时，如果运行的线程少于 corePoolSize，则创建新线程来处理请求，即使其他辅助线程是空闲的。如果运行的线程多于 corePoolSize 而少于 maximumPoolSize，则仅当队列满时才创建新线程。如果将 maximumPoolSize 设置为基本的无界值（如 Integer.MAX_VALUE），则允许线程池适应任意数量的并发任务。在大多数情况下，核心池和最大池的大小仅基于构造参数来设置。Mycat 设置的 corePoolSize 和 maximumPoolSize 相同，所以它是一个固定大小的线程池，由如下代码可以看到核心池和最大池来自同一个参数 size。

```
public class NameableExecutor extends ThreadPoolExecutor {
    protected String name;
    public NameableExecutor(String name, int size, BlockingQueue<Runnable> queue,
ThreadFactory factory) {
        super(size, size, Long.MAX_VALUE, TimeUnit.NANOSECONDS, queue, factory);
        this.name = name;
    }
    ……
}
```

从以上代码中可以看到有一个参数是 factory，这是 Mycat 线程池在创建的过程中自定义的线程工厂。当然也可以使用 JDK 默认的工厂，但是有时候我们需要对每一个线程命名、添加唯一的识别 ID 号，设置是否守护线程及判断归属于哪个线程组，所以 Mycat 使用了自定义的工厂给每个线程“穿上美丽的外套”。

当最大线程池（maximumPoolSize）和工作队列容量已经饱和时，新提交的任务将被拒绝。在 JDK 中有如下几种拒绝策略。

- 遭到拒绝时将抛出运行时异常 RejectedExecutionException。
- 直接调用 run 方法并且阻塞执行。
- 直接丢弃后来的任务。
- 丢弃在队列中队首的任务。

除了上面的 4 种拒绝策略，用户还可以自定义拒绝策略，在 Mycat 1.5 中使用的拒绝策略比较简单，使用了上面的第 1 种拒绝策略，也是 JDK 默认的拒绝策略。

Mycat AIO 使用了 withFixedThreadPool 创建线程池，这是 JAVA NIO 包中自带的创建线程池的方法，但它在本质上是使用 Executors.newFixedThreadPool()创建线程池，而 Executors.newFixedThreadPool()在本质上又是使用 ThreadPoolExecutor 创建一个核心池与最大池固定的线程池。

6.4 Mycat 内存管理及缓存架构与实现

这里讨论的内存管理指的是 Mycat 缓冲区管理，众所周知设置缓冲区的唯一目的是提高系统的性能。缓冲区通常是将部分常用或者需要预取的数据存放在缓冲池中以便系统直接访问，避免使用磁盘 I/O 访问磁盘数据，而 I/O 操作是相当耗时的，所以合理利用缓冲池的空间可以大大提高系统的性能。

6.4.1　Mycat 内存管理

Mycat 的缓冲区采用的是 java.nio.ByteBuffer，由 BufferPool 类统一管理，缓存区分为直接缓冲区和非直接缓冲区，简单地理解非直接缓冲区是 JVM 分配的内存，直接缓冲使用的是操作系统的内存，对 JVM 内存没有影响。直接缓冲区是 I/O 的最佳选择，但可能比创建非直接缓冲区要花费更多的成本。直接缓冲区使用的内存是通过调用本地操作系统方面的代码分配的，ByteBuffer.allocateDirect()创建直接缓冲区，通过 isDirect()方法的返回值可以判断是否为直接缓冲区。

1. Mycat 缓冲池的组成

缓冲池的最小单位是 chunk，默认的 chunk 大小（DEFAULT_BUFFER_CHUNK_SIZE）为 4096 字节，BufferPool 的总大小为 DEFAULT_BUFFER_CHUNK_SIZE×processors×1000，其中 processors 为处理器的数量。对 I/O 进程而言，它们共享一个缓冲池。缓冲池有两种类型：本地缓存线程缓冲区和其他缓冲区，其中本地缓存线程指的是线程名以“$_”开头的线程，分配 buffer 时优先获取 ThreadLocalPool 中的 buffer，没有命中时会获取 BufferPool 中的 buffer。

```
        public static final String LOCAL_BUF_THREAD_PREX = "$_";
        private  final ThreadLocalBufferPool  localBufferPool;

        private final int chunkSize;
        private final ConcurrentLinkedQueue<ByteBuffer> items = new
ConcurrentLinkedQueue<ByteBuffer>();
        private long sharedOptsCount;
        //private volatile int newCreated;
           private AtomicInteger newCreated = new AtomicInteger(0);
        private final long threadLocalCount;
        private final long capactiy;
        private long totalBytes = 0;
        private long totalCounts = 0;
```

各变量代表的含义如下。

- chunkSize：每个 chunk 的大小。
- capacity：chunk 的个数，计算方式为 BufferPool 的总大小/chunkSize。
- items：ByteBuffer 队列，初始大小为 capacity，其中每个 ByteBuffer 由 ByteBuffer.*createDirectBuffer* 创建，也就是说这个队列里都是直接缓冲区。

- threadLocalCount：本地线程数量，由 capacity 的某个比例计算得出，看起来相当于每个处理器分到的 chunk 个数。
- localBufferPool：本地线程缓冲区，类型为继承了 ThreadLocal<BufferQueue>的 ThreadLocalBufferPool，在 BufferQueue 中包含了类似的 ByteBuffer 链表 items，其容量为 threadLocalCount。

2. 分配 Mycat 缓冲池

分配缓冲池时可以指定大小，也可以用默认值，它们分别由以下两种方法分配。

- allocate()：先检测是否为本地线程，当执行线程为本地缓存线程时，localBufferPool 取出一个可用的 buffer。如果不是，则从 ConcurrentLinkedQueue 队列中取出一个 buffer 进行分配，如果队列没有可用的 buffer，则创建一个直接缓冲区。
- allocate(size)：如果用户指定的 size 不大于 chunkSize，则调用 allocate()进行分配；反之则调用 createTempBuffer(size)创建临时非直接缓冲区。

分配缓冲池的代码片段如下：

```
public ByteBuffer allocate() {
        ByteBuffer node = null;
        if (isLocalCacheThread()) {
            // allocate from threadlocal
            node = localBufferPool.get().poll();
            if (node != null) {
                return node;
            }
        }
        node = items.poll();
        if (node == null) {
            //newCreated++;
            newCreated.incrementAndGet();
            node = this.createDirectBuffer(chunkSize);
        }
        return node;
    }
```

3. Mycat 缓冲池回收

回收时先判断 buffer 是否有效，有如下情况时缓冲池不回收。

- 不是直接缓冲区。
- buffer 是空的。

- buffer 的容量大于 chunkSize。

在满足了回收条件后进行 buffer clear 操作，实际上是将以下三个参数归位。

- position = 0。
- limit = capacity。
- mark = -1。

重置后判断 buffer 是否属于本地缓存线程，若 localBufferPool 还有空余容量，则将其放入，否则直接放入 items 中。

6.4.2 Mycat 缓存架构与实现

使用缓存的好处在于为某些特定场景节省二次计算的开销，省掉了路由计算的开销，直接从相应的缓存中获取结果。

1. 缓存框架的选择

Mycat 支持 ehcache、mapdb、leveldb 缓存，可过配置文件 cacheservice.properties 决定使用哪种缓存框架。

配置参考如下：

```
#used for Mycat cache service conf
factory.encache=org.opencloudb.cache.impl.EnchachePooFactory
#key is pool name ,value is type,max size, expire seconds
pool.SQLRouteCache=encache,10000,1800
pool.ER_SQL2PARENTID=encache,1000,1800
layedpool.TableID2DataNodeCache=encache,10000,18000
layedpool.TableID2DataNodeCache.TESTDB_ORDERS=50000,18000
```

在配置中可以指定各类缓存统一的类型、大小、过期时间，也可以指定缓存框架，其中 factory.encache 参数指定了由哪个缓存类实现 Mycat 缓存的接口（CachePool），还有 org.opencloudb.cache.impl.LevelDBCachePooFactory、org.opencloudb.cache.impl.MapDBCachePooFactory 两个选项，它们分别对应不同的缓存框架。

这个步骤是在缓存初始化时完成的，由 CacheService 类的 init()方法完成。

2. 缓存内容

Mycat 有路由缓存、表主键到 datanode 缓存、ER 关系缓存。

- **路由缓存**即 SQLRouteCache，根据 SQL 语句查找路由信息的缓存，为 CachePool 类型，它的 key 为虚拟库名+SQL 语句，value 为路由信息 RouteResultSe，该缓存只针对 select 语句，如果执行了之前已经执行过的某个 SQL 语句（缓存命中），那么路由信息就不需要重复计算了，直接从缓存中获取。由 org.opencloudb.route.RouteService 类中的 route()方法实现路由信息缓存的读取，在获取前都会首先通过 schema+SQL 查询缓存该路由信息是否存在，如果存在则直接取出，请参考 RouteService 的 route()方法中有关于此缓存的相关代码。
- **表主键到 dataNode 缓存**的目的在于当分片字段与主键字段不同时，直接通过主键值查询是无法定位具体分片的（只能全分片下发），所以设置之后就可以利用主键值查找到分片名，缓存的 key 是 id 的值，value 是节点名，请参考 MultiNodeQueryHandler 的 rowResponse()的代码。
- **ER 关系缓存**在 ER 分片时使用，而且在 insert 查询中才会使用缓存，当子表插入数据时根据父子关联字段确定子表分片，下次可以直接从缓存中获取所在的分片，key 为虚拟库名+SQL 语句，value 为 dataNode 名。

另外可以通过 9066 端口查看缓存信息，命令如下。

```
MySQL> show @@cache;
```

6.5 Mycat 连接池架构与实现

客户端请求数据库的时候，需要向数据库获得数据库连接，而数据库建立连接通常需要消耗相对较大的资源，用完之后关闭也需要相当大的开销。假设应用系统有 20 万的访问量，那么数据库服务器就需要创建、关闭 20 万次连接，这无形中造成数据库的资源浪费，且容易导致数据库服务器内存溢出、宕机等。

为了解决这些问题，可以采用数据库连接池技术。连接池技术的核心思想是连接复用，通过预先建立一个一定数据量的连接池及对池子的连接进行分配、释放、闲置管理等策略，使得该连接池中的连接可以高效、安全地复用，避免了数据库连接频繁建立、关闭的开销。采用连接池技术需要考虑的问题有并发问题、事务问题、连接创建、释放、闲置管理、关闭等。

6.5.1 Mycat 连接池

这里讨论的连接池主要是 Mycat 后端连接池，也就是 Mycat 后端与各数据库节点之间的连接架构。

Mycat 启动时首先初始化一个最小的连接池，也就是将一定数量的连接放在池子中，无论这些连接有没有被使用，它将一直留在池子中，初始时具体放多少连接到池子中由 dataHost 标签 minCon 的值决定，minCon 值默认是 10。Mycat 的最大连接数由 maxCon 决定，也就是说数据库的连接不能超过 maxCon，当池子的活动连接数达到最大的连接数时，其他请求将被加入队列中等待。

1. Mycat 连接池的创建

Mycat 按照每个 dataHost 创建一个连接池，根据 schema.xml 文件的配置取得最小的连接数 minCon，取得 minCon 后用一个 for 循环逐一进行初始化。在初始化每个连接时需要判断用户选择的是 JDBC 还是原生的 MySQL 协议，如果选择 MySQL 协议，则还需要判断用户选择的是 AIO 还是 NIO，创建成功后把该连接添加到连接池的队列中。在创建过程中，Mycat 启用一个线程池去完成这个工作，这个线程就是我们前面介绍过的 BusinessExecutor。

2. Mycat 连接池的分配

分配连接就是从连接池队列中取出一条连接，Mycat 的连接分配也不例外，不过在取出连接之前需要做一些工作。在取出一条连接时，Mycat 需要根据负载均衡的类型选择不同的数据源，因为连接与数据源绑在一起，所以 Mycat 需要知道读写的是哪些数据源，才能分配相应的连接。负载均衡类型由 balance 参数决定，主要判断用户是否开启了负载均衡、是否全部开启负载均衡等，详细情况请参考 dataHost 标签 balance 属性的说明。

取出数据源后，Mycat 还需要知道用户读写的是哪个 schema，因为一个 schema 对应一个连接队列，这个队列保存在 ConMap 中，这是一个 map 数据结构，schema 是 key，value 是连接队列。

根据 schema 取出相应的连接队列后，判断是否超出最大的连接限制，选择一条进行自动提交或手动提交连接。经过这些判断后，Mycat 最终以 poll 方式从队列中取出一条连接。

3. Mycat 连接池的释放

按理来说连接池的释放是用完便返回给连接池的，但是在释放前 Mycat 需要做一系列的操作。

Mycat 首先解除路由结果节点与连接的绑定，然后需要判断该连接是否属于自动提交，如果属于自动提交，则在检查连接的同步状态后将连接归还队列，如果不属于自动提交并且连接关闭，则需要回滚事务。

4. Mycat 连接池的管理与维护

Mycat 的任务 dataNodeConHeartBeatCheck 对所有数据源连接进行定期管理，它从 conMap

中取出所有的连接队列进行空闲、超时检查，dataNodeIdleCheckPeriod 参数决定了检查的间隔，默认时间间隔是 300 秒。

6.5.2 Mycat 连接池架构及代码实现

Mycat 启动时首先获得各个 dataHost 的 PhysicalDBPool，分别对各个 host 初始化，取得最小连接数 minCon 属性的值，再获取这个 dataHost 的 datasource，根据 datasource 的 getConnection 创建连接，初始创建连接数等于 minCon 的值，getConnection 先是从已有的连接队列中取，如果没有它就创建；如果没有连接，则 createNewConnection 方法从线程池 BusinessExecutor 分配线程并创建连接。

PhysicalDatasource 的 createNewConnection 方法是个抽象方法，它分别被 MySQLDataSource 和 JDBCDatasource 继承，这意味着后端有两种选择：一是选择原生 MySQL 协议；二是选择 JDBC。在 Mycat 中后端实现了 NIO 和 AIO，所以走原生 MySQL 协议的通信效率更高。当选择走 MySQL 协议时，MySQLDataSource 继承了 createNewConnection，实际上调用了 MySQLConnectionFactory 的 make 方法创建连接对象，创建成功后队列连接数加 1。如图 6-11 所示。

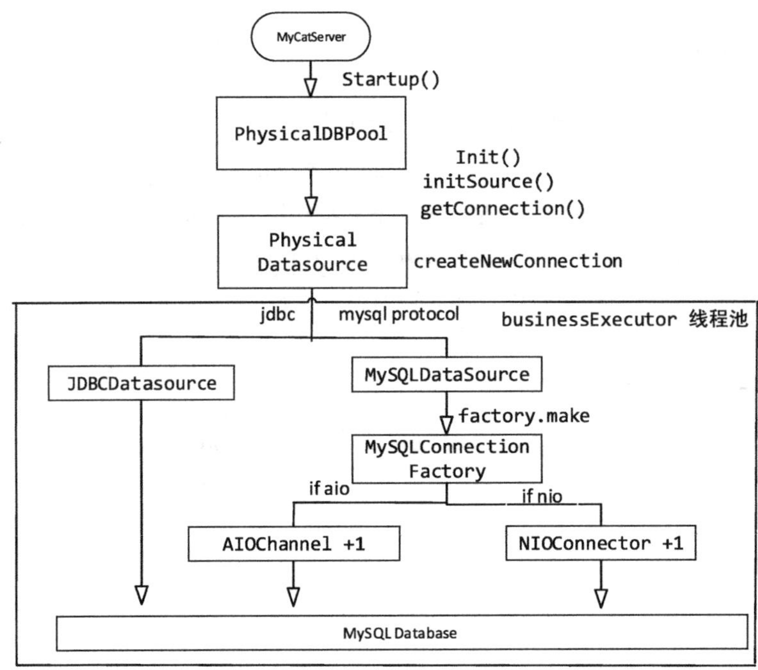

图 6-11

实现连接池的类在 org.opencloudb.backend 包下，如下所述。

- ConQueue 是一个连接池队列，有自动提交、手动提交两种类型。
- ConMap 存储着 schema 的 ConQueue，也就是说一个 schema 对应一个连接队列。
- PhysicalDatasource 数据源对象是一个抽象类，用于获取数据源连接。
- PhysicalDBPool 连接池对象。

另外，timer 线程负责对连接进行管理和维护，以及释放超时连接和闲连接处理等。

连接池的创建代码入口如下：

```
MycatServer.java
for (PhysicalDBPool node : dataHosts.values()) {
            String index = dnIndexProperties.getProperty(node.getHostName(),
                   "0");
            if (!"0".equals(index)) {
                LOGGER.info("init datahost: " + node.getHostName()
                       + "  to use datasource index:" + index);
            }
            node.init(Integer.valueOf(index));
            node.startHeartbeat();
}
```

连接池的维护代码入口如下：

```
MycatServer.java
// 数据节点定时连接空闲超时检查任务
    private TimerTask dataNodeConHeartBeatCheck(final long heartPeriod) {
        方法内容省略
}
```

6.6 Mycat 主从切换架构与实现

为了确保数据库产品的稳定性，很多数据库拥有双机热备功能，也就是第 1 台数据库服务器对外提供增删改业务；第 2 台数据库服务器主要进行读操作。它的原理就是让主数据库（master）处理事务性增、改、删操作（insert、update、delete），而从数据库（slave）处理 select 查询操作，利用数据库的同步机制将 master 的数据同步到 slave 上，而允许主从数据同步存在时间差。

实现数据库的读写分离时，人们关心的是当 master 宕机后 slave 承载的业务如何切换到 master 继续提供服务，以及 slave 宕机后如何将 master 切换到 slave 上。如何实现切换很简单，手动切换数据连接就行了，但手动切换一般比较被动而且不是运维工作的首选，所以本节重点讲解 Mycat 如何实现自动切换。

6.6.1 Mycat 主从切换概述

Mycat 的读写分离依赖于 MySQL 的主从同步，也就是说 Mycat 没有实现数据的主从同步功能，但实现了自动切换功能。

在 Mycat 中实现了以下三种类型的自动切换。

- 自动切换。
- 基于 MySQL 主从同步状态的切换。
- 基于 MySQL Galera Cluster 的切换（适用于集群，Mycat 1.4.1 以上版本支持）。

1. 自动切换

自动切换是 Mycat 主从切换的默认配置，当主机或从机宕机后，Mycat 自动切换到可用的服务器上。假设写服务器为 M，读服务器为 S，则

- 正常时，写 M 读 S；
- 当 M 挂了之后，读写 S；恢复 M 后，写 S 读 M；
- 当 S 挂了之后：读写 M；恢复 S 后，写 M 读 S。

自动切换的缺点是当 M、S 都没有宕机，但由于网络或种种原因，M 的数据没有及时同步到 S 上时，应用程序可能读到的是旧数据。这也是读写分离机制普遍遇到的问题。基于这个问题，Mycat 提出了第 2 种方案。

参考配置如下：

```
<dataHost name="localhost1" maxCon="1000" minCon="10" balance="1"
writeType="0" dbType="MySQL" dbDriver="native">
<heartbeat>select user()</heartbeat>
<!-- can have multi write hosts -->
<writeHost host="hostM1" url="localhost:3306" user="root"
password="123456">
<!-- can have multi read hosts -->
<readHost host="hostS1" url="localhost2:3306" user="root"
password="123456"
weight="1" />
</writeHost>
</dataHost>
```

或者

```
<dataHost name="localhost1" maxCon="1000" minCon="10" balance="1"
writeType="0" dbType="MySQL" dbDriver="native">
<heartbeat>select user()</heartbeat>
<!-- can have multi write hosts -->
<writeHost host="hostM1" url="localhost:3306" user="root"
password="123456">
</writeHost>
<writeHost host="hostS1" url="localhost:3307" user="root"
password="123456">
</writeHost>
</dataHost>
```

2. 基于 MySQL 主从同步延迟的切换

这种切换与自动切换不同的是，Mycat 检测到主从数据延迟时，会自动切换到拥有最新数据的 MySQL 服务器上，防止读到很久以前的数据。原理是通过 Mycat 的心跳检测 show slave status 中的 Seconds_Behind_Master、Slave_IO_Running、Slave_SQL_Running 三个字段，来确定当前主从同步的状态和 Seconds_Behind_Master 主从复制时延。Seconds_Behind_Master 为 0 时表示主从复制良好，没有延时差异；不为零时表示主从已经出现延时，数字越大表示从库落后主库越多。但是 Seconds_Behind_Master 只有在高速网络并且网络良好的状态下，才能正确地体现本身的意义。

MySQL 主从复制的原理是 slave 通过 I/O thread 从 master 上将 binlog 抽取到本地，然后通过 SQL thread 将 relaylog 重做，而 Seconds_Behind_Master 表示本地 relaylog 中未被执行完的那部分的差值。如果拉过来的日志全部被执行完，那么这个值为 0；如果没有被执行完，则表示有差异。但是在网络拥挤或通信环境不佳的情况下，主库的日志没能及时抽取到 slave，这时即使我们看到 Seconds_Behind_Master 为 0，主从也会有延迟，读到的数据也有可能为旧数据。所以利用 Mycat 实现主从同步延迟切换时，一定要保证网络处于高速稳定的环境下。

参考配置如下：

```
<dataHost name="localhost1" maxCon="1000" minCon="10" balance="0"
writeType="0" dbType="MySQL" dbDriver="native" switchType="2"
slaveThreshold="100">
<heartbeat>show slave status </heartbeat>
<!-- can have multi write hosts -->
<writeHost host="hostM1" url="localhost:3306" user="root"
password="123456">
</writeHost>
<writeHost host="hostS1" url="localhost:3316" user="root"
password="123456" />
</dataHost>
```

3. 基于 MySQL Galary Cluster 切换

Galera Cluster for MySQL 是一个基于同步复制多个主节点的 MySQL 集群解决方案，它具有高可用、可扩展性强、易于使用等特点，可以在任意节点读写，能够自动管理各个节点，可以移除有故障的节点，也可以自动加入新节点，实现了真正行级别的并发复制，客户端连接它就像连接普通单机 MySQL 一样。关于它的安装和说明请参考 http://galeracluster.com/products/。

在目前的开源中间件中，只有 Mycat 完美地支持了 Galera Cluster for MySQL 集群模式。

Mycat 从 1.4.1 版本开始支持 MySQL 集群模式，让读更加安全、可靠。心跳检查语句配置为 show status like 'wsrep%'，配置如下：

```
<dataHost name="localhost1" maxCon="1000" minCon="10" balance="0"
writeType="0" dbType="MySQL" dbDriver="native" switchType="3" >
<heartbeat> show status like 'wsrep%' </heartbeat>
<writeHost host="hostM1" url="localhost:3306"
user="root"password="123456">
</writeHost>
<writeHost
host="hostS1"url="localhost:3316"user="root"password="123456" ></writeHost>
</dataHost>
```

6.6.2 Mycat 主从切换的实现

数据节点的心跳任务监听各节点的健康状况，由 timer 线程执行；心跳分为由 JDBC 发起的心跳和由 MySQL 原生协议发起的心跳两种，这里主要介绍由 MySQL 原生协议发起的心跳。

为了避免同一时段内有重复的心跳任务执行，MySQL 原生协议发起的心跳首先检查心跳是否正在被执行。如果没有检查到，则会发起一个 SQL JOB 线程发送心跳的查询语句，接下来的任务由 SQLJOB 线程完成。SQL JOB 发送一条 SQL 语句，根据响应的结果判断心跳是否成功。如果响应失败，则记录失败记录，超过五次就切换数据源，如果响应成功，则需要判断用户配置的是哪种类型的切换。

- 如果是基于延迟的切换，则判断结果集中的字段 Slave_IO_Running、Slave_SQL_Running 是否都为 yes，以及 Seconds_Behind_Master 是否小于 slaveThreshold 的值，三者不能全部满足就切换。
- 如果是基于 MySQL Galera Cluster 的切换，则判断结果集中的字段 wsrep_connected、wsrep_ready 是否皆为 NO，以及 wsrep_cluster_status 是否为 Primary，三者不能全部满足就切换。

以上只是满足切换条件的判断，那么 Mycat 又如何实现切换呢？这里需要介绍一个关键的变量。Mycat 在初始化时把可用的数据源都存于 PhysicalDatasource 的数组中，在获取数据源时根据当前的活动索引号 activedIndex 从数组中取出一个可用的数据源，假设 activedIndex 为 0，那么把 activedIndex 改为 1 即可实现切换，这就是切换的主要思路。

但 Mycat 实现切换要比上面的思路复杂得多。如果切换前 Mycat 判断 switchType 为-1，则表示不自动切换，切换函数结束。取出下一个数据，为了保证它是一个健康的数据源，需要判断心跳连接是否正常，如果不正常则再取下一个，由此循环。如果下一个数据源的心跳正常，那么就把 activedIndex 指向它，然后重新初始化这个数据源。当原来配置的 MySQL 写节点宕机恢复后，它只能作为一个从节点跟随一个新的主节点，主从需要重新配置，在线下手工完成这些动作后再加入 Mycat。

图 6-12 描述了它的主要流程。

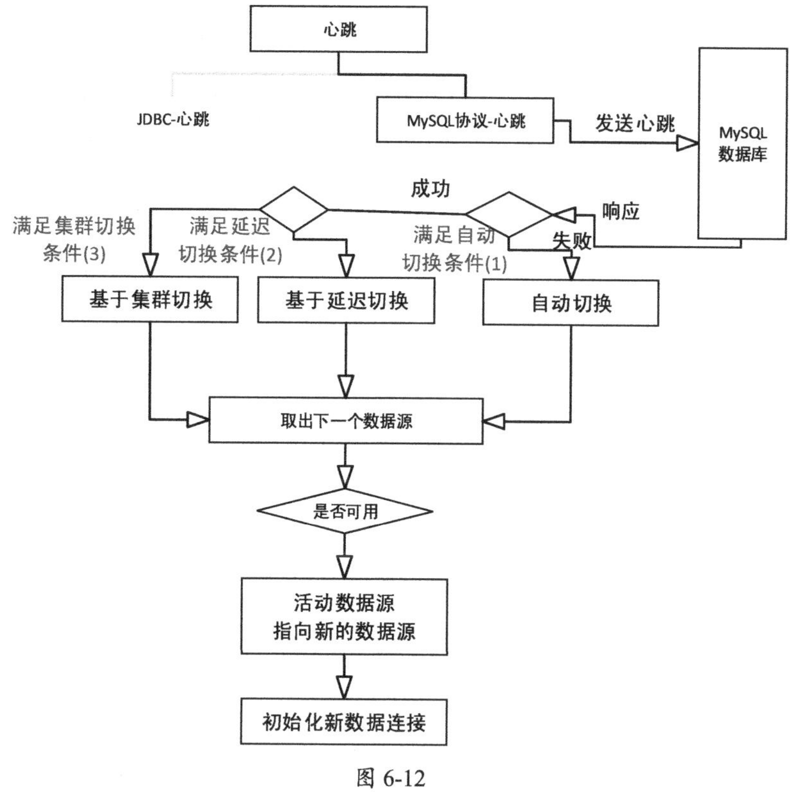

图 6-12

切换的核心代码如下。

（1）发送心跳后对结果集的判断代码如下：

MySQLDetector.java

```
@Override
   public void onResult(SQLQueryResult<Map<String, String>> result) {
 if (result.isSuccess()) {
        ......
 }else{
     heartbeat.setResult(MySQLHeartbeat.ERROR_STATUS, this,  null);
}
```

（2）数据源切换代码如下：

```
PhysicalDBPool.java
   public boolean switchSource(int newIndex, boolean isAlarm, String reason) {

   ......省略部分代码
          int current = activedIndex;
          if (current != newIndex) {
          // switch index
             activedIndex = newIndex;
          // init again
          this.init(activedIndex);
   ......
}
```

第 7 章 Mycat 核心技术分析

本章主要介绍 Mycat 核心技术。在了解 Mycat 分布式事务、路由解析、跨库实现并对业务深入理解后，结合 Mycat 的特点，在设计上就可以规避 Mycat 支持得不好的地方，提升系统的性能和稳定性。

7.1 Mycat 分布式事务的实现

随着并发量、数据量越来越大及业务已经细化到不能再按照业务划分，我们不得不使用分布式数据库提高系统的性能。在分布式系统中，各个节点在物理上都是相对独立的，每个节点上的数据操作都可以满足 ACID。但是，各独立节点之间无法知道其他节点事务的执行情况，如果想让多台机器中的数据保存一致，就必须保证所有节点上的数据操作要么全部执行成功，要么全部不执行，比较常规的解决方法是引入“协调者”来统一调度所有节点的执行。

7.1.1 XA 规范

X/Open 组织（即现在的 Open Group）定义了分布式事务处理模型。X/Open DTP 模型（1994）包括应用程序（AP）、事务管理器（TM）、资源管理器（RM）、通信资源管理器（CRM）四部分。事务管理器（TM）是交易中间件，资源管理器（RM）是数据库，通信资源管理器（CRM）是消息中间件。通常把一个数据库内部的事务处理看作本地事务，而分布式事务处理的对象是全局事务。全局事务是指在分布式事务处理环境中，多个数据库可能需要共同完成一个工作，这个工作就是一个全局事务。在一个事务中可能更新几个不同的数据库，此时一个数据库对自己内部所做操作的提交不仅需要本身的操作成功，还需要全局事务相关的其他数据库的操作成功。如果任一数据库的任一操作失败，则参与此事务的所有数据库所做的所有操作都必须回滚。XA 就是 X/Open DTP 定义的交易中间件与数据库之间的接口规范（即接口函数），交易中间件用它来通知数据库事务的开始、结束、提交、回滚等，XA 接口函数由数据库厂商提供，根据这一思想衍生出二阶段提交协议和三阶段提交协议。

7.1.2 二阶段提交

所谓的两个阶段是指准备阶段和提交阶段。

准备阶段指事务协调者（事务管理器）向每个参与者（资源管理器）发送准备消息，每个参与者要么直接返回失败消息（如权限验证失败），要么在本地执行事务，写本地的 redo 和 undo 日志但不提交，可以进一步将准备阶段分为以下三步。

（1）协调者节点向所有参与者节点询问是否可以执行提交操作（vote），并开始等待各参与者节点的响应。

（2）参与者节点执行询问发起为止的所有事务操作，并将 undo 信息和 redo 信息写入日志。

（3）各参与者节点响应协调者节点发起的询问。如果参与者节点的事务操作实际执行成功，则它返回一个“同意”消息；如果参与者节点的事务操作实际执行失败，则它返回一个“中止”消息。

提交阶段指如果协调者收到了参与者的失败消息或者超时，则直接向每个参与者发送回滚（Rollback）消息，否则发送提交（Commit）消息，参与者根据协调者的指令执行提交或者回滚操作，释放所有事务在处理过程中使用的锁资源。

二阶段提交所存在的缺点如下。

（1）同步阻塞问题，在执行过程中所有参与节点都是事务阻塞型的，当参与者占用公共资源时，其他第三方节点访问公共资源时不得不处于阻塞状态。

（2）单点故障，由于协调者的重要性，一旦协调者发生故障，则参与者会一直阻塞下去。

（3）数据不一致，在二阶段提交的第 2 个阶段中，当协调者向参与者发送 commit 请求之后发生了局部网络异常或者在发送 commit 请求的过程中协调者发生了故障，则会导致只有一部分参与者接收到了 commit 请求，而在这部分参与者在接收到 commit 请求之后就会执行 commit 操作，其他部分未接收到 commit 请求的机器则无法执行事务提交，于是整个分布式系统便出现了数据不一致的现象。

由于二阶段提交存在诸如同步阻塞、单点问题、数据不一致、宕机等缺陷，所以，研究者们在二阶段提交的基础上做了改进，提出了三阶段提交。

7.1.3 三阶段提交

三阶段提交（Three-phase commit，3PC），也叫作三阶段提交协议（Three-phase commit protocol），是二阶段提交（2PC）的改进版本。三阶段提交把二阶段提交的准备阶段再次一分为二，这样三阶段提交就有 CanCommit、PreCommit、DoCommit 三个阶段。

（1）CanCommit 阶段：三阶段提交的 CanCommit 阶段其实和二阶段提交的准备阶段很像，协调者向参与者发送 commit 请求，参与者如果可以提交就返回 Yes 响应，否则返回 No 响应。

（2）PreCommit 阶段：协调者根据参与者的反应情况来决定是否可以记录事务的 PreCommit 操作。根据响应情况，有以下两种可能。

- 假如协调者从所有参与者那里获得的反馈都是 Yes 响应，则执行事务。
- 假如有任何一个参与者向协调者发送了 No 响应，或者等待超时之后协调者都没有接到参与者的响应，则执行事务的中断。

（3）DoCommit 阶段：该阶段进行真正的事务提交，也可以分为执行提交、中断事务两种执行情况。

执行提交的过程如下。

- 协调者接收到参与者发送的 ACK 响应后，将从预提交状态进入提交状态，并向所有参与者发送 doCommit 请求。
- 事务提交参与者接收到 doCommit 请求之后，执行正式的事务提交，并在完成事务提交之后释放所有的事务资源。
- 事务提交完之后，向协调者发送 ACK 响应。
- 协调者接收到所有参与者的 ACK 响应之后，完成事务。

中断事务的过程如下。

- 协调者向所有参与者发送 abort 请求。
- 参与者接收到 abort 请求之后，利用其在第 2 个阶段记录的 undo 信息来执行事务的回滚操作，并在完成回滚之后释放所有的事务资源。
- 参与者完成事务回滚之后，向协调者发送 ACK 消息。
- 协调者接收到参与者反馈的 ACK 消息之后，执行事务的中断。

7.1.4 Mycat 中分布式事务的实现

Mycat 在 1.6 版本以后已经完全支持 XA 分布式强事务类型了，先通过一个简单的示例来了解 Mycat 中 XA 的用法。

用户应用侧（AP）的使用流程如下：

（1）set autocommit=0

在应用层需要设置事务不能自动提交；

（2）set xa=on

在 SQL 中设置 XA 为开启状态；

（3）执行 SQL

insert into travelrecord(id,name) values(1,'N'),(6000000,'A'),(321,'D'),(13400000,'C'),(59,'E');

（4）commit 或者 rollback

对事务进行提交（提交成功或者回滚异常）。

完整的流程图如图 7-1 所示。

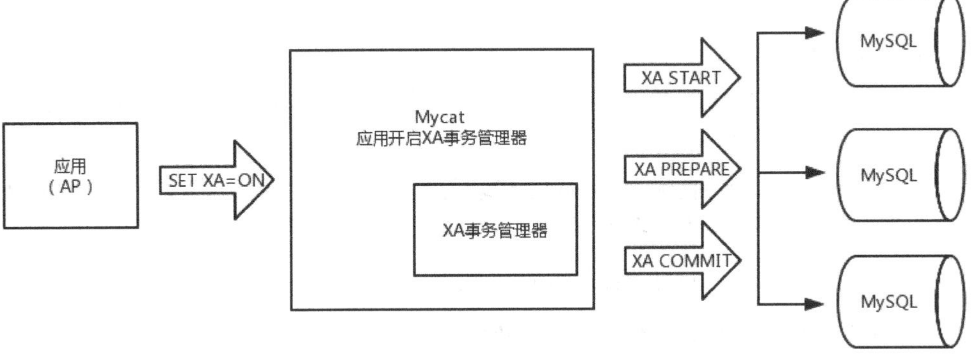

图 7-1

Mycat 内部实现侧的实现流程如下：

（1）set autocommit=0

将 MysqlConnection 中的 autocommit 设置为 false；

（2）set xa=on

在 Mycat 中开启 XA 事务管理器，用 MycatServer.getInstance().genXATXID()生成 XID，用

XA START XID 命令进行 XA 事务开始标记，继续拼装 SQL 业务（Mycat 会将上面的 insert 数据分片到不同的节点上），拼装 XA END XID，XA PREPARE XID 最后进行 1pc 提交并记录日志到 tm.log 中，如果 1pc 阶段有异常，则直接回滚事务 XA ROLLBACK xid。

（3）在多节点 MySQL 中全部进行 2pc 提交（XA COMMIT），提交成功后，事务结束；如果有异常，则对事务进行重新提交或者回滚。

Mycat 中的 XA 分布式事务的异常处理流程如下：

（1）一阶段 commit 异常：如果 1pc 提交任意一个 mysql 节点无法提交或者异常，则全部节点的事务进行回滚，抛出异常给应用侧事务回滚。

（2）Mycat Crash Recovery

Mycat 崩溃以后，根据 tm.log 事务日志再进行重启恢复，mycat 启动后执行事务日志查找各个节点中已经 prepared 的 XA 事务，进行 commit 或者 rollback。

1. 相关类说明

通过用户应用侧发送 set xa = on ; SQL 开启 Mycat 内部 XA 事务管理器的功能，事务管理器将对 MySQL 数据库进行 XA 方式的事务管理，具体事务管理功能的实现代码如下：

- MySQLConnection：数据库连接。
- NonBlockingSession：用户连接 Session。
- MultiNodeCoordinator：协调者。
- CommitNodeHandler：分片提交处理。
- RollbackNodeHandler：分片回滚处理。

2. 代码解析

XA 事务启动的源码如下：

```
public class MySQLConnection extends BackendAIOConnection {
    //设置开启事务
    private void getAutocommitCommand(StringBuilder sb, boolean autoCommit) {
        if (autoCommit) {
            sb.append("SET autocommit=1;");
        } else {
            sb.append("SET autocommit=0;");
        }
    }
     public void execute(RouteResultsetNode rrn, ServerConnection sc,
```

```
            boolean autocommit) throws UnsupportedEncodingException {
        if (!modifiedSQLExecuted && rrn.isModifySQL()) {
            modifiedSQLExecuted = true;
        }
        //获取当前事务 ID
        String xaTXID = sc.getSession2().getXaTXID();
        synAndDoExecute(xaTXID, rrn, sc.getCharsetIndex(), sc.getTxIsolation(),
                autocommit);
    }
    ……
```

……//省略此处代码，建议读者参考 GitHub 仓库的 MyCAT-Server 项目的 MySQLConnection.java 源码

}

用户应用侧设置手动提交以后，Mycat 会在当前连接中加入

```
SET autocommit=0;
```

将该语句加入到 StringBuffer 中，等待提交到数据库。

用户连接 Session 的源码如下：

```
public class NonBlockingSession implements Session {
    ……
```

……//省略此处代码，建议读者参考 GitHub 仓库的 MyCAT-Server 项目的 NonBlockingSession.java 源码

}

SET XA = ON ;语句分析

用户应用侧发送该语句到 Mycat 中，由 SQL 语句解析器解析后交由 SetHandle 进行处理

```
c.getSession2().setXATXEnabled (true);
```

调用 NonBlockSession 中的 setXATXEnabled 方法设置 XA 开关启动，并生成 XID,代码如下：

```
public void setXATXEnabled(boolean xaTXEnabled) {

    LOGGER.info("XA Transaction enabled ,con " + this.getSource());
    if (xaTXEnabled && this.xaTXID == null) {
        xaTXID = genXATXID();
    }
}
```

另外，NonBlockSession 会接收来自于用户应用侧的 commit，调用 commit 方法进行处理事务提交的逻辑。

在 commit()方法中，首先会 check 节点个数，一个节点和多个节点分为不同的处理过程，这里只讲下多个节点的处理方法 checkDistriTransaxAndExecute();

该方法会对多个节点的事务进行提交。

协调者的源码如下：

```
public class MultiNodeCoordinator implements ResponseHandler {
    ……
```

……//省略此处代码，建议读者参考 GitHub 仓库 MyCAT-Server 项目的 MultiNodeCoordinator.java 源码

}

在 NonBlockSession 的 checkDistriTransaxAndExecute()方法中，NonBlockSession 会话类会调用专门进行多节点协同的 MultiNodeCoordinator 类进行具体的处理，在 MultiNodeCoordinator 类中，executeBatchNodeCmd 方法加入 XA 1PC 提交的处理，代码片段如下：

```
for (RouteResultsetNode rrn : session.getTargetKeys()) {
……
    if (mysqlCon.getXaStatus() == TxState.TX_STARTED_STATE)
    {
      //recovery Log
    participantLogEntry[started] = new
    ParticipantLogEntry(xaTxId,conn.getHost(),0,conn.getSchema(),((MySQLConne
    ction) conn).getXaStatus());
      String[] cmds = new String[]{"XA END " + xaTxId,
          "XA PREPARE " + xaTxId};
      if (LOGGER.isDebugEnabled()) {
        LOGGER.debug("Start execute the batch cmd : "+ cmds[0] + ";" +
    cmds[1]+","+
              "current connection:"+conn.getHost()+":"+conn.getPort());
      }
      mysqlCon.execBatchCmd(cmds);
    }
……
}
```

在 MultiNodeCoordinator 类的 okResponse 方法中，则进行 2pc 的事务提交

```
    MySQLConnection mysqlCon = (MySQLConnection) conn;
    switch (mysqlCon.getXaStatus())
    {
      case TxState.TX_STARTED_STATE:
        if (mysqlCon.batchCmdFinished())
        {
          String xaTxId = session.getXaTXID();
          String cmd = "XA COMMIT " + xaTxId;
          if (LOGGER.isDebugEnabled()) {
            LOGGER.debug("Start execute the cmd :"+cmd+",current host:"+
                mysqlCon.getHost()+":"+mysqlCon.getPort());
          }
          //recovery log
          CoordinatorLogEntry coordinatorLogEntry =
    inMemoryRepository.get(xaTxId);
          for(int i=0; i<coordinatorLogEntry.participants.length;i++){
            LOGGER.debug("[In Memory
    CoordinatorLogEntry]"+coordinatorLogEntry.participants[i]);

    if(coordinatorLogEntry.participants[i].resourceName.equals(conn.getSchema
    ())){
              coordinatorLogEntry.participants[i].txState =
    TxState.TX_PREPARED_STATE;
            }
          }
          inMemoryRepository.put(session.getXaTXID(),coordinatorLogEntry);
    fileRepository.writeCheckpoint(inMemoryRepository.getAllCoordinatorLogEnt
    ries());
```

```
            //send commit
            mysqlCon.setXaStatus(TxState.TX_PREPARED_STATE);
            mysqlCon.execCmd(cmd);
        }
        return;
    ......
    }
```

分片事务提交处理的源码如下：

```
public class CommitNodeHandler implements ResponseHandler {
    //结束 XA
    public void commit(BackendConnection conn) {
         ......
......//省略此处代码，建议读者参考 GitHub 仓库 MyCAT-Server 项目的 CommitNodeHandler.java 源码
    }
    //提交 XA
    @Override
    public void okResponse(byte[] ok, BackendConnection conn) {
    ......
......//省略此处代码，建议读者参考 GitHub 仓库的 MyCAT-Server 项目的 CommitNodeHandler.java 源码
}
```

在 Mycat 中同样支持单节点 MySQL 数据库的 XA 事务处理，在 CommitNodeHandler 类中就是对单节点的 XA 二阶段处理，处理方式与 MultiNodeCoordinator 类同，通过 commit 方法进行 1pc 的提交，而通过 okResponse 的方法进行 2pc 阶段的事务提交。

分片事务回滚处理的源码如下：

```
public class RollbackNodeHandler extends MultiNodeHandler {
    ......
......//省略此处代码，建议读者参考 GitHub 仓库的 MyCAT-Server 项目的 RollbackNodeHandler.java 源码

}
```

在 RollbackNodeHandler 的 rollback 方法中加入了对 XA 事务的 rollback 处理，用户应用侧发起的 rollback 会在这个方法中进行处理。

```
for (final RouteResultsetNode node : session.getTargetKeys()) {
......
    //support the XA rollback
    MySQLConnection mysqlCon = (MySQLConnection) conn;
    if(session.getXaTXID()!=null) {
       String xaTxId = session.getXaTXID();
       mysqlCon.execCmd("XA END " + xaTxId + ";");
```

```
        mysqlCon.execCmd("XA ROLLBACK " + xaTxId + ";");
    }else {
        conn.rollback();
    }
......
}
```

同样，该方法会对所有的 MySQL 数据库节点发起 xa rollback 指令。

7.2　Mycat SQL 路由的实现

Mycat SQL 的路由是和 SQL 解析组件息息相关的，SQL 路由模块是 Mycat 数据库中间件最重要的模块之一，使用 Mycat 主要是为了分库分表，而分库分表的核心就是路由，在理解了路由模块的实现后就能明白分库分表这四个字的真正含义了。

7.2.1　路由的作用

我们先来看看图 7-2。

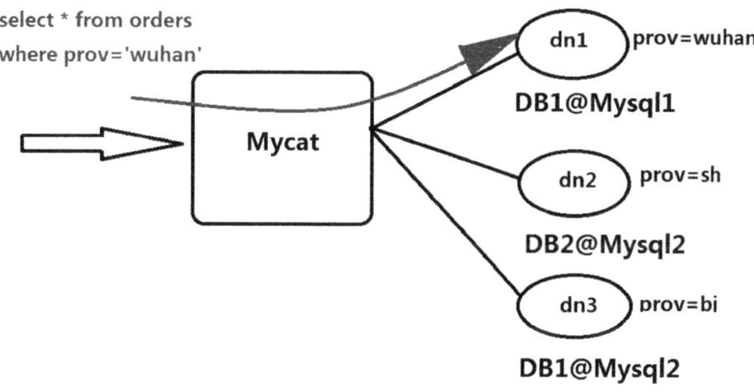

图 7-2

从图 7-2 来看，Mycat 在接收到应用系统发来的查询语句后，将其发送到后端连接的 MySQL 数据库服务器去执行，但是后端连接了三个数据库服务器，具体查询哪一台数据库服务器呢？这个就是 SQL 路由需要实现的功能。事件源的注册和事件的回调都由模式中的事件分享器完成。

SQL 路由要能保证 SQL 语句路由到相应范围的节点，这句话的意思是如果一个表有三个分片，也就是数据存储在三个节点上，则路由到两个节点时，查询出来的数据是缺失的；如果路由到四个节点，则会浪费资源，并且降低系统的性能和可能路由到一个不存在此表的节点。当然，路由到相应范围的节点的效率也不是最高的，所以设计好分片节点是在保证数据正确的情况下性能最高的分片方案。例如：orders 配置了 dn1、dn2、dn3 三个节点，Mycat 配置了 dn1、dn2、dn3、dn4 四个节点，查询语句 select * from orders 的正确范围是转发到 dn1、dn2、dn3 三个节点，如果只转发到一个或两个节点，则结果集不正确，如果转发到 dn1、dn2、dn3、dn4 四个节点，则报 table orders not exists。select * from orders where pro= 'wuhan' 语句只能在 dn1 节点查询到，而在 dn2、dn3 节点查询不到，如果同时转发到 dn1、dn2、dn3 三个节点，则会使性能较低。若要针对这个语句提高性能，则需要把 pro 设为分片字段，并且按照这个字段的第 1 个字母分片（假设 a..j 为 dn3，k..s 为 dn2，t..z 为 dn1），按这个规则分片，就只会路由到 dn1 这个节点，select * from orders where pro= 'sh' 只会路由到 dn2，select * from orders where pro= 'bi' 只会路由到 dn3 节点，当然具体的分片方案应该按照业务来设计，这样才更合理、更高效。

7.2.2 SQL 解析器

Mycat 在 1.3 版本以前默认使用 Fdbparser 这个 foundationdb 出品的开源 SQL 解析器，在 2015 年被苹果收购后，从开源变为闭源了。Mycat 在 1.4 版本中移除了这个解析器相关的代码，如果没有被收购，那么 Fdbparser 解析器也会被取代。我们接着来测试 SQL 解析器的性能。

目前采用 Java 开发的开源解析器有 JSQLParser（https://GitHub.com/JSQLParser/JSQLParser）和 Druidparser（https://GitHub.com/alibaba/druid/wiki/SQL-Parser），我们对 Fdbparser、JSQLParser、Druidparser 这三种解析器做个性能对比，对同一条 SQL 语句使用 3 种解析器解析出 ast 语法树（这是编译原理上的说法，SQL 解析式就是解析器自定义的 statement 类型），执行 10 万次、100 万次的时间对比。

```
package demo.test;
import java.io.StringReader;
import java.sql.SQLSyntaxErrorException;
import net.sf.jsqlparser.JSQLParserException;
import net.sf.jsqlparser.parser.CCJSqlParser;
import net.sf.jsqlparser.parser.CCJSqlParserManager;
import net.sf.jsqlparser.statement.Statement;
import org.opencloudb.parser.SQLParserDelegate;
import com.alibaba.druid.sql.ast.SQLStatement;
import com.alibaba.druid.sql.dialect.MySQL.parser.MySQLStatementParser;
import com.foundationdb.sql.parser.QueryTreeNode;
import com.foundationdb.sql.parser.SQLParser;
import com.foundationdb.sql.parser.SQLParserFeature;
```

```
    public class TestParser {
        public static void main(String[] args) {
            String sql = "insert into employee(id,name,sharding_id) values(5,
'wdw',10010)";
            int count = 1000000;
            long start = System.currentTimeMillis();
            System.out.println(start);
            try {
                for(int i = 0; i < count; i++) {
                    SQLParser parser = new SQLParser();

parser.getFeatures().add(SQLParserFeature.DOUBLE_QUOTED_STRING);
                    parser.getFeatures().add(SQLParserFeature.MySQL_HINTS);
                    parser.getFeatures().add(SQLParserFeature.MySQL_INTERVAL);
                    // fix 位操作符号解析问题 add by micmiu
                    parser.getFeatures().add(SQLParserFeature.INFIX_BIT_OPERATORS);
                    QueryTreeNode ast =parser.parseStatement(sql);
                    // QueryTreeNode ast = SQLParserDelegate.parse(sql,"utf-8" );
                }

            } catch (Exception e) {
                // TODO Auto-generated catch block
                e.printStackTrace();
            }
            long end = System.currentTimeMillis();
            System.out.println(count + "times parse,fdb cost:" + (end - start) + "ms");

            start = end;
            try {
                for(int i = 0; i < count; i++) {
                    //Statements stmt = CCJSqlParserUtil.parseStatements(sql);
                    Statement stmt =new CCJSqlParserManager().parse(new
StringReader(sql));
                }
            } catch (JSQLParserException e) {
                // TODO Auto-generated catch block
                e.printStackTrace();
            }
            end = System.currentTimeMillis();
            System.out.println(count + "times parse,JSQLParser cost:" + (end - start)
+ "ms");

            start = end;
            for(int i = 0; i < count; i++) {
                MySQLStatementParser parser = new MySQLStatementParser(sql);
                SQLStatement statement = parser.parseStatement();
            }
```

```
            end = System.currentTimeMillis();
            System.out.println(count + "times parse ,druid cost:" + (end - start) +
"ms");
        }
    }
```

输出结果如下：

```
1000000times parse,fdb cost:24468ms
1000000times parse,JSQLParser cost:11469ms
1000000times parse ,druid cost:1454ms
```

通过上面的代码测试，执行 100 万次查询，Druid 比 Fdbparser 快了 16 倍，比 JSQLParser 快了近 8 倍。从这里我们可以得出一个选择，目前 Mycat 采用 Druid 作为 SQL 解析器，测试数据表明采用 Druid 比采用 Fdbparser 的整体性能提高了 20%以上。

7.2.3 路由计算

通过 MySQL 可视化工具连接到 Mycat，并在可视化工具执行一条 SQL 语句时，将 SQL 语句通过网络传输发送到 Mycat 中，Mycat 首先会解析协议，然后调用 ServerQueryHandler.query 方法。Query 方法的定义非常简单，只需要传入参数为字符串类型的 SQL 语句就可以识别 EXPLAIN、SET、SHOW、SELECT 等命令。识别 SQL 语句后通过调用相应的 Handler 实现类执行，而查询、插入语句则会调用 ServerConnection.execute 方法，在此方法中为了获取 Schema 的配置信息，调用了 routeEndExecuteSQL 方法，执行 routeEndExecuteSQL 后才真正完成了路由计算和 SQL 执行。

路由计算的入口方法为 org.opencloudb.route.RouteService 类中的 route 方法。RouteResultset 是路由结果集，最后调用 session.execute 执行，流程图如图 7-3 所示。

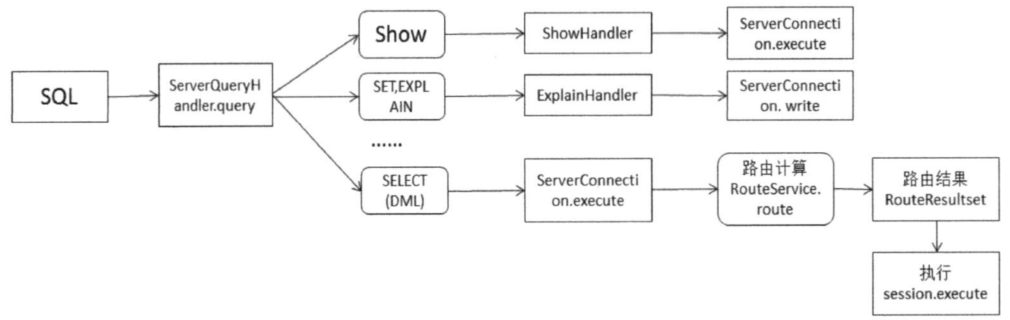

图 7-3

1. 路由接口

先来看看路由接口的定义：

```
public RouteResultset route(SystemConfig sysconf, SchemaConfig schema, int
sqlType, String stmt, String charset, ServerConnection sc) throws
SQLNonTransientException{…}
```

路由接口中包含 SystemConfig、SchemaConfig、sqlType、charset、ServerConnection 等其他输入参数，但对于路由计算来说，这些参数都不是最主要的参数。例如 SystemConfig 和 SchemaConfig 这两个参数完全可以不用传入，我们可以直接用其他方式获取，例如：

```
SystemConfig sysconf =MycatServer.getInstance().getConfig().getSystem();
SchemaConfig schema = MycatServer.getInstance().getConfig().getSchemas().
get(sc.getSchema());
```

这些参数可以是一些次要的参数，对路由计算本身是比较次要的，但是对其他流程有用。另外一个需要传入这些参数的原因是路由计算的流程比较复杂，耗时比较长，要经过很多方法的调用，如果在每个方法中都通过曲折的途径去计算和获取这些参数，则也是一种性能损耗。

2. 路由解析流程

路由解析的总体流程图如图 7-4 所示。

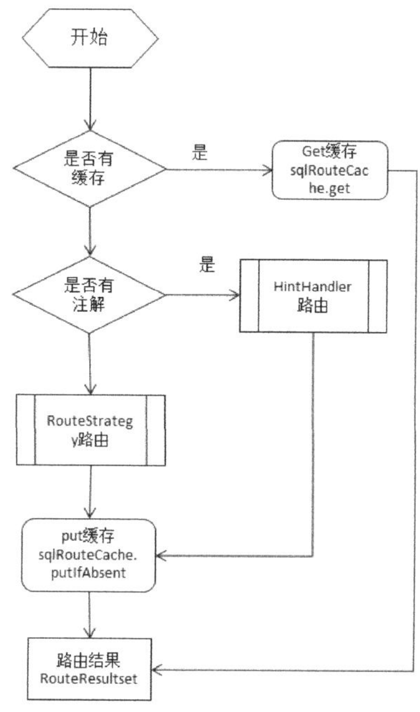

图 7-4

在进入 RouteService 中的 route 方法后，第 1 步是判断 SQL 的路由缓存是否命中，如果命中了，则直接返回路由结果，因为路由过程的计算比较耗时间和资源，所以把 SQL 语句的路由结果缓存起来，对性能有很大的提升；第 2 步是判断是否有注解，如果有，则进入 HintHandler.route 路由子流程。Mycat 支持两种路由注解，格式分别是/*!Mycat:……*/和新注解格式/*#Mycat:……*/，分别有 sql、schema、datanode、catlet 四种类型；第 3 步是直接进入 RouteStrategy 路由子流程。路由解析采用了策略模式，每种解析器实现了一种路由策略，非常方便做自定义扩展，目前只实现了 Druid 解析的 MySQL 策略，其他的暂时不支持。而路由策略工厂类 RouteStrategyFactory 原来支持 Fdbparser 和 Druidparser 这两种解析策略，Mycat 在 1.4 版本的时候移除了对 Fdbparser 的支持，目前只支持 Druidparser 解析策略。

Mycat 路由策略类如图 7-5 所示。

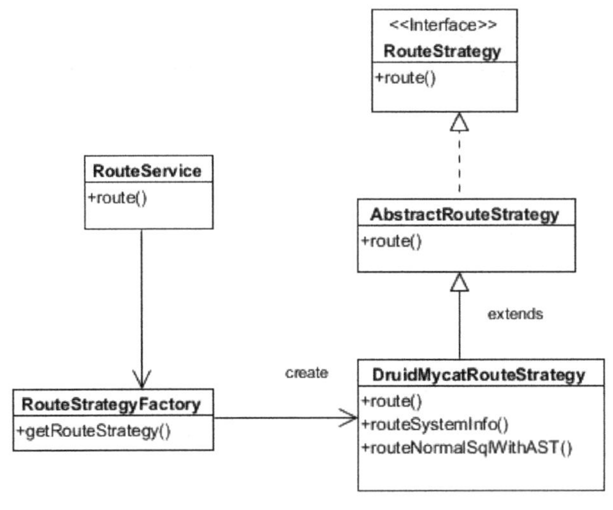

图 7-5

RouteStrategy 路由子流程图如图 7-6 所示。

RouteStrategy 流程封装在 AbstractRouteStrategy 类的 route 方法中，相当于策略的模板方法。在这个方法内部可以先处理一些路由之前的逻辑，比如在 insert 的时候要插入自动编号的 ID 或者 ER 分片子表插入的处理，具体内容可以查看 AbstractRouteStrategy.beforeRouteProcess（路由前的处理过程）；然后可以处理 SQL 的拦截器，通过拦截器可以记录 SQL 语句的操作，实现对 SQL 语句执行的监控；之后才是对 SQL 的解析。这里 SQL 解析有三个分支：（1）如果是没有分片的表，则直接进入单节点的路由（RouterUtil.routeToSingleNode）；（2）如果有分片表，则进入系统命令 routeSystemInfo 的解析方法；（3）如果分支 2 解析出来的路由结果集为空，则进入 AST 语法树解析的方法 routeNormalSqlWithAST。

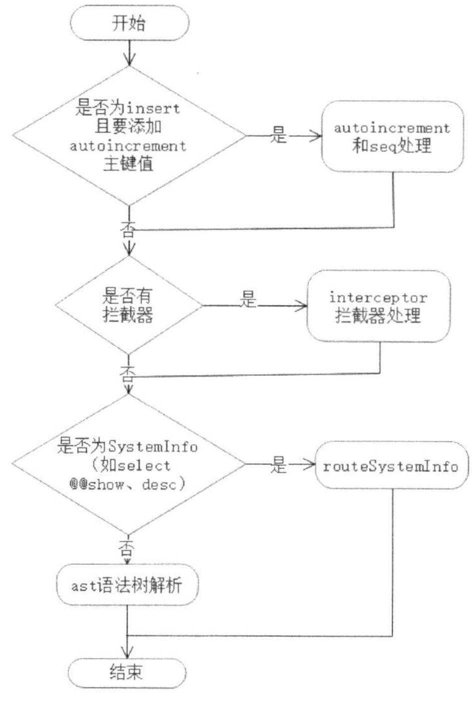

图 7-6

DruidMycatRouteStrategy 重载了 routeSystemInfo 和 routeNormalSqlWithAST 方法，routeNormalSqlWithAST 就是 DruidMycatRouteStrategy 的 AST 语法树解析流程，如图 7-7 所示。

通过流程我们可以看到会先判断是否支持多数据库类型，目前支持的数据库有 MySQL、Oracle、SQLServer、DB2、PostgreSQL，代码如下：

```
if (schema.isNeedSupportMultiDBType()) {
            parser = new MycatStatementParser(stmt);
        } else {
            parser = new MySQLStatementParser(stmt);
            }
```

如果支持，则创建 MycatStatementParser，否则创建 MySQLStatementParser。它们有什么区别呢？经过对代码进行简单分析，我们发现 MycatStatementParser 继承了 MySQLStatementParser 并做了简单扩展，用来支持 load data 的指令，因为 Druid 默认提供的 load data 解析有 Bug，所以这里替换为自己的解析实现。

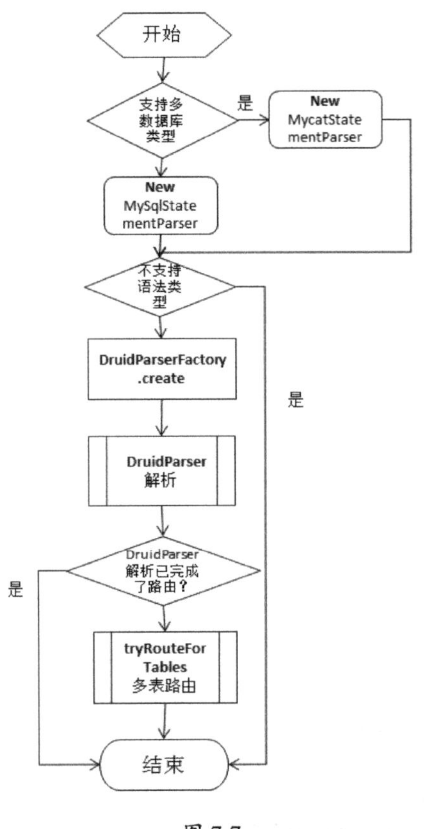

图 7-7

我们再来看看 DruidParser 工厂类 DruidParserFactory，创建 DruidParser 的定义是 public static DruidParser create(SchemaConfig schema,SQLStatement statement,SchemaStatVisitor visitor)，在此方法内部判断 SQLStatement 的类型是 SQLSelectStatement、MySQLInsertStatement、MySQLDeleteStatement、MySQLCreateTableStatement、MySQLUpdateStatement、MySQLAlterTableStatement，则分别采用相应的 DruidSelectParser、DruidInsertParser、DruidUpdateParser 等去解析。这里只有 SQLSelectStatement 类型比较特殊，它还是会先检查是否支持多数据库类型，如果支持，则先解析出表名，然后判断表所在的数据库类型，再根据数据库类型创建相应数据库的解析类 DruidSelectOracleParser、DruidSelectDb2Parser、DruidSelectSqlServerParser、DruidSelectPostgresqlParser，默认是 DruidSelectParser，支持 MySQL 数据库。它们之间的区别不大，只是针对不同的数据库支持相应的分页语法。比如支持 Oracle 的三层嵌套和 row_number 两种分页语法及 rownum 控制最大条数的语法；支持 SOLServer 的 row_number 和 row_number 与 top 结合两种分页；支持 top 限制最大条数；支持 DB2 的 row_number 分页和 fetch first rows only 语法。

在上述 AST 解析流程中我们可以看到有两个子流程，分别是 DruidParser 解析和多表路由 tryRouteForTables。

DruidParser 解析指的是利用 ast 语法树，把 Druid 解析器解析出来的表名、条件表达式、字段列表、值列表等信息组成 SQLStatement 对象，通过 DefaultDruidParser 类中的 parser 方法对 SQLStatement 对象进行路由计算。如图 7-7 所示。

DefaultDruidParser 类中的解析流程为：首先调用 visitor 方法进行解析，然后调用 statementParse 方法进行解析，最后调用 changeSql 方法进行 SQL 语句改写。在解析过程中通过把 RouteResultset 作为参数传递路由信息，通过调用 RouteResultset.isFinishedRoute()方法判断是否完成了路由，如果完成了，则 AST 语法树解析结束，否则进入多表路由 tryRouteForTables 子流程。

详细的 Druid 语法树解析相关类如图 7-8 所示。

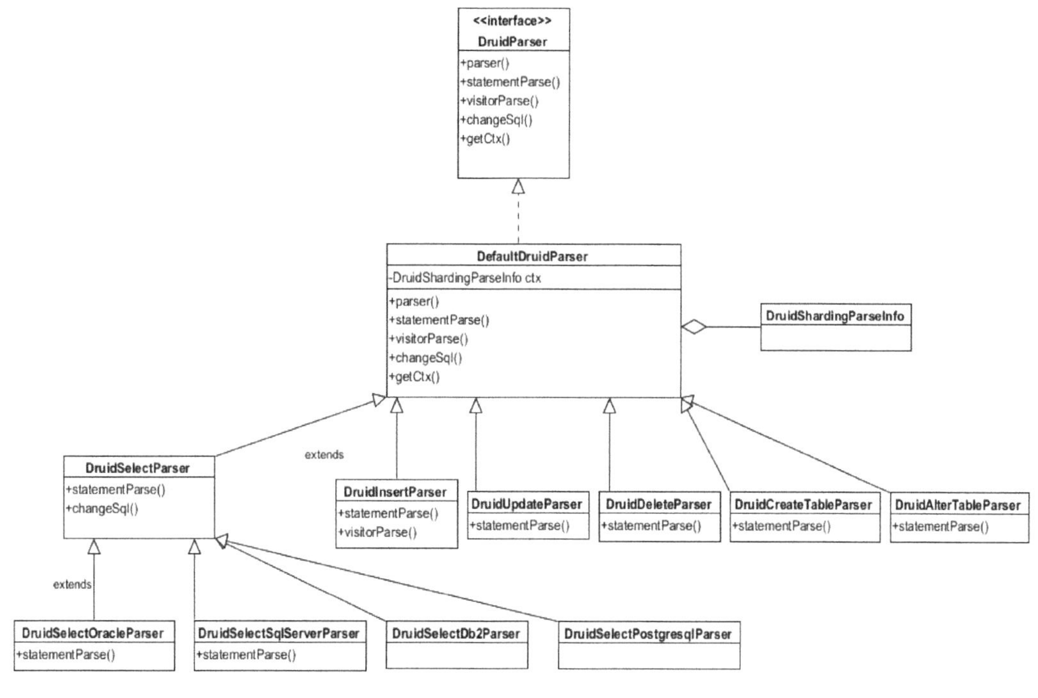

图 7-8

DruidParser 接口默认的实现类是 DefaultDruidParser。我们看一下 DruidParser 的接口方法，如表 7-1 所示。

表 7-1

方 法	说 明
parser	解析的入口方法
visitorParse	通过 visitor 解析，可以很方便地获取表名、条件、字段列表、值列表等，对各种语句的 statement 都适用
statementParse	statement 方式解析。子类覆盖该方法一般是将 SQLStatement 转型后再解析（如转型为 MySQLInsertStatement）
changeSql	该方法用来改写 SQL。如 select 语句加 limit，insert 语句加自增长值等，主要是为了规范代码结构
getCtx	获取解析结果，返回 DruidShardingParseInfo 对象。该对象包含解析到的表名列表、条件列表等信息，用于后续计算路由

由图 7-8 我们可以知道，针对不同类型的 SQL 有不同的实现：查询操作由 DruidSelectParser 实现；插入操作由 DruidInsertParser 实现；更新操作由 DruidUpdateParser 实现；删除表、创建表、修改表分别由 DruidDeleteParser、DruidCreateTableParser、DruidAlterTableParser 实现。继承自 DefaultDruidParser 的类基本上重载 statementParse 方法，而且 DruidSelectParser 默认支持 MySQL 查询语句。为了支持 DruidInsertParser，DruidUpdateParser 等其他解析类大部分重载了 statementParse 方法。为了兼容其他数据库的特殊分页语法，继承 DruidSelectParser 的类有 DruidSelectOracleParser、DruidSelectDb2Parser、DruidSelectSqlServerParser、DruidSelectPostgresqlParser，分别支持 Oracle、DB2、SQL Server、PostgreSQL。

最后只剩下一个多表路由 tryRouteForTables 的解析了，这个方法在 RouterUtil.java 类中，其流程图如图 7-9 所示。

首先判断是否为分片表，如果不是分片表，则判断为默认节点（在 Mycat 1.5 中新加的功能）。可以在 schema.xml 中配置一个默认节点，例如<schema name="TESTDB" checkSQLschema="false" sqlMaxLimit="1000" dataNode="dn1">，这里配置了默认节点 dn1，所有的未分片表都路由到 dn1 节点中，实现代码片段如下：

```
    public static RouteResultset tryRouteForTables(SchemaConfig schema,
DruidShardingParseInfo ctx,
                RouteCalculateUnit routeUnit, RouteResultset rrs, boolean isSelect,
LayerCachePool cachePool)
                        throws SQLNonTransientException {

            List<String> tables = ctx.getTables();
            if(schema.isNoSharding()||(tables.size() >= 1&&isNoSharding(schema,
tables.get(0)))) {
                return routeToSingleNode(rrs, schema.getDataNode(), ctx.getSql());
            }
```

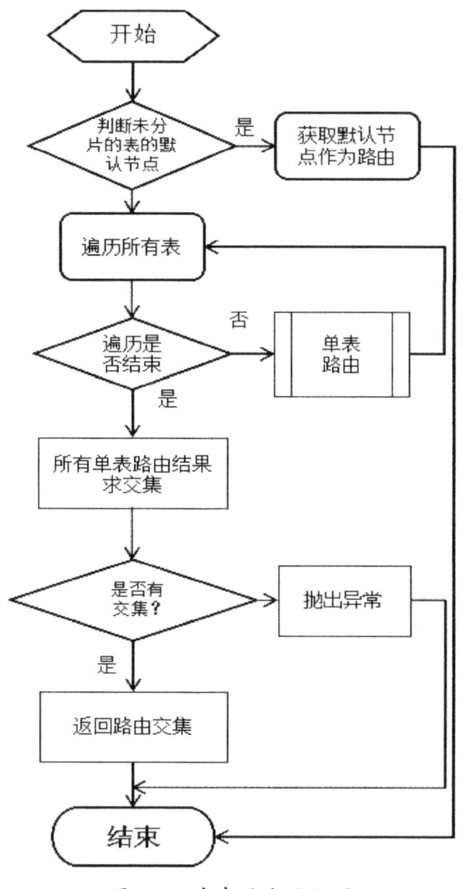

图 7-9　多表路由流程图

然后遍历所有的表，如果没有结束，则针对单表进入单表路由并保存路由结果，在遍历完成后判断单表路由结果的交集，如果有结果则返回，否则抛出异常。

单表路由也是首先判断是不是未分片且有默认的节点配置，代码如下：

```
if (isNoSharding(schema, tableName)) {
        return routeToSingleNode(rrs, schema.getDataNode(), ctx.getSql());
    }
```

如果判断是全局表，则再次判断是否是 select 语句，如果是 select 语句，则直接访问任意一个节点即可，因为全局表的数据都是一致的，否则就要返回所有的 DataNode 节点，因为更新和删除操作需要全部的 DataNode 同步改变。最后判断是否是分片表，若不是分片表，则可以直接返回路由结果，否则还要根据分片规则和条件语句判断来获取正确的路由结果。如图 7-10 所示。

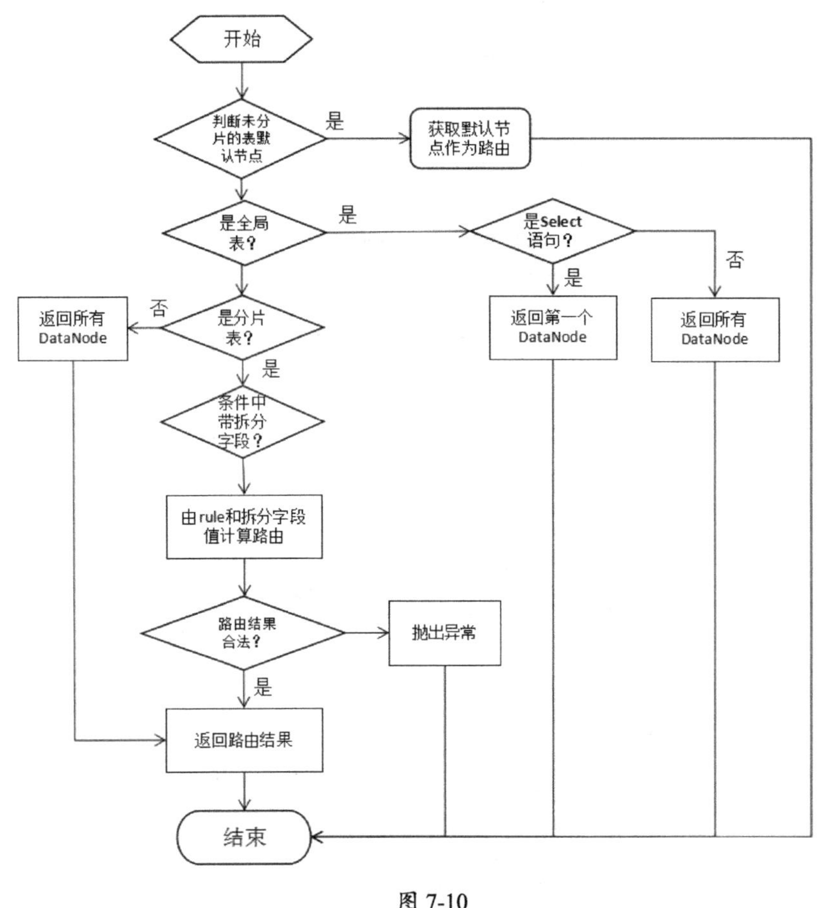

图 7-10

7.3 Mycat 跨库 Join 的实现

Join 操作是关系型数据库中最常用的一个特性，也是关系型数据库必备的功能，然而在分布式环境中，它是一个 S 级的难点，目前在市场上没有完美的解决方案。目前市场的关系型数据库对分布式 Join 的支持其实有限，还有各种各样的限制。

Mycat 对跨库 Join 的支持有四种实现方式，分别是全局表、ER 分片、catlet（HBT）和 ShareJoin，下面我们详细讲解这几种实现方式。

7.3.1　全局表

任何一个企业级的系统都会有很多基础信息表，它们的数据结构相对比较简单，类似于字典表，它们和业务表之间有关系，但是这种关系肯定不会是主从关系，大多数是一种属性关系，相当于配置信息，比如省份、城市、区域等。

当业务的数据量达到一定的规模，我们需要对业务表进行分片时，业务表和相关的基础信息之间的关系就变得非常麻烦，而这些基础信息表有如下一些特点。

- 数据的变化不频繁。
- 数据量的规模也比较小，在一般情况下为几万条记录，很少超过几十万条记录。

鉴于这些特点，Mycat 定义了一种特殊的表，叫作全局表。我们看看全局表是如何进行插入操作的，如图 7-11 所示。

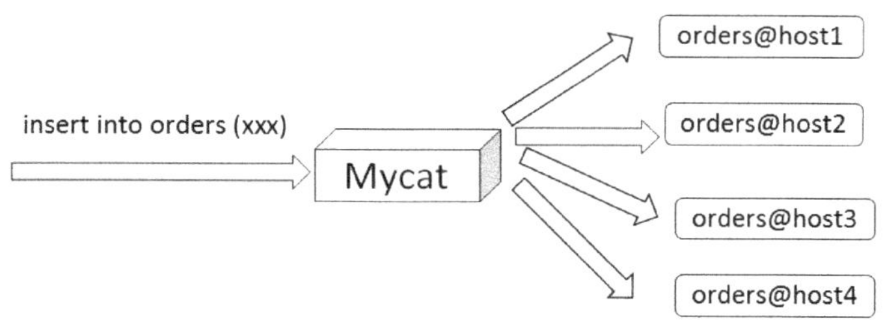

图 7-11

可以看出，在全局表中进行插入操作时，每个节点同时并发插入和更新数据，而在读取数据的时候，因为任意节点的数据都是相同的，所以任意节点都可以读，这样在提升读性能的同时解决了跨节点 Join 的效率。

全局表的重要特性如下。

- 全局表的插入、更新、删除操作会实时地在所有节点同步执行，保持各个分片的数据的一致性。
- 全局表的查询操作会从任意节点执行，因为所有节点的数据都一致。
- 全局表可以和任意表进行 Join 操作。

将基础信息表和字典表定义成全局表，从业务上来看，业务表分片后，不会影响它们之间的 Join 操作，这是解决跨库 Join 的一种实现，通过全局表再加上 E-R 关系的分片规则，Mycat 可以满足大部分企业级的应用系统。

7.3.2 ER 分片

数据库的 ER 关系模型是目前应用系统中经常使用到的，Mycat 参考了 NewSQL 领域的新秀 Foundation DB 的设计思路。Foundation DB 创新性地提出了 Table Group 的概念，子表的存储位置依赖于主表，并且在物理上紧邻存放，因此彻底解决了 Join 的效率和性能问题，并根据这一思路，提出了基于 E-R 关系的数据分片策略，子表的记录与所关联的父表记录存放在同一个数据分片上。

存在关联关系的父、子表在数据插入的过程中，子表会被 Mycat 路由到与其相关的父表记录的节点上，从而使父、子表的 Join 查询可以下推到各个数据库节点完成，这是最高效的跨节点 Join 处理技术，也是 Mycat 的首创，如图 7-12 所示。

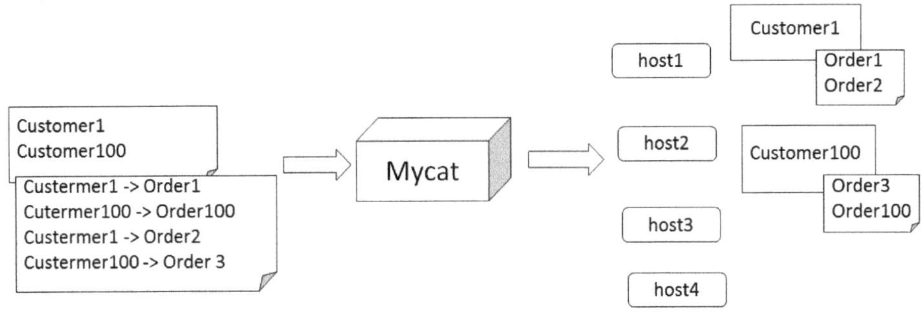

图 7-12 ER 分片示意图

Customer 表采用了按范围分片的策略，每个范围内的 Customer 在一个分片上，Customer1 和 Customer100 分片在 host1、host2 上。从图 7-12 可以知道，Customer1 的 Order1 和 Order2 订单在 host1 分片上，Customer100 的 Order3 和 Order100 订单在 host2 分片上，所以 orders 表的记录依赖父表进行分片，两个表的关联关系为 orders.customer_id=customer.id。

按照这样的分片规则，host1 分片上的 customer 与 orders 就可以进行局部的 Join 操作，在 host2 分片上也一样，就是在各自的分片上先做 Join，然后在 Mycat 内部合并两个节点的数据，这样就完成了跨分片的 Join。试想一下，每个分片上 orders 表有 1000 万条，所以 10 个分片就有 1 亿条，基于 E-R 映射的数据分片模式基本上解决了比较多的应用所面临的问题。

但是有很多业务存在多层 E-R 关系，对于这种情况，其实有种非常简单的方法，就是用空间换时间的概念，简单地说就是多层 E-R 关系如第二、三层的 E-R 关系，通过增加冗余字段来解决这个问题。比如，在上面的例子中还有个 Orders 的子表 Order_items，只要这个表新增一个字段 customer_id 就可以了，也就是把第二、三层的 E-R 关系转化为第一层的 E-R 关系进行分片。

7.3.3　catlet

catlet 是 Mycat 为了解决跨分片 Join 提出的一种创新思路，也叫作人工智能（HBT）。解决跨分片 Join 没有完美的方案，不但需要考虑分片数据获取的问题，还要考虑到性能问题，性能太低也是不可取的。Mycat 参考数据库中存储过程的实现方式，提供了类似的方案，用户可以根据系统提供的 API 接口实现跨分片 Join。HBT 的实现模板可以在 org.opencloudb.sqlengine 包路径下找到。

- catlet 接口：需要按照业务逻辑自己实现。
- EngineCtx：核心类 HBT 的执行引擎。
- AllJobFinishedListener：全部 Job 完成事件的侦听器。
- BatchSQLJob：批量任务的执行器。
- SQLJob：SQL 执行的任务类。
- SQLJobHandler：SQL 任务执行后获取数据的 handler。
- OneRawSQLQueryResultHandler：Raw 查询返回的结果类。
- SQLQueryResult：SQL 查询返回的数据类。
- SQLQueryResultListener：SQL 查询返回的数据侦听器。

在采用这个方案开发的时候，必须实现 catlet 的两种方法，代码如下：

```
import org.opencloudb.cache.LayerCachePool;
import org.opencloudb.config.model.SchemaConfig;
import org.opencloudb.config.model.SystemConfig;
import org.opencloudb.server.ServerConnection;
public interface Catlet {
    /*
     * EngineCtx 执行 SQL 并给客户端返回结果集
     */
    void processSQL(String sql, EngineCtx ctx);
    //路由的方法，传递系统配置和 schema 配置等
    void route(SystemConfig sysConfig, SchemaConfig schema,
            int sqlType, String realSQL, String charset, ServerConnection sc,
            LayerCachePool cachePool) ;

}
```

执行流程如图 7-13 所示。

在图 7-13 中可以看到有 route 和 processSQL 的方法，并且按顺序执行，route 路由的方法可以通过传递进来的配置信息，按照业务逻辑重新录用，路由结果可用在 processSQL 的方法中，

processSQL 是真正执行 SQL 语句的过程。

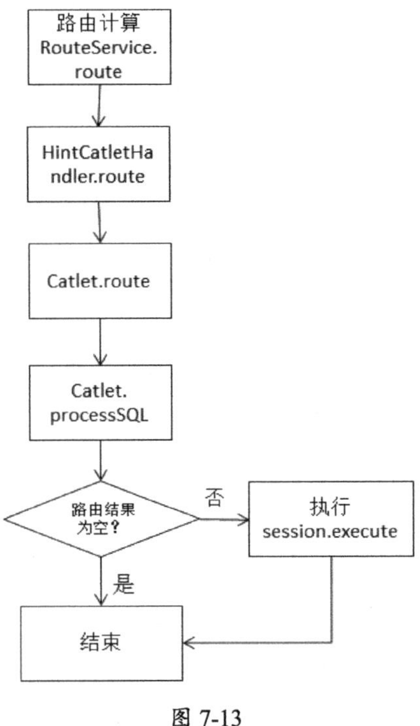

图 7-13

EngineCtx 是整个 HBT 的执行引擎，通过 executeNativeSQLSequnceJob 和 executeNativeSQLParallJob 执行 SQL 语句，参数 dataNodes 是需要执行的节点，它们的区别是：executeNativeSQLSequnceJob 是顺序执行的任务，BatchSQLJob 执行器中的任务执行完后才执行新的任务；executeNativeSQLParallJob 的方法会马上执行，都是放入 businessExecutor 线程池中执行，HBT 也可以认为是个 SQL 执行框架，完全以异步的模式执行。

catlet 编写完成并编译通过以后，必须放在 Mycat_home/catlets 目录下，系统会动态加载相关 Class（需要按照 Java Class 的目录结构存放，比如 demo\catlets\XXXCatlet.class，目前支持 jar 文件），而且每隔 1 分钟扫描一次文件是否更新，若更新则自动重新加载，因此无须重启服务。

7.3.4 ShareJoin

ShareJoin 是一个简单的跨分片 Join，基于 HBT 的方式实现。目前支持两个表的 Join，原理就是解析 SQL 语句，拆分成单表的 SQL 语句执行，然后把各个节点的数据汇集。

相关类图如图 7-14 所示。

- JoinParser：SQL 语句的解析。
- TableFilter：保存解析后的各个子表。
- ShareJoin：实现语句、字段、记录拆分管理。
- ShareDBJoinHandler：第 1 个表执行后获取数据的 handler。
- ShareRowOutPutDataHandler：最后一个表执行后获取数据的 handler。

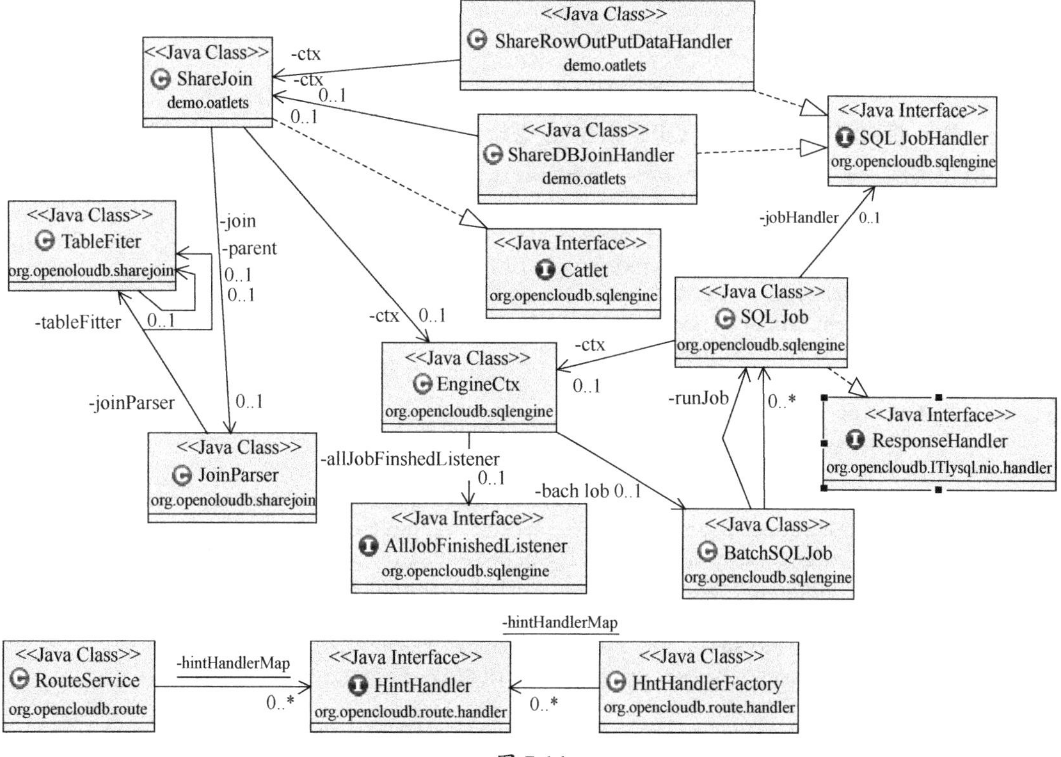

图 7-14

如下所述的主表和子表分别是拆分出的第 1 条 SQL 语句和第 2 条 SQL 语句中的表。

```
    public class ShareJoin implements Catlet {
    private EngineCtx ctx;  //HBT 的执行引擎
    private RouteResultset rrs ;//路由结果集
    private JoinParser joinParser;//Join 解析器
    private Map<String, byte[]> rows = new ConcurrentHashMap<String, byte[]>();//
存记录的结果集       private Map<String,String> ids = new
ConcurrentHashMap<String,String>();//join 字段的值
    private List<byte[]> fields; //主表的字段
    private ArrayList<byte[]> allfields;//所有的字段
```

```
    private boolean isMfield=false; //已经获取主表的字段了
    private int mjob=0;    //job 的任务数
    private int maxjob=0;  //最大的任务数
    private int joinindex=0;//关联 join 表字段的位置
    private int sendField=0; //输出 field 的标志
    private boolean childRoute=false;//是否重新路由标志   //重新路由使用
    private SystemConfig sysConfig;
    private SchemaConfig schema;
    private int sqltype;
    private String charset;
    private ServerConnection sc;
    private LayerCachePool cachePool;
```

第 1 步，获取路由的配置信息和原始 SQL 语句，Join 解析器（joinParser）解析原始语句：

```
    public void route(SystemConfig sysConfig, SchemaConfig schema, int sqlType,
            String realSQL, String charset, ServerConnection sc,
            LayerCachePool cachePool) {
        int rs = ServerParse.parse(realSQL);
        this.sqltype = rs & 0xff;
        this.sysConfig = sysConfig;
        this.schema = schema;
        this.charset = charset;
        this.sc = sc;
        this.cachePool = cachePool;
        try {
            MySQLStatementParser parser = new MySQLStatementParser(realSQL);
            SQLStatement statement = parser.parseStatement();
            if (statement instanceof SQLSelectStatement) {
                SQLSelectStatement st = (SQLSelectStatement) statement;
                SQLSelectQuery sqlSelectQuery = st.getSelect().getQuery();
                if (sqlSelectQuery instanceof MySQLSelectQueryBlock) {
                    MySQLSelectQueryBlock MySQLSelectQuery =
(MySQLSelectQueryBlock) st
                            .getSelect().getQuery();
                    joinParser = new JoinParser(MySQLSelectQuery, realSQL);
                    joinParser.parser();
                }
            }
        } catch (Exception e) {
        }
    }
```

第 2 步，执行 SQL 语句：

```
    public void processSQL(String sql, EngineCtx ctx) {
        String ssql=joinParser.getSql();//拆分的第 1 条 SQL 语句
        getRoute(ssql);//对第 1 条 SQL 语句重新路由
        RouteResultsetNode[] nodes = rrs.getNodes();//获取路由节点
```

```
        if (nodes == null || nodes.length == 0 || nodes[0].getName() == null
                || nodes[0].getName().equals("")) {
            ctx.getSession().getSource().writeErrMessage(ErrorCode.ER_NO_DB_ERROR,
                    "No dataNode found ,please check tables defined in schema:"
            + ctx.getSession().getSource().getSchema());
        return;
        }
        this.ctx=ctx;
        String[] dataNodes =getDataNodes();
        maxjob=dataNodes.length;//节点数就是最大的任务数
        ShareDBJoinHandler joinHandler = new
ShareDBJoinHandler(this,joinParser.getJoinLkey()); //多个节点执行第 1 条 SQL 语句
        ctx.executeNativeSQLSequnceJob(dataNodes, ssql, joinHandler);
        EngineCtx.LOGGER.info("Catlet exec:"+getDataNode(getDataNodes())+" sql:"
+ssql); //所有任务完成的侦听器
        ctx.setAllJobFinishedListener(new AllJobFinishedListener()
{       @Override
            public void onAllJobFinished(EngineCtx ctx) {
            ctx.writeEof();
            EngineCtx.LOGGER.info("发送数据 OK");
            }
         });
        }
      }
```

第 3 步，Join 第 1 条 SQL 语句的字段列表，每个节点的表结构一样，只需要获取一次：

```
public void putDBFields(List<byte[]> mFields){
    if (!isMfield){
        fields=mFields;
    }
}
```

第 4 步，Join 第 1 条 SQL 语句的记录结果集：

```
public void putDBRow(String id,String nid, byte[] rowData,int findex){
rows.put(id, rowData);
ids.put(id, nid);
joinindex=findex;
//ids.offer(nid);
 int batchSize = 999;     // 满 1000 条，发送一个查询请求
if (ids.size() > batchSize) {
 createQryJob(batchSize);
}
}
```

第 5 步，Join 第 1 条 SQL 语句的节点任务完成：

```
public void endJobInput(String dataNode, boolean failed){
      mjob++;
```

```
        if (mjob>=maxjob){
          createQryJob(Integer.MAX_VALUE);
          ctx.endJobInput();
        }
       // EngineCtx.LOGGER.info("完成"+mjob+":" + dataNode+" failed:"+failed);
    }
```

第 6 步，创建第 2 次查询任务：

```
private void createQryJob(int batchSize) {
        int count = 0;
        Map<String, byte[]> batchRows = new ConcurrentHashMap<String, byte[]>();
        String theId = null;
        StringBuilder sb = new StringBuilder().append('(');
        String svalue="";
        for(Map.Entry<String,String> e: ids.entrySet() ){
            theId=e.getKey();
            batchRows.put(theId, rows.remove(theId));
            if (!svalue.equals(e.getValue())){
                if(joinKeyType == Fields.FIELD_TYPE_VAR_STRING
                        || joinKeyType == Fields.FIELD_TYPE_STRING){
                          // joinkey 为 varchar
                        sb.append("'").append(e.getValue()).append("'").append(',');
                          // ('digdeep','yuanfang')
                }else{ // 默认 joinkey 为 int/long
                    sb.append(e.getValue()).append(','); // (1,2,3)
                }
            }
            svalue=e.getValue();
            if (count++ > batchSize) {
                break;
            }
        }
        /*
        while ((theId = ids.poll()) != null) {
            batchRows.put(theId, rows.remove(theId));
            sb.append(theId).append(',');
            if (count++ > batchSize) {
                break;
            }
        }
        */
        if (count == 0) {
            return;
        }
        jointTableIsData=true;
        sb.deleteCharAt(sb.length() - 1).append(')');
        String sql = String.format(joinParser.getChildSQL(), sb);
```

```
            //if (!childRoute){
              getRoute(sql);
             //childRoute=true;
            //}
            ctx.executeNativeSQLParallJob(getDataNodes(),sql, new
ShareRowOutPutDataHandler(this,fields,joinindex,joinParser.getJoinRkey(),
batchRows));
            EngineCtx.LOGGER.info("SQLParallJob:"+getDataNode(getDataNodes())+"
sql:" + sql);
        }
    class ShareRowOutPutDataHandler implements SQLJobHandler{
        //获取上面第 5 步 job 执行后字段列表的事件：
    @Override
        public void onHeader(String dataNode, byte[] header, List<byte[]> fields) {
            this.fields = fields;
            ctx.putDBFields(fields);
        }
    @Override
        public boolean onRowData(String dataNode, byte[] rowData) {
            int fid=this.ctx.getFieldIndex(fields,joinkey);
            String id = ResultSetUtil.getColumnValAsString(rowData, fields, 0);
            //主键，默认 id
            String nid = ResultSetUtil.getColumnValAsString(rowData, fields, fid);
            // 放入结果集
            //rows.put(id, rowData);
            ctx.putDBRow(id,nid, rowData,fid);
            return false;
        }
    //处理完成，标记方法：
    @Override
        public void finished(String dataNode, boolean failed) {
            ctx.endJobInput(dataNode,failed);
        }
    }
```

执行第 2 条 SQL 语句：

```
    class ShareRowOutPutDataHandler implements SQLJobHandler {
    ......
    ......//此处省略部分源码，建议读者参考 GitHub 上 mycat server 项目的
ShareRowOutPutDataHandler 类的源码

        @Override
        public void finished(String dataNode, boolean failed) {
        //    EngineCtx.LOGGER.info("完成 2:" + dataNode+" failed:"+failed);
        }
    }
```

7.4 Mycat 数据汇聚和排序的实现

通过 Mycat 实现数据汇聚和排序，不仅可以减少各分片和客户端之间的数据传输 I/O，也可以帮助开发者从复杂的数据处理中解放出来，从而专注于开发业务代码。数据汇聚和排序在 Query 语句中使用得比较频繁，由于汇聚和排序都涉及数据排序操作，所以把它们放在同一小节中讲解。

7.4.1 数据排序

在 MySQL 中有两种排序方式：一种利用有序索引获取有序数据，另一种通过相应的排序算法将获取到的数据在内存中进行排序。在 Mycat 中数据排序采用堆排序法对多个分片返回有序数据，并在合并、排序后再返回到客户端，执行流程如图 7-15 所示。

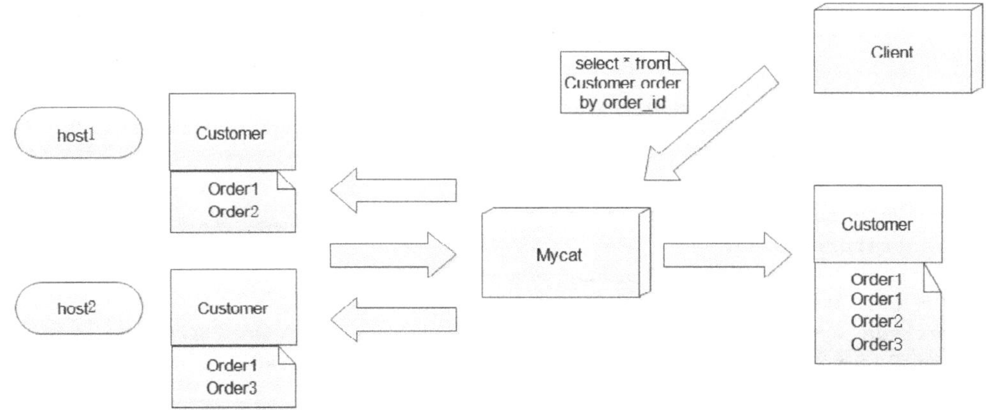

图 7-15

1. 相关类说明

- RowDataSorter：数据排序。
- RowDataCmp：字段对比。
- HeapItf：堆排序接口。
- MaxHeap：最大堆排序。

2. 代码解析

以下是排序算法的调用入口及数据排序处理的实现类：

```
public class DataMergeService implements Runnable {
……
……//此处省略部分源码，建议读者参考 GitHub 上 mycat server 项目的 DataMergeService 类的源码
}
```

堆排序 MaxHeap 的代码实现如下：

```
public class MaxHeap implements HeapItf {
private RowDataCmp cmp;//order by 字段对比
private List<RowDataPacket> data;//数据列表

public MaxHeap(RowDataCmp cmp, int size) {
    this.cmp = cmp;
    this.data = new ArrayList<>();
}

@Override
public void buildHeap() {
    int len = data.size();
    for (int i = len / 2 - 1; i >= 0; i--) {
        heapifyRecursive(i, len);
    }
}
// 递归版本
protected void heapifyRecursive(int i, int size) {
    int l = left(i);
    int r = right(i);
    int max = i;
    if (l < size && cmp.compare(data.get(l), data.get(i)) > 0)
        max = l;
    if (r < size && cmp.compare(data.get(r), data.get(max)) > 0)
        max = r;
    if (i == max)
        return;
    swap(i, max);
    heapifyRecursive(max, size);
}
private int right(int i) {
    return (i + 1) << 1;
}

private int left(int i) {
    return ((i + 1) << 1) - 1;
```

```
    }

//交换数据
    private void swap(int i, int j) {
        RowDataPacket tmp = data.get(i);
        RowDataPacket elementAt = data.get(j);
        data.set(i, elementAt);
        data.set(j, tmp);
    }

    @Override
    public RowDataPacket getRoot() {
        return data.get(0);
    }

    @Override
    public void setRoot(RowDataPacket root) {
        data.set(0, root);
        heapifyRecursive(0, data.size());
    }

    @Override
    public List<RowDataPacket> getData() {
        return data;
    }

    @Override
    public void add(RowDataPacket row) {
        data.add(row);
    }

    @Override
    public boolean addIfRequired(RowDataPacket row) {
        // 淘汰堆里最小的数据
        RowDataPacket root = getRoot();
        if (cmp.compare(row, root) < 0) {
            setRoot(row);
            return true;
        }
        return false;
    }

    @Override
    public void heapSort(int size) {
        final int total = data.size();
        // 容错处理
        if (size <= 0 || size > total) {
```

```
            size = total;
        }
        final int min = size == total ? 0 : (total - size - 1);

        // 末尾与头交换，交换后调整最大堆
        for (int i = total - 1; i > min; i--) {
            swap(0, i);
            heapifyRecursive(0, i);
        }
    }

}
```

字段的排序代码实现如下：

```
public class RowDataCmp implements Comparator<RowDataPacket> {
    ……
……//此处省略部分源码，建议读者参考 GitHub 上 mycat server 项目的 RowDataCmp.java 源码

}
```

7.4.2 数据汇聚

group by 实际上也需要对数据进行排序，与 order by 相比 group by 多了排序后的数据分组，在分组的同时可以使用聚合函数 count、sum、max、min、avg。在 MySQL 中 group by 有三种实现方式：第 1 种是利用松散（Loose）索引扫描实现；第 2 种是使用紧凑（Tight）索引扫描实现；第 3 种是使用临时表实现。当所有分片返回已经汇聚好的数据时，Mycat 对返回的所有数据进行合并、汇聚后再返回到客户端，执行流程如图 7-16 所示。

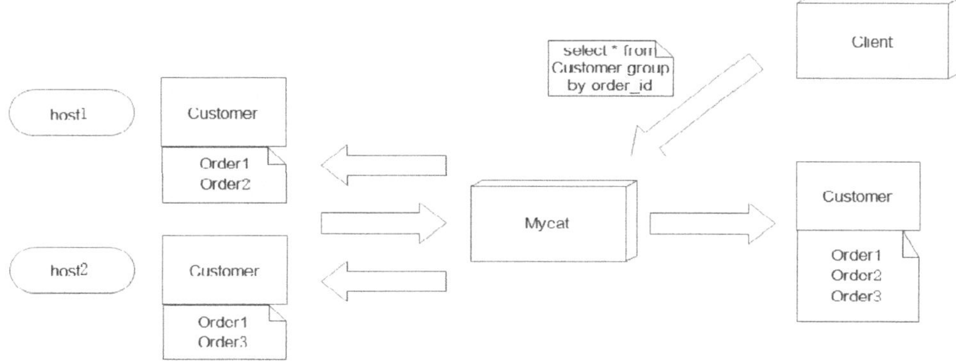

图 7-16

1. 相关类说明

- MergeCol：聚合方法。
- ColMeta：聚合方法及聚合类型。
- RowDataPacketGrouper：数据汇聚类。

2. 代码解析

以下是数据汇聚的调用入口及数据返回：

```
public class DataMergeService implements Runnable {
……
……//此处省略部分源码，建议读者参考 GitHub 上 mycat server 项目的 DataMergeService.java 源码
}
```

数据汇聚代码实现如下：

```
public class RowDataPacketGrouper {
……
……//此处省略部分源码，建议读者参考 GitHub 上 mycat server 项目的 RowDataPacketGrouper.java 源码

}
```

第 8 章

Mycat 多数据库支持原理与实现

前几章介绍了 Mycat 的设计原理、相关设计和实现，本章主要介绍 Mycat 对多数据库的支持，主要包括 MySQL 通信协议的实现、PostgreSQL 协议的实现及对 JDBC 的支持。

8.1 MySQL 协议在 Mycat 中的实现

8.1.1 MySQL 协议概述

1. 概述

MySQL 协议处于应用层之下、TCP/IP 网络层之上，在 MySQL 客户端和 MySQL 服务器之间使用。它包含了连接器（Connector/C、连接器/J 等）、MySQL 代理和主从复制服务器之间的通信，并支持 SSL 加密、传输数据的压缩、连接与身份验证及数据交互等。其中，握手认证阶段和命令执行阶段是 MySQL 协议中的两个重要阶段。下面主要以 Mycat 4.1 之后的版本作为参考进行分析。

2. 握手认证阶段

握手认证阶段是客户端连接服务器的必经之路，在客户端与服务端完成 TCP 的三次握手以后，服务端会向客户端发送一个初始化握手包，握手包中包含了协议版本、MySQLServer 版本、线程 ID、服务器的权能标志和字符集等信息。

客户端在接收到服务端的初始化握手包之后，会发送身份验证包给服务端（AuthPacket），此时，用户名、密码等信息就包含在该包中。

服务器接收到客户的登录验证包之后，需要进行逻辑校验，校验该登录信息是否正确，例如用户名、密码和验证字符串等信息。如果信息都符合，则返回一个 OKPacket，表示登录成功，否则返回 ERR_Packet，表示拒绝。如图 8-1 所示。

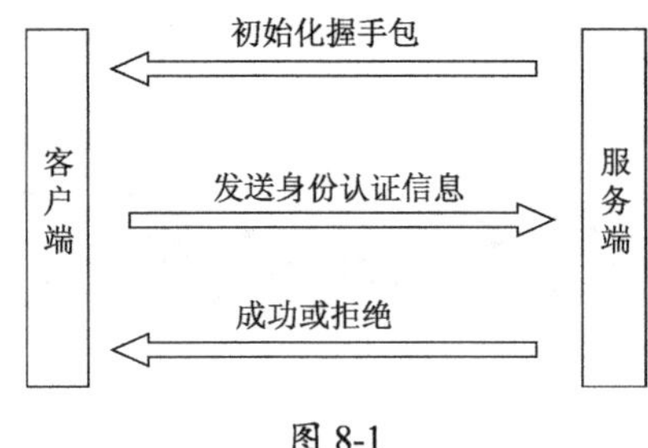

图 8-1

在 TCP 通道建立的基础上，将握手阶段的流程总结如下。

- 从服务器到客户端：握手初始化包。
- 从客户端到服务器：客户端认证包。
- 从服务器到客户端：OK 包、错误（Error）包。

对握手认证阶段的报文分析如下。

1）初始化握手包

报文分为消息头和消息体两部分，其中消息头占用固定的 4 个字节，消息体的长度由消息头中的长度字段决定，如图 8-2 所示。

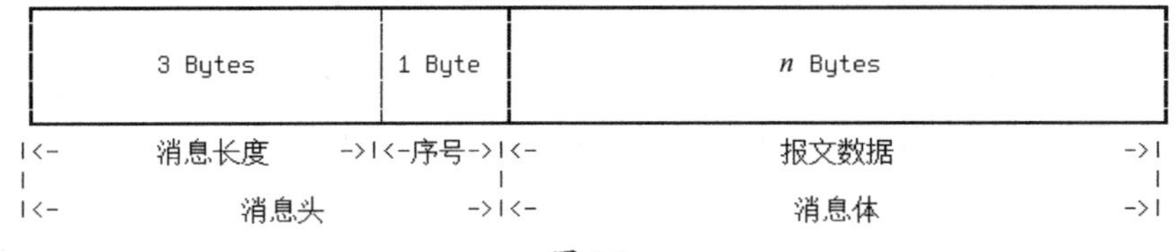

图 8-2

下面举例说明握手包的格式，Packet Length 是包的长度，Packet Number 是包的序号，我们将二者称为消息头；Server Greeting 是消息体，消息体中包含了协议版本、MySQLServer 版本、线程 ID 和字符集等信息。

```
MySQLProtocol
    Packet Length: 77
    Packet Number: 0
    Server Greeting
        Protocol: 10
        Version: 5.7.4-m14
```

```
Thread ID: 1
Salt: 5Ry.v)d@
Server Capabilities: 0xf7ff
Server Language: latin1 COLLATE latin1_swedish_ci (8)
Server Status: 0x0002
Extended Server Capabilities: 0x80ff
Authentication Plugin Length: 21
Unused: 00000000000000000000
Salt: PS2<xg|Y*b2n
Authentication Plugin: MySQL_native_password
```

通过以上分析我们知道了握手包的结构，消息头固定是 3+1 个字节，那么消息体的每个字段之间占位多少个字节并采用什么标志来识别各个字段呢？如表 8-1 所示，第 1 个字段是协议版本占位 1 个字节；第 2 个字段是服务器版本占 *N* 个字节，遇 null 为结束；第 3 个字段为线程 ID，其他字段都这样排序。

表 8-1

字 段 名 称	字　　节
protocol_version	1
server_version	N（Null 为分隔符）
thread_id	4
scramble_buff	8
(filler) always 0x00	1
server_capabilities	2
server_language	1
server_status	2
(filler) always 0x00 …	13
rest of scramble_buff (4.1)	13

我们知道在以 NULL 结束的字段中，如果字符串本身包含了 NULL，那么该字符串传达的信息就有可能丢失了，在 MySQL 协议中如何保证信息的完整性呢？这时就要借助消息体中的 Packet Length 字段了，Packet Length 告诉消息体后面还有多少长度。

2）登录认证包

客户端在接收到服务器端发来的初始握手包后，向服务端发出认证请求，该请求包中包含了以下信息：

```
MySQLProtocol
   Packet Length: 199
   Packet Number: 1
   Login Request
```

```
Client Capabilities: 0xa68d
Extended Client Capabilities: 0x007f
MAX Packet: 16777216
Charset: Unknown (28)
Username: root
Password: f22bc6e9ec3e6a0f6925b5ff988ca824d0b6cb62
Schema: db1
Client Auth Plugin: MySQL_native_password
Connection Attributes
```

可以看到在上面的例子中，包的长度是 199，序号是 1，Login Request 是消息体部分，第 1 个字段是客户端权能标志，用于与客户端协商通信方式，标志位的含义与握手初始化报文中的相同。客户端收到服务器发来的初始化报文后，会对服务器发送的权能标志进行修改，保留自身所支持的功能，然后将权能标志返回给服务器，从而保证服务器与客户端通信的兼容性。

MAX Packet 是最大的消息长度，是客户端发送请求报文时所支持的消息的最大长度值。Charset 是字符集编号，Username、Password 分别是用户名和密码。各字段的名称与字节数如表 8-2 所示。

表 8-2

字　节	字段名称
4	客户端权能标志
4	最大消息长度
1	字符编码的编号
23	填充字符，一般为 0x00
n（Null-Terminated String）	user
n（Length Coded Binary）	scramble_buff (1 + *x* bytes)
n（Null-Terminated String）	databasename (可选)

3）OK 包或 ERROR 包

服务器接收到客户的登录验证包之后，如果通过认证，则返回一个 OKPacket，如果未通过认证，则返回一个 ERROR 包。

一个 OK 包的例子如下：

```
MySQLProtocol
    Packet Length: 8
    Packet Number: 2
    Affected Rows: 0
    Server Status: 0x0002
    Warnings: 0
    Message:
```

OK 包的字段说明如表 8-3 所示。

表 8-3

字　节	说　明
1	OK 报文，值恒为 0x00
1～9（Length Coded Binary）	受影响行数（Length Coded Binary）
1～9（Length Coded Binary）	索引的 ID 值（Length Coded Binary）
2	服务器状态
2	告警计数
n	服务器消息（字符串到达消息尾部时结束，无结束符，可选）

ERROR 包的字段说明如表 8-4 所示。

表 8-4

字　节	字 段 说 明
1	Error 报文，值恒为 0xFF
2	错误编号（小字节序）
1	服务器状态标志，恒为“#”字符
5	服务器状态（5 个字符）
n	服务器消息

一个错误包的例子如下：

```
MySQLProtocol
    Packet Length: 75
    Packet Number: 2
    Error Code: 1045
    SQL state: 28000
    Error message: Access denied for user 'root'@'192.168.58.1' (using password: YES)
```

3. 命令执行阶段

在握手认证阶段通过并完成以后，客户端可以向服务器发送各种命令来请求数据，此阶段的流程是：命令请求→返回结果集。

1）命令包

```
MySQLProtocol
    Packet Length: 16
    Packet Number: 0
    Request Command Query
        Command: Query (3)
        Statement: select * from a
```

包体只有两个字段：一个是命令，为一个字节；另一个是命令参数，为 N 个字节。在上面的例子中，命令 3 表示查询命令，这是用得最多的命令，因为 insert、update、delete、select 都属于命令 3。各命令的代码含义可参考官方文档 http://dev.MySQL.com/doc/internals/en/text-protocol.html。

2）结果集包

一个结果集包由一系列的包组成，包含了包头、查询返回的列包、查询返回的行包和 EOF 包。结果集包比较复杂，下面以一个简单的例子来说明结果集包：

```
MySQL> select * from a;
+------+------+-------------+
| id   | name | create_date |
+------+------+-------------+
| 1000 | 李工 | 2016-03-10  |
| 1002 | 陈工 | 2016-04-10  |
| 1004 | 周工 | 2016-04-10  |
+------+------+-------------+
3 rows in set (0.00 sec)
```

上面是一条简单的查询语句，结果集一共返回了 9 个包，结果集返回的包计算公式为：$n+m+3$，其中 n 为行数，m 为列数。比如上面查询了 3 行数据和 3 列数据，那么返回的结果集包为 3+3+3=9 个。这 9 个包的结构如表 8-5 所示。

表 8-5

结果集包头		
字段包 1	字段包 2	字段包 3
EOF 包		
行包 1	行包 2	行包 3
EOF 包		

现在我们分别分析结果集包头、字段包、EOF 包和行包的结构。

结果集包头共有两个字段：field_count 描述了列的数目，extra 为可选项，如表 8-6 所示。

表 8-6

字 节 数	字 段 说 明
1-9	field_count
1-9	Extra(可选)

一个结果集包头的例子如下：

```
MySQLProtocol
    Packet Length: 1
```

```
    Packet Number: 1
    Number of fields: 3
```

列包用于描述结果集中的字段信息，紧跟在结果集包头之后，一个列（字段）对应一个包，其格式如表 8-7 所示。

表 8-7

字 节 数	字 段 说 明
n（带长度标识字符串）	catalog
n（带长度标识字符串）	db
n（带长度标识字符串）	table
n（带长度标识字符串）	org_table
n（带长度标识字符串）	name
n（带长度标识字符串）	org_name
1	filler
2	charsetNumber
4	length
1	type
2	flags
1	decimals
2	filler，通常为 0x00
n（Length Coded Binary）	default

“带长度标识字符串”的意思是这个字段由两部分组成：长度定义部分和内容部分。长度定义部分规定了内容的长度，MySQL 规定了各类数据库、表等名称不能超过 256 个字符。

一个字段包的例子如下：

```
MySQLProtocol
    Packet Length: 31
    Packet Number: 2
    Catalog: def
    Database: db1
    Table: a
    Original table: a
    Name: id
    Original name: id
    Charset number: binary COLLATE binary (63)
    Length: 11
    Type: FIELD_TYPE_LONG (3)
    Flags: 0x5003
    Decimals: 0
```

EOF 结构用于标识 Field 和 Row Data 的结束，在预处理语句中，EOF 也被用来标识参数的结束，其结构如表 8-8 所示。

表 8-8

字　节	说　明
1	EOF 值（0xFE）
2	告警计数
2	状态标志位

EOF 值永远为 0XFE，告警计数为服务器的告警数量，在所有数据都发送给客户端后该值才有效。状态标志位包含类似 SERVER_MORE_RESULTS_EXISTS 这样的标志位。

一个 EOF 包的例子如下：

```
MySQLProtocol
    Packet Length: 5
    Packet Number: 5
    EOF marker: 254
    Warnings: 0
    Server Status: 0x0022
```

行包与列包中间有一个 EOF 包间隔，MySQL 以每行数据一个包的形式包含在结果集中，不管一行有多少个字段，这些字段的值都组成一行数据。它的格式非常简单，如表 8-9 所示。

表 8-9

字　节	说　明
n	字段值（Length Coded String）

下面是一个行包的例子，数据经过抓包软件做了解析：

```
MySQLProtocol
    Packet Length: 21
    Packet Number: 6
    text: 1000
    text: \357\277\275\357\277\275\357\277\275\357\277\275
    text: 2016-03-10
```

4. 关于报文标识符

1）字符串（以 NULL 结尾，Null-Terminated String）

字符串长度不固定，当遇到 NULL（0x00）字符时结束。

2）二进制数据（长度编码，Length Coded String）

当字符串中包含了 NULL 字符，无法标识一个字段时，就要用到长度标识的字符串了。带长度标识的字符串（Length Coded String）由两部分组成：长度定义部分和内容部分。数据的长度不固定，长度值由数据前的 1～9 个字节决定，其中长度值所占的字节数不定，字节数由第 1 个字节决定，如表 8-10 所示。

表 8-10

第 1 个字节值	后续字节数	长度值说明
0～250	0	第 1 个字节的值即数据的真实长度
251	0	后面两个字节的值，空数据，数据的真实长度为 0
252	2	后面两个字节的值
253	3	后面 3 个字节的值
254	8	后面 8 个字节的值

3）字符串（长度编码，Length Coded Binary）

字符串的长度不固定，无 NULL（0x00）结束符，编码方式与上面的 Length Coded String 相同。

8.1.2　Mycat 的 MySQL 协议实现

1. MySQL 协议之 I/O 框架实现

在 Mycat 中同时实现了 NIO 和 AIO，通过配置可以选择 NIO 或 AIO。NIO 对应的设计模式是 Reactor，AIO 对应的设计模式是 Proactor。AIO 需要 JDK 1.7 的支持。

Proactor 主要体现在对于回调的处理，它的性能更高，能够处理耗时长的并发场景。Proactor 的实现逻辑复杂；依赖操作系统对异步的支持，目前实现纯异步操作的操作系统较少，比较优秀的如 Windows IOCP，但由于其 Windows 系统用于服务器的局限性，目前应用范围较小；而 UNIX/Linux 系统对纯异步的支持有限，应用事件驱动的主流还是通过 select/epoll 来实现的，这也是 Mycat 保留 NIO 的原因，默认启动 NIO。

在 Mycat 中实现异步的 I/O 框架的细节请参考 7.2 节。本章重点介绍 TCP 协议之上的 MySQL 协议实现。

2. MySQL 协议之握手认证实现

通过 7.2 节我们已经分析了 I/O 架构及 NIO、AIO 的实现，这里我们讨论在实现 AIO 的基础上 MySQL 通信协议在握手阶段的实现。

Mycat Server 在启动阶段已经选择好采用 NIO 还是 AIO，因此建立 I/O 通道后，Mycat 服务端一直等待着客户的连接，当有连接到来的时候，Mycat 首先发送握手包。我们知道握手包一共有 9 个字段，包头有两个字段。包头部分定义为一个抽象类 MySQLPacket，包含了两个重要属性：包的长度（packetLength）和包的序号（packet Id）。以 Mycat 1.5 为例，它的 MySQL 协议版本为 10，服务器版本为 5.5.8-Mycat-1.5-RELEASE-20160321172631。发送握手包的源码片段如下（详见 org.opencloudb.net.FrontendConnection）：

```
// 发送握手数据包
        HandshakePacket hs = new HandshakePacket();
        hs.packetId = 0;
        hs.protocolVersion = Versions.PROTOCOL_VERSION;
        hs.serverVersion = Versions.SERVER_VERSION;
        hs.threadId = id;
        hs.seed = rand1;
        hs.serverCapabilities = getServerCapabilities();
        hs.serverCharsetIndex = (byte) (charsetIndex & 0xff);
        hs.serverStatus = 2;
        hs.restOfScrambleBuff = rand2;
        hs.write(this);
```

握手包序号由于是 Server 端发送的第 1 个 MySQL 包，所以握手包的序号一般定义为 0，Server Capabilities 包含了服务器处理能力标识及是否使用压缩协议，允许在函数名后使用空格。所有函数名可以预留字；允许使用关闭连接之前的不活动交互超时的描述，而不是等待超时秒数都在这个标识中描述。初始化完握手包后通过 FrontendConnection 写回客户端。

客户端接收握手包后，紧接着向服务器发起一个认证包，认证包主要包含以下信息：

```
public long clientFlags;
public long maxPacketSize;
public int charsetIndex;
public byte[] extra;// from FILLER(23)
public String user;
public byte[] password;
public String database;
```

客户端发送的认证包转由 FrontendAuthenticator 的 Handle 来处理，这个阶段要做的事情主要是拆包，检查用户名、密码的合法性，检查连接数是否超出。部分源码如下：

```
if (!checkUser(auth.user, source.getHost())) {
  ……
}
if (!checkPassword(auth.password, auth.user)) {
   ……
}
if ( isDegrade( auth.user ) ) {
  ……
}
```

认证通过后，发送 OK 包：

```
source.write(source.writeToBuffer(AUTH_OK, buffer));
```

紧接着处理前端的 Handler 转到命令处理的 FrontendCommandHandler：

```
source.setHandler(new FrontendCommandHandler(source));
```

此时进入了命令执行阶段。

3. MySQL 协议之命令执行的实现

命令执行阶段就是我们熟悉的 SQL 命令和 SQL 语句执行的阶段，这个阶段首先要做的事情就是对客户端发来的包进行拆包，并判断命令的类型，然后解析 SQL 语句，执行相应的命令，最后把执行结果封装在结果集的包中，返回给客户端。本节不讨论 SQL 解析，也不讨论 SQL 路由和 SQL 命令的实际执行，只关心客户端与服务端通信协议的实现。

1）拆包的实现

在 MySQL 中用一系列数字代表命令的类型。如下面的代码片段所示，我们熟悉的 COM_QUERY 用 3 表示，客户端只需要返回 3 即可表示这是一个 select、update、delete、insert 类型的命令。

```
org.opencloudb.net.MySQL.MySQLPacket:
public static final byte COM_SLEEP = 0;
public static final byte COM_QUIT = 1;
public static final byte COM_INIT_DB = 2;
public static final byte COM_QUERY = 3;
public static final byte COM_FIELD_LIST = 4;
……
```

从客户端发来的命令包先由 ServerLoadDataInfileHandler 的 handle 加载数据、填充数据，随后用一个 switch 语句判断命令的类型，如下面的代码片段所示。Mycat 实现了 9 种类型的命令，比如执行 SQL 语句、退出、Ping、预处理和心跳等。

```
switch (data[4])
        {
            case MySQLPacket.COM_INIT_DB:
                ……
            case MySQLPacket.COM_QUERY:
                ……
            case MySQLPacket.COM_PING:
                ……
            case MySQLPacket.COM_QUIT:
                ……
            case MySQLPacket.COM_PROCESS_KILL:
                ……
            case MySQLPacket.COM_STMT_PREPARE:
                ……
```

```
            case MySQLPacket.COM_STMT_EXECUTE:
                ……
            case MySQLPacket.COM_STMT_CLOSE:
              ……
            case MySQLPacket.COM_HEARTBEAT:
                ……
            default:
                    ……
    }
```

通过以上代码可知，不同的命令由不同的方法执行，执行结果打包返回，包装在 MySQL 协议的结果集包中。命令的执行还要经过 SQL 解析、路由计算、分发和缓存等复杂的技术处理环节。

2）返回结果集

结果集包由包头、列包、行包、EOF 包组成。打包的顺序是：包头→列包→EOF 包→行包。假设查询结果有 3 行 3 列记录，则根据计算公式 $n+m+3$ 将返回 9 个包。由 ouputResultSet 方法负责将查询结果打包并返回给客户端。

取得结果集的字段数，写入包头：

```
ResultSetHeaderPacket headerPkg = new ResultSetHeaderPacket();
headerPkg.fieldCount = fieldPks.size();
headerPkg.packetId = ++packetId;
```

写入字段包，一个字段一个包，由之迭代：

```
ist<byte[]> fields = new ArrayList<byte[]>(fieldPks.size());
Iterator<FieldPacket> itor = fieldPks.iterator();
while (itor.hasNext()) {
   FieldPacket curField = itor.next();
   curField.packetId = ++packetId;
   byteBuf = curField.write(byteBuf, sc, false);
   byteBuf.flip();
   byte[] field = new byte[byteBuf.limit()];
   byteBuf.get(field);
   byteBuf.clear();
   fields.add(field);
   itor.remove();
}
```

写入 EOF 包，EOF 包在列包与行包的中间位置，起分隔作用：

```
EOFPacket eofPckg = new EOFPacket();
eofPckg.packetId = ++packetId;
byteBuf = eofPckg.write(byteBuf, sc, false);
```

写入数据行包，这是数据的内容包，一行数据一个包，多节点合并结果后再返回给客户端：

```
// output row
while (rs.next()) {
    RowDataPacket curRow = new RowDataPacket(colunmCount);
    for(int i = 0; i < colunmCount; i++) {
        int j = i + 1;
        curRow.add(StringUtil.encode(rs.getString(j),
        sc.getCharset()));
    }
    curRow.packetId = ++packetId;
    byteBuf = curRow.write(byteBuf, sc, false);
    byteBuf.flip();
    byte[] row = new byte[byteBuf.limit()];
    byteBuf.get(row);
    byteBuf.clear();
    this.respHandler.rowResponse(row, this);
}
```

最后写入一个 EOF 包，打包结束。以上代码请参考 org.opencloudb.jdbc.ouputResultSet。

8.2　PostgreSQL 协议在 Mycat 中的实现

由于 PostgreSQL 在目前的生产环境中用户比较多，所以 Mycat 基于 PostgreSQL 的前后端交互协议实现了和 PostgreSQL 的通信功能，支持 PostgerSQL 主从和数据分片的功能。在前端客户端看来，PostgreSQL 是透明的（部分不兼容的 SQL 语句除外），前端应用可以把 PostgreSQL 当作 MySQL 来使用。

8.2.1　PostgreSQL 介绍

PostgreSQL 是一个免费的开源的对象关系型数据库，是由加州大学伯克利分校计算机系写的 POSTGRES 软件包发展而来的，经过 15 年的打磨和改进，是当前声誉很好的开源关系型数据库之一。由于其独特的背景，PostgerSQL 有着浓厚的学院风（不是贬义词，在国外学院派中有严格、严谨、专注的美誉），有一部分与数学相关的功能，例如索引的执行计划由专注于该领域十几年的教授参与开发，代码质量非常高。

PostgreSQL 的当前版本是 8.5.1。

1. 应用领域

有不少优秀企业使用了 PostgreSQL 作为自己的存储方案，金山、腾讯、百度有不少项目也使用了 PostgreSQL，甚至不少创业公司的最初的对象关系数据库使用了 PostgreSQL。

PostgreSQL 由于其独特的优势，在数据挖掘、地理位置应用方面有非常不错的市场份额。不少地理应用如地理位置计算、地理范围查询（查询附近人）等利用 PostgreSQL 就比较容易实现。不少公司将 PostgreSQL 作为线上应用数据的备库，用于做报表和数据分析的相关应用，所以 PostgreSQL 和 MySQL 混用的情况在一些大型公司比较常见。

2. 实现意义

由于不少用户反映自己的很多应用采用了 PostgreSQL 作为 BI 使用的备库，所以 Mycat 也有心尝试 MySQL 以外的数据库作为 Mycat 后端的支持，于是就有了 PostgreSQL 的支持计划，并反复对 PostgreSQL 通信协议进行研究，为后续对其提供支撑做参照。

当前 PostgreSQL 的相关实现源码在 https://github.com/MycatApache/Mycat-Server 的 io.Mycat.backend.postgresql 子包下。

8.2.2 PostgreSQL 协议

PostgreSQL 前后端的交互是基于 TCP 连接的，采用命令应答式交互，客户端每发一个命令请求，服务端便返回一个或者多个数据包，通信过程分为三个阶段，如图 8-3 所示。

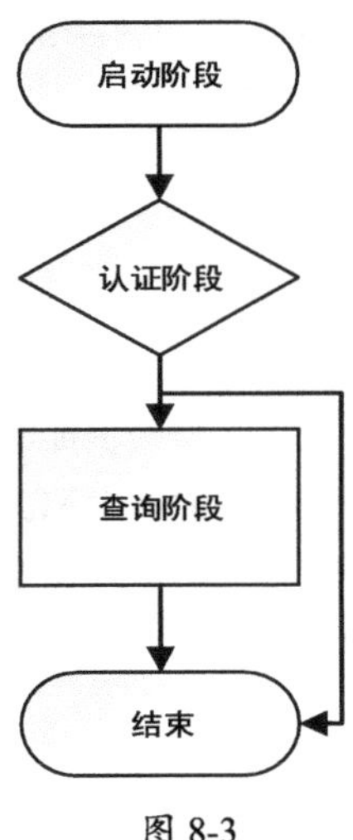

图 8-3

启动阶段由客户端开启一个 TCP 连接，主动发送一个启动包给 PostgreSQL（为了方便描述，之后的“后台”指的都是 PostgreSQL 数据服务器）。启动包中包含：客户端名称、协议版本、用户名和环境参数等。

后台根据客户端发送的启动包会返回一个认证请求的包，客户端则按要求发送认证凭证数据包；认证通过后则进入查询阶段，到了这个阶段，后台就可以响应客户端的 SQL 查询请求了；当客户端给服务器发送一个关闭数据包时，这一次会话结束。

前后端交互的数据包都有统一的格式（启动包除外），如图 8-4 所示。

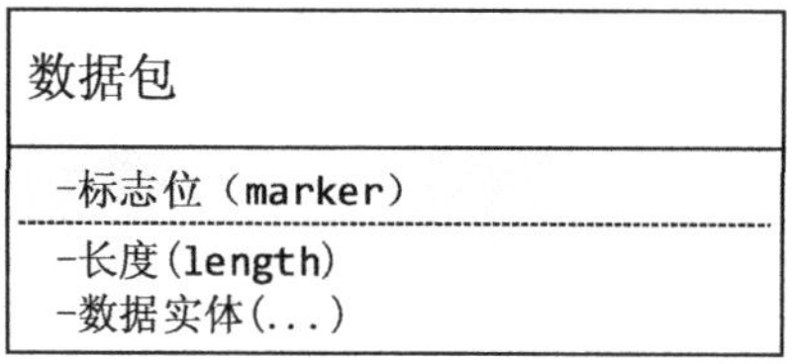

数据包
-标志位（marker）
-长度(length) -数据实体(...)

图 8-4

每个数据包都有一个英文字母或者数字字符标识（区分大小写）。数据包的详细结构请见附录。

数据包的标志位和意义如表 8-11 所示。

表 8-11

类　型	标 志 位	包　名	说　明
后台响应	'R'	AuthenticationPacket	服务器给前端的认证结果包，当启动包发给服务器后，服务器会返回
	'E'	ErrorResponse	请求处理出错
	'K'	BackendKeyData	本次会话的密钥
	'S'	ParameterStatus	参数状态，当用户设置某些环节的参数时，会返回此数据包
	'Z'	ReadyForQuery	后台已经初始化成功，等待接收查询请求
	'N'	NoticeResponse	参数有更改便会通知客户端
	'C'	CommandComplete	查询请求处理完成
	'T'	RowDescription	查询请求结果集列描述
	'D'	DataRow	查询请求结果集行数据
	'I'	EmptyQueryResponse	空语句请求响应
	'G'	CopyInResponse	复制进数据请求响应
	'H'	CopyOutResponse	复制出数据请求响应
	'1'	ParseComplete	扩展查询协议，SQL 语句预编译成功
	'2'	BindComplete	预编译语句变量绑定成功
客户端请求	无	StartupMessage	启动数据包
	'p'	PasswordMessage	认证密码数据包
	'Q'	Query	简单的查询语句

续表

类　型	标 志 位	包　名	说　明
客户端请求	'X'	Terminate	结束会话请求
	'P'	Parse	预编译解析请求
	'B'	Bind	绑定预编译语句的变量
	无	CancelRequest	取消请求，例如查询长时间没有响应时，即可发送此请求
	'C'	Close	关闭一个预编译语句请求
	'f'	CopyFail	复制失败请求
	'D'	Describe	预编译语句描述
	'E'	Execute	执行预编译语句
	'H'	Flush	完成预编译查询
	'F'	FunctionCall	函数调用
	无	SSLRequest	启用 SSL 通信
	'S'	Sync	同步请求

基于 Mycat 的需要，我们对部分协议进行了实现，基本满足了 Mycat 的要求。

1. 认证协议

初始化和认证是息息相关的，在这个阶段出现任何错误时，PostgreSQL 都会断掉前端的 Socket 连接，连接后 1 分钟内未给后端发启动数据包时，PostgreSQL 也会将连接断开，当然这是 PostgreSQL 安全保护的一部分。

PostgreSQL 和 MySQL 通信协议的最大不同之处在于，第 1 次连上 Socket 时，MySQL 会主动发送一个启动包给前端，然后前端根据 MySQL 发送过来的数据包发送相应的认证凭证。PostgreSQL 则不一样，客户端一旦连上就要马上给启动包，告知 PostgreSQL 客户端的名称、版本、系统参数、字符集等，否则一旦超出服务器设置的时间，服务器就会主动断开 TCP 连接。

当服务器接收到启动包后，会返回一个认证包（AuthenticationPacket，见图 8-5）或者错误包，一旦返回错误，则 TCP 连接会被断开（这一点在整个认证过程中都一样）。

```
AuthenticationPacket
-标志位（marker）= 'R' [1byte]
-长度（length）[4byte]
-认证类型（authType）[4byte]
-盐值（salt） [0byte/4byte/2byte]
```

图 8-5

认证包有 6 种类型，如下所述。

1）Ok(0)

代表认证成功，一般 ReadyForQuery 可以使用这个连接操作，但是等到 ReadyForQuery 还是有可能出现错误。

2）KerberosV5(2)

现在前端必须与服务器进行一次 Kerberos V5 认证对话。

3）CleartextPassword(3)

使用明文认证，收到此消息后需要将用户名、密码构造成一个认证请求包并发送给 PostgreSQL。

4）CryptPassword(4)

密码使用 Crypt 加密，加密时要加入 CryptPassword 数据包里面的验证。

5）MD5Password(5)

密码使用 MD5 加密，加密时要加入 MD5Password 数据包里面带的两个 byte 的哈希验证。

6）SCMCredential(6)

这个响应只对那些支持 SCM 信任消息的本地 UNIX 域连接出现。前端必须发出一条 SCM 信任消息，然后发送一个数据字节（数据字节的内容并不会被注意，它的作用只是确保服务器等待了足够长的时间来接受信任信息）。

此时需要根据 AuthenticationPacket 的类型发送对应的认证令牌 PasswordMessage，PostgreSQL 会返回 AuthenticationPacket 或者 ErrorResponse。如果返回的是 AuthenticationPacket 而且类型是 Ok(0)认证通过，则紧接着会返回多个 ParameterStatus 和一个 ReadyForQuery，BackendKeyData 接收到 ReadyForQuery 就代表后端可以响应前端应用的查询请求了。如图 8-6 所示。

在任何阶段都有可能返回一个 NoticeResponse，此数据包是一些警告信息和后端改变配置时的通知信息。

2. 简单查询

简单查询是指向 PostgreSQL 发送可执行的 SQL 命令语句，为非编译型的 SQL 语句，例如：

```
SELECT * FROM T_COOLLF WHERE USER = 'REST' ;
UPDATE TD_COOLLF SET USES = 'XOO' where 1=1
```

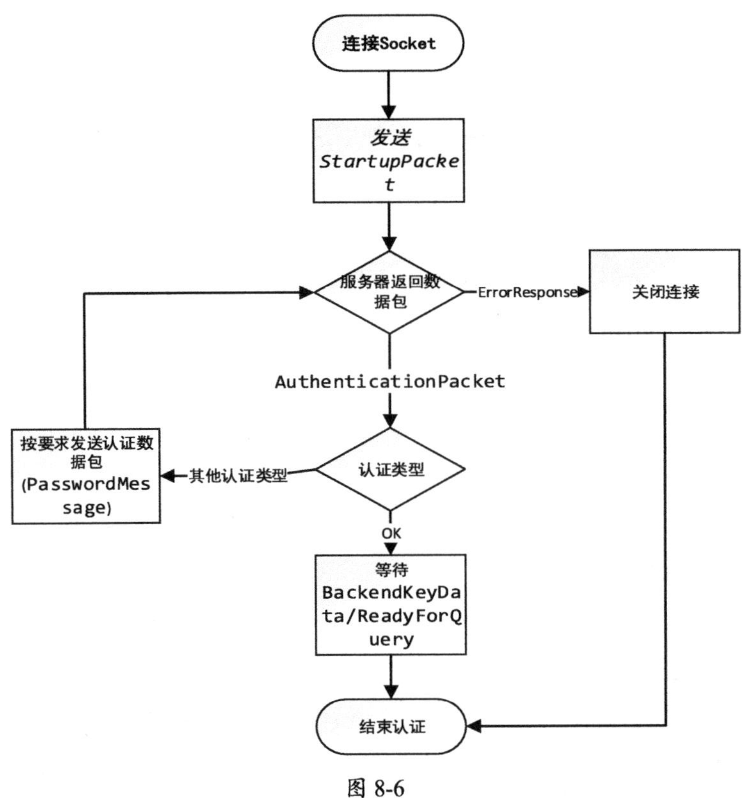

图 8-6

一次性可以传多条语句（但 Mycat 解析层和路由层已经进行了封装，此处一般是简单的 SQL 语句及一些事务控制、字符同步的 SQL 语句）。

将可执行的 SQL 封装成 Query，每个 Query 请求最终都会返回一个 CommandComplete 或者 ErrorResponse 数据包，然后通过 ReadyForQuery 进入下一个循环。

如果是处理成功的查询语句，则在 CommandComplete 之前会有一个 RowDescription 和 0 或多个 DataRow 查询结果，DataRow 代表行记录。

如果 Query 传入的是一个空字符串查询，则在 CommandComplete 之前返回的是 EmptyQueryResponse。

简单查询的流程如图 8-7 所示。

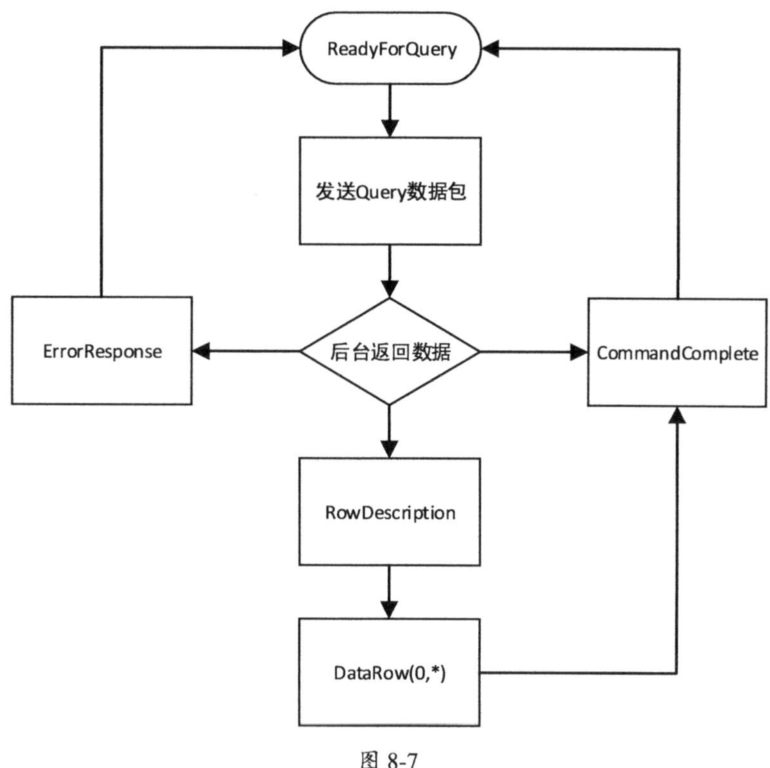

图 8-7

3. 关闭流程

客户端发送一个 Terminate 数据包给 PostgreSQL，后端就会把连接断开，此时客户端的 TCP 连接将不可用。如果需要再次进行查询，则需要重新开始一个连接并进行认证。

8.2.3　PostgreSQL 实现

在 Mycat 中，PostgreSQL 协议和 MySQL 协议的实现相似，只是在启动时会主动发送启动数据包，而 MySQL 是开启 TCP 连接后等待后端发送一个认证的要求数据包。

本实例最初是写了一个基于 Socket 通信的测试程序来直接连接 PostgreSQL 服务端，实现和调试数据包的组装和解析，代码在 io.Mycat.backend.postgresql.PostgresqlKnightriders 中可以找到。

```
List<String[]> paramList = new ArrayList<String[]>();
   String user = "postgres";
   String password = "coollf";
```

```
        String database = "Mycat";
        String appName = "Mycat-Server";
        paramList.add(new String[] { "user", user });
        paramList.add(new String[] { "database", database });
        paramList.add(new String[] { "client_encoding", "UTF8" });
        paramList.add(new String[] { "DateStyle", "ISO" });
        paramList.add(new String[] { "TimeZone", createPostgresTimeZone() });
        paramList.add(new String[] { "extra_float_digits", "3" });
        paramList.add(new String[] { "application_name", appName });
        boolean nio = false;
        try {
            Socket socket = new Socket("localhost", 5432);
            .......
```

通过测试程序实现验证数据包和交互流程，此过程参考了 PostgreSQL JDBC 驱动的实现，明白大体流程后，开始将测试代码在 Mycat 中进行集成和调试。

1. 嵌入点

对一个大型工程进行扩展，首先要找到嵌入点。例如对于一个 Web 工程，我们首先要到 web.xml 里去看看配置了一些什么内容，Mycat 实现同样通过查看 Mycat 的主配置文件，我们会发现每个后端数据库连接都是一个 dataHost 节点。如下所示是一个 MySQL 后端物理库的连接配置：

```
    <dataHost name="localhost1" maxCon="500" minCon="10" balance="0" dbType=
"MySQL" dbDriver="native">
        <heartbeat>select user()</heartbeat>
        <!-- can have multi write hosts -->
        <writeHost host="hostM1" url="localhost:3306" user="root" password=
"123456">
        <!-- can have multi read hosts -->
        <!-- <readHost host="hostS1" url="localhost:3307" user="root" password=
"123456"   /> -->
        </writeHost>
    </dataHost>
```

dbType="MySQL"和 dbDriver="native"标识了数据库的类型和驱动。我们通过更改 dbType="PostgreSQL"和 dbDriver="native"来支持 PostgreSQL 数据库。

在查找后，我们知道这个配置的作用点在 io.Mycat.server.config.loader.ConfigInitializer.createDataSource 方法上。

通过查看代码，我们会发现每个后端数据库的连接配置都对应了一个 PhysicalDatasource，即每个后端逻辑库对应了一个 PhysicalDatasource 对象。

要实现 PostgreSQL 协议，就要实现一个自定义的 PhysicalDatasource 对象。

```
public class PostgreSQLDataSource extends PhysicalDatasource {

    public PostgreSQLDataSource(DBHostConfig config, DataHostConfig hostConfig,
            boolean isReadNode) {
        super(config, hostConfig, isReadNode);
        // TODO Auto-generated constructor stub
    }

    @Override
    public DBHeartbeat createHeartBeat() {
        // TODO Auto-generated method stub
        return null;
    }

    @Override
    public void createNewConnection(ResponseHandler handler, String schema)
            throws IOException {
        // TODO Auto-generated method stub

    }

}
```

可以通过改写 io.Mycat.server.config.loader.ConfigInitializer.createDataSource 方法的相关代码，让 Mycat 支持新协议，并在 PostgreSQLDataSource 类的两个要实现的方法上写一些测试代码并设置断点。重新启动并进入我们设置的断点，第一步寻找嵌入点的工作就算完成了。

2. 实现后端连接

通过上一节我们找到了 Mycat 后端物理库连接配置的关键嵌入点。通过调试和查看代码，可以看出 PhysicalDatasource 主要提供两个服务：创建 DBHeartbeat 对象和创建后端连接 BackendConnection 对象。

由于此时 DBHeartbeat 已经有了 MySQL 的实现，所以我们暂时可以使用 MySQL 的 DBHeartbeat 来充数。此时我们的重点便是实现一个 PostgreSQL 的后端连接。参照 MySQL 的后端连接实现，我们可以继承 io.Mycat.net.Connection 类来减少开发的复杂度，通过参考 MySQL 的实现，我们可以知道要实现的核心方法如下。

- io.Mycat.backend.postgresql.PostgreSQLBackendConnection.query：心跳和验证性查询。
- io.Mycat.backend.postgresql.PostgreSQLBackendConnection.execute：咨询 Mycat 客户端发送的查询请求处理（路由到这个库的请求）。
- io.Mycat.backend.postgresql.PostgreSQLBackendConnection.commit：咨询 Mycat 客户端

发送的提交事务请求处理。

- io.Mycat.backend.postgresql.PostgreSQLBackendConnection.rollback：咨询 Mycat 客户端发送的回滚事务请求处理。
- io.Mycat.backend.postgresql.PostgreSQLBackendConnection.quit：Mycat 出错后，主动关闭连接。

后台逻辑库和 Mycat 之间的连接被抽象成一个 NIO 连接，由 Mycat 内部的 NIO 框架来处理，这时我们会发现每一类连接都会有一个自己的 NIOHandler 类来处理由服务器发送的数据包。

```
/**
 * NIOHandler 是无状态的，多个连接共享一个，用于处理连接的事件，各方法需要不阻塞，尽快返回结果
 *
 * @author wuzh
 */
public interface NIOHandler<T extends Connection> {

    /**
     * 连接建立成功的通知事件
     *
     * @param con
     *            当前连接
     */
    public void onConnected(T con) throws IOException;

    /**
     * 连接失败
     *
     * @param con
     *            失败的连接
     * @param e
     *            连接异常
     */
    public void onConnectFailed(T con, Throwable e);

    /**
     * 连接关闭通知
     * @param con
     * @throws IOException
     */
    public void onClosed(T con,String reason);

    /**
     * 收到数据需要处理
     *
```

```
     * @param con
     *            当前连接
     * @param data
     *            收到的数据包
     * @param readedLength    数据包的长度
     */
    void handle(T con, ByteBuffer data,int start,int readedLength);
}
```

NIOHandler 抽象了连接的几种情况分别是：连接完成、连接失败、连接关闭和对方有数据发送过来。然后写测试代码，设置断点验证流程。

此时我们会发现与 PostgreSQL 之间的连接总是被掐断，而我们写的测试代码是没有问题的。出现这种情况的原因是启动流程的客户端要主动发起一个 StartupMessage 数据包给 PostgreSQL。

通过 io.Mycat.backend.postgresql.PostgreSQLBackendConnectionHandler.onConnected 方法向 PostgreSQL 后台数据库发送一个 StartupMessage 数据包，发现连接依然会被掐断。经过多次断点调试，onConnected 确认被执行了，但是对应的数据也被写入了 NIO 的写列队。

经过调试发现，ServerSocket 程序从 I/O 流中读不到数据，那么只能说明 Mycat 没有发送数据到 PostgreSQL 数据库。通过调试 NIO 的相关代码，发现在调用 onConnected 方法时 Selector 已经被置为读模式，直到读到数据才会被置为写模式，才会开始调用对应的 channel 写入数据。由于 Mycat 中的 NIO 是基于 Mycat 通信协议流程设计的，所以没有考虑到一旦连接成功就向后台发送数据的情况。基于此我们在连接成功的地方发送一个数据包以达到我们的目标。

我们在 io.Mycat.net.NIOConnector.run 方法上加入了一段发送 StartupMessage 数据包的代码 io.Mycat.backend.postgresql.PostgreSQLBackendConnectionHandler.handle，此时 Mycat 将接收到 PostgreSQL 发送过来的数据。

3. 处理响应

我们在上一节成功接收到了 PostgreSQL 发送过来的数据，可以通过之前的测试代码解析数据包，就可以知道服务器发送过来的是错误请求还是认证请求。

由于认证流程较长，所以未认证成功之前便认为连接正在建立，只有认证成功后才开始处理业务数据。我们根据连接状态将 PostgreSQLBackendConnectionHandler 中的 handle 方法的任务拆解如下：

```
    /***
     * 进行连接处理
     *
     * @param con
```

```
         * @param buf
         * @param start
         * @param readedLength
         */
        private void doConnecting(PostgreSQLBackendConnection con, ByteBuffer buf,
int start, int readedLength)

        /***
         * 进行业务处理
         *
         * @param con
         * @param buf
         * @param start
         * @param readedLength
         */
        private void doHandleBusinessMsg(PostgreSQLBackendConnection con,
ByteBuffer buf, int start, int readedLength)
```

到目前为止，我们基本上可以和 PostgreSQL 进行通信了。偶尔由于断点调试时间较长使连接被掐断，但是认证完成之后就不会出现这种尴尬了。一旦完成认证，就算查询出错了，也会发送一个 ErrorResponse 数据包。

简单的 select 查询语句的返回略微复杂，所以我们把 SQL 查询结果返回的数据包做了封装，SelectResponse 先处理 RowDescription 和 DataRow 的数据包并保存起来，然后等到处理 CommandComplete 数据包时再统一处理，以减少逻辑的复杂度。

到现在为止，我们的 Mycat 和 PostgreSQL 的连接还没实现事务、心跳，也会经常报错，经过调试发现是因为很多 MySQL 查询语句不能在 PostgreSQL 上使用，包括我们在 MySQL 中用来开始事务的语句：

```
SET autocommit=0
```

处理完这些烦琐的事情后，Mycat 的 PostgreSQL 协议功能就基本上可用了。后续的改进和提高需要大家的参与和协助。

8.3 Mycat 对 JDBC 支持的实现

Mycat 最开始的设计只考虑支持 MySQL 数据库，在 Mycat 应用越来越广泛的时候，各种各样的需求逐渐出现了，很多企业存在多种数据库并存的现象，他们需要一款中间件来支持多种不同类型的数据库，这样可以极大地减少维护的工作量，还便于构建一套跨多种应用的数据分析系统。

如图 8-8 所示，客户端通过 MySQLProtocal 或 JDBC 连接 Mycat，而 Mycat NIO 和 MySQL 数据库交互，通过 JDBC 和 Oracle、SQL Server、Postgresql 交互，这里可以把 Mycat 当作数据库路由器，异构多种数据源，支持大型企业的复杂业务环境。

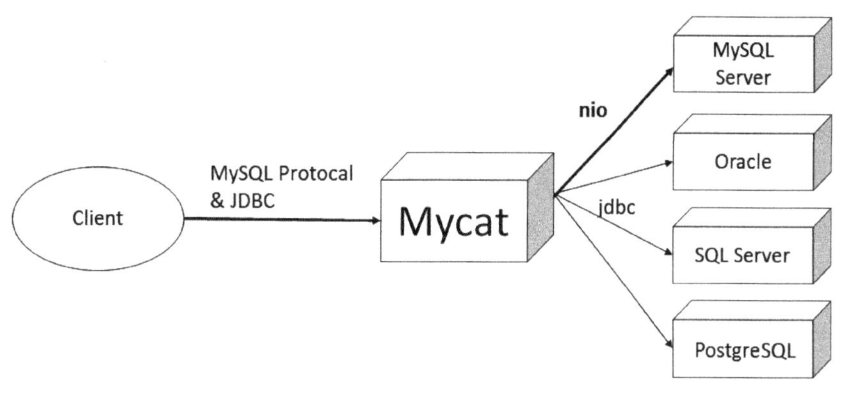

图 8-8

Mycat 从 1.3 版本开始实现了对 JDBC 的支持，这个功能的实现使 Mycat 支持市场上的大部分关系型数据库，Mycat 从对单一 MySQL 数据库的支持，变为支持目前的众多数据库，包括 Oracle、SQL Server、PostgreSQL、MongoDB、SequoiaDB 等 NoSQL 数据库。对于客户端来说不论 Mycat 后端采用哪种数据库，客户端和 Mycat 的通信是相同的，和具体的后端数据库无关，相当于数据库是透明的。Mycat 会把其他数据库模拟成 MySQLServer 来使用，Mycat 本身是通过 JDBC 访问其他数据库的，所以需要把相应数据库的 JDBC 驱动的 jar 包放入 Mycat 的 lib 下。

Mycat 在支持其他数据库的基础上，还保持着原数据库的一些特性，比如 Oracle 特有的分页 rownum 仍然可以使用 Oracle 语法，SQL Server 的 top 语法也可以使用。

8.3.1 Oracle 配置

配置支持 Oracle 数据库时，需要先修改 Mycat 目录 conf 的 schema.xml 配置文件中的以下 dataHost 配置：

```
    <dataHost name="oracle1" maxCon="1000" minCon="1" balance="0" writeType="0"
    dbType="oracle" dbDriver="jdbc">
    <heartbeat>select 1 from dual</heartbeat>
    <connectionInitSql>alter session set nls_date_format='yyyy-mm-dd hh24:mi:ss'
</connectionInitSql>
    <writeHost host="hostM1" url="jdbc:oracle:thin:@127.0.0.1:1521:nange" user=
"base"
       password="123456" > </writeHost>
    </dataHost>
```

- dbDriver 一定为 jdbc，dbType 代表数据库的类型，可以为 Oracle 等数据库。
- url 是 jdbc 连接的地址，和一般开发 Java Web 的 jdbc.url 的地址一致，user、password 分别是用户名和密码。
- heartbeat 是心跳包的查询语句。
- select 1 from dual 连接 Oracle 的初始化语句，connectionIniSql 初始化本次会话的日期显示格式。
- 需要 ojdbc14-x.jar 包（也支持其他版本）。
- 其他配置同 table、dataNode 标签。

Mycat 支持 Oracle 的三层嵌套和 row_number 的两种分页语法及 rownum 控制最大条数的语法，还可以使用 limit 语法分页。Mycat 在判断后端的具体数据库后，会转化成相应数据库的分页语法。

8.3.2 SQL Server 配置

配置支持 SQL Server 数据库时，需要先修改 Mycat 目录 conf 的 schema.xml 配置文件中的以下 dataHost 配置：

```
    <dataHost name="sqlserver1" maxCon="1000" minCon="1" balance="0" writeType="0"
dbType="sqlserver" dbDriver="jdbc">
<heartbeat></heartbeat>
<connectionInitSql></connectionInitSql>
<writeHost host="hostM1" url="jdbc:sqlserver://localhost:1433" user="sa"
password="sa" >
</writeHost>
</dataHost>
```

- dbDriver 一定为 jdbc，dbType 代表数据库的类型，可以为 SQL Server、Oracle、MongoDB。
- url 是 jdbc 连接的地址，和一般开发 Java Web 的 jdbc.url 地址一致，user、password 分别是用户名和密码。
- heartbeat 是心跳包的查询语句，可以为空。
- connectionInitSql 是连接 SQL Server 的初始化语句，可以为空。
- 需要 mssqljdbc*.jar 包（也支持其他版本）。
- 如果需要支持多个数据库，则可以不用指定数据库名，在 dataNode 中指定。

Mycat 支持 row_number 和 row_number 与 top 结合这两种分页语法，另外支持 top 限制最大条数，还支持 limit 语法自动转换成 SQL Server 的分页语法。

8.3.3 MongoDB 配置

配置支持 MongoDB 数据库时，需要先修改 Mycat 目录 conf 的 schema.xml 配置文件中的以下 dataHost 配置：

```
<dataHost name="jdbchost" maxCon="1000" minCon="1" balance="0"
 writeType="0" dbType="mongodb" dbDriver="jdbc">
<heartbeat>select user()</heartbeat>
<writeHost host="hostM" url="mongodb://192.168.0.99/test"
user="admin" password="123456" ></writeHost> </dataHost>
```

- dbDriver 一定为 jdbc，dbType 代表数据库的类型，可以为 MongoDB、Oracle，通过这个配置可以支持其他数据库。
- url 是 jdbc 连接的地址，和一般开发 Java Web 的 jdbc.url 地址一致。
- user、password 分别是用户名和密码，可以是任意值，目前不支持 MongoDB 配置用户名和密码。
- heartbeat 是心跳包的查询语句，可以为空。
- 如果需要支持多个 MongoDB 数据库，则可以不用指定数据库名，在 dataNode 中指定。

实现原理为通过实现标准的 JDBC 接口，在 JDBC 驱动包中调用 MongoDB 提供的 Java API 实现对 MongoDB 的操作。

（1）解析 SQL 语句（Druid SQL Parser 为 SQL 解析器）。

（2）转化为 MongoDB API。

（3）发送到 MongoDB 服务端。

（4）获取结果转化为 ResultSet。

支持的语法如下。

创建表（注意在 MongoDB 中不用创建表也可以使用）：

```
create table people (name varchar(30),age int,sex int,diqu
varchar(20),lev int);
```

在插入数据的时候必须有字段名，否则会提示错误：

```
insert into people (name,age,sex,diqu,lev) values('mongo',22,1,'sz',1);
insert into people values('mongo',22,1,'sz',1);
ERROR 3009 (HY000):java.lang.RuntimeException:number of values and columns have
to match
```

查询插入的数据：

```
mysql> select * from people where name='mongo';
+--------------------------+-------+------+------+------+------+
| _id                      | name  | age  | sex  | diqu | lev  |
+--------------------------+-------+------+------+------+------+
| 54a21dbd4001d690588ffe32 | mongo |   22 |    1 | sz   |    1 |
+--------------------------+-------+------+------+------+------+
1 row in set (0.10 sec)
```

更新语句：

```
mysql> update people set age =23 where name='mongo';
Query OK, 1 row affected (0.05 sec)

mysql> select * from people where name='mongo';
+--------------------------+-------+------+------+------+------+
| _id                      | name  | age  | sex  | diqu | lev  |
+--------------------------+-------+------+------+------+------+
| 54a21dbd4001d690588ffe32 | mongo |   23 |    1 | sz   |    1 |
+--------------------------+-------+------+------+------+------+
1 row in set (0.00 sec)
```

支持多种形式的 select 查询语句，如下所述。

（1）支持带“*”号的查询：

```
mysql> select * from people where name='mongo';
+--------------------------+-------+------+------+------+------+
| _id                      | name  | age  | sex  | diqu | lev  |
+--------------------------+-------+------+------+------+------+
| 54a21dbd4001d690588ffe32 | mongo |   23 |    1 | sz   |    1 |
+--------------------------+-------+------+------+------+------+
1 row in set (0.00 sec)
```

（2）支持对指定字段名的查询：

```
mysql> select name,age from people where name='mongo';
+-------+------+
| name  | age  |
+-------+------+
| mongo |   23 |
+-------+------+
1 row in set (0.00 sec)
```

对于 select 语句的 where 条件支持如下。

（1）支持大于表达式：

```
mysql> select name,age from people where age>23;
+--------------------------+--------+------+
| _id                      | name   | age  |
```

```
+--------------------------+--------+------+
| 549eb4dc40018c3cf9748d65 | feng   |   30 |
| 549eb4dc40018c3cf9748d66 | jifeng |   30 |
| 549eb4dc40018c3cf9748d67 | zhou   |   32 |
+--------------------------+--------+------+
3 rows in set (0.00 sec)
```

（2）支持小于表达式：

```
mysql> select name,age from people where age<23;
+--------------------------+------+------+
| _id                      | name | age  |
+--------------------------+------+------+
| 549eb4dc40018c3cf9748d68 | sohu |   21 |
+--------------------------+------+------+
1 row in set (0.00 sec)
```

（3）支持小于等于表达式：

```
mysql> select name,age from people where age<=23;
+--------------------------+-------+------+
| _id                  | name  | age  |
+--------------------------+-------+------+
| 54a21dbd4001d690588ffe32 | mongo |   23 |
| 549eb4dc40018c3cf9748d68 | sohu  |   21 |
+--------------------------+-------+------+
2 rows in set (0.00 sec)
mysql> select _id, name,age from people where age>=23;
+--------------------------+--------+------+
| _id                  | name   | age  |
+--------------------------+--------+------+
| 54a21dbd4001d690588ffe32 | mongo  |   23 |
| 549eb4dc40018c3cf9748d65 | feng   |   30 |
| 549eb4dc40018c3cf9748d66 | jifeng |   30 |
| 549eb4dc40018c3cf9748d67 | zhou   |   32 |
+--------------------------+--------+------+
4 rows in set (0.00 sec)
```

（4）支持大于等于表达式：

```
mysql> select name,age from people where age>=23;
+--------------------------+--------+------+
| _id                      | name   | age  |
+--------------------------+--------+------+
| 54a21dbd4001d690588ffe32 | mongo  |   23 |
| 549eb4dc40018c3cf9748d65 | feng   |   30 |
| 549eb4dc40018c3cf9748d66 | jifeng |   30 |
| 549eb4dc40018c3cf9748d67 | zhou   |   32 |
+--------------------------+--------+------+
4 rows in set (0.00 sec)
```

（5）支持不等于表达式：

```
mysql> select  name,age from people where age<>23;
+--------------------------+--------+------+
| _id                      | name   | age  |
+--------------------------+--------+------+
| 549eb4dc40018c3cf9748d65 | feng   |   30 |
| 549eb4dc40018c3cf9748d66 | jifeng |   30 |
| 549eb4dc40018c3cf9748d67 | zhou   |   32 |
| 549eb4dc40018c3cf9748d68 | sohu   |   21 |
+--------------------------+--------+------+
4 rows in set (0.00 sec)
```

（6）支持 and 表达式：

```
mysql> select name,age from people where age>30 and lev=2;
+--------------------------+------+------+
| _id                      | name | age  |
+--------------------------+------+------+
| 549eb4dc40018c3cf9748d67 | zhou |   32 |
+--------------------------+------+------+
1 row in set (0.00 sec)
```

（7）支持 and 表示范围：

```
mysql> select name,age from people where age>18 and age<30;
+--------------------------+-------+------+
| _id                      | name  | age  |
+--------------------------+-------+------+
| 54a21dbd4001d690588ffe32 | mongo |   23 |
| 549eb4dc40018c3cf9748d68 | sohu  |   21 |
+--------------------------+-------+------+
2 rows in set (0.00 sec)
```

（8）支持 or 表达式：

```
mysql> select  name,age from people where age>30 or diqu<>'gz';
+--------------------------+-------+------+
| _id                      | name  | age  |
+--------------------------+-------+------+
| 54a21dbd4001d690588ffe32 | mongo |   23 |
| 549eb4dc40018c3cf9748d67 | zhou  |   32 |
+--------------------------+-------+------+
2 rows in set (0.00 sec)
```

（9）支持 and 和 or 混合条件表达式：

```
mysql>select name,age from people where age>30 or (diqu='gz' and lev=1);
+--------------------------+------+------+
| _id                      | name | age  |
+--------------------------+------+------+
```

```
| 549eb4dc40018c3cf9748d65 | feng |   30 |
| 549eb4dc40018c3cf9748d67 | zhou |   32 |
| 549eb4dc40018c3cf9748d68 | sohu |   21 |
+--------------------------+------+------+
3 rows in set (0.00 sec)
```

支持排序，包括多字段排序、Limit 等，如下所述。

（1）支持升降序：

```
mysql> select * from people order by age;
+--------------------------+--------+------+------+------+------+
| _id                      | name   | age  | sex  | diqu | lev  |
+--------------------------+--------+------+------+------+------+
| 549eb4dc40018c3cf9748d68 | sohu   |   21 |    2 | gz   |    1 |
| 54a21dbd4001d690588ffe32 | mongo  |   23 |    1 | sz   |    1 |
| 549eb4dc40018c3cf9748d65 | feng   |   30 |    1 | gz   |    1 |
| 549eb4dc40018c3cf9748d66 | jifeng |   30 |    1 | gz   |    2 |
| 549eb4dc40018c3cf9748d67 | zhou   |   32 |    1 | gz   |    2 |
+--------------------------+--------+------+------+------+------+
5 rows in set (0.00 sec)

mysql> select * from people order by age desc;
+--------------------------+--------+------+------+------+------+
| _id                      | name   | age  | sex  | diqu | lev  |
+--------------------------+--------+------+------+------+------+
| 549eb4dc40018c3cf9748d67 | zhou   |   32 |    1 | gz   |    2 |
| 549eb4dc40018c3cf9748d65 | feng   |   30 |    1 | gz   |    1 |
| 549eb4dc40018c3cf9748d66 | jifeng |   30 |    1 | gz   |    2 |
| 54a21dbd4001d690588ffe32 | mongo  |   23 |    1 | sz   |    1 |
| 549eb4dc40018c3cf9748d68 | sohu   |   21 |    2 | gz   |    1 |
+--------------------------+--------+------+------+------+------+
5 rows in set (0.00 sec)
```

（2）支持 limit：

```
mysql> select * from people limit 3;
+--------------------------+--------+------+------+------+------+
| _id                      | name   | age  | sex  | diqu | lev  |
+--------------------------+--------+------+------+------+------+
| 54a21dbd4001d690588ffe32 | mongo  |   23 |    1 | sz   |    1 |
| 549eb4dc40018c3cf9748d65 | feng   |   30 |    1 | gz   |    1 |
| 549eb4dc40018c3cf9748d66 | jifeng |   30 |    1 | gz   |    2 |
+--------------------------+--------+------+------+------+------+
3 rows in set (0.00 sec)
```

（3）支持 delete 删除语句：

```
mysql> delete from people where name='zz';
Query OK, 1 row affected (0.04 sec)
```

（4）支持删除表的 drop 语句：

```
drop table people;
```

8.3.4 源码分析

jdbc 的这部分代码相对于 Mycat 来说是比较简单的，主要有 4 个类文件：JDBCDatasource、JDBCConnection、JDBCHeartbeat、ShowVariables，它们分别是 jdbc 物理数据源、jdbc 连接类、jdbc 心跳类和 Show 命令支持。

JDBCDatasource 类文件的代码如下：

```
package org.opencloudb.jdbc;
……
……
public class JDBCDatasource extends PhysicalDatasource {
……
……//省略此处代码，建议读者参考 GitHub 仓库 MyCAT-Server 项目的 JDBCDatasource.java 源码

}
```

JDBCConnection 主要执行 SQL 语句，然后把执行结果返回给 mpp（SQL 合并引擎，Mycat 处理多节点结果集排序、分组、分页），需要实现 ResponseHandler 的接口。

在执行 SQL 语句的过程，JDBCConnection 先判断是 select、show 语句还是 ddl 语句。

- 如果是 show 语句，并且不是 MySQL 数据库，则执行 ShowVariables.execute，构造 MySQL 的固定信息包。
- 如果是 SELECT CONNECTION_ID()语句，则执行 ShowVariables.justReturnValue，也是构造 MySQL 的固定信息包。
- 如果是 select 语句，则执行并且返回结果数据集。
- 如果是 ddl 语句，则执行并且返回 OkPacket。

在线程中执行 executeSQL 的方法，并把线程放入 Mycat Server 的线程池中执行，据测试结果显示，这比不用线程方式执行 SQL 语句的效率提高了 20%～30%。

以上过程建议读者参考 GitHub 仓库 MyCAT-Server 项目的 JDBCConnection.java 源码。